Occupational Stress in the
Service Professions

Occupational Stress in the Service Professions

Maureen F Dollard
University of South Australia

Anthony H Winefield
University of South Australia

and

Helen R Winefield
University of Adelaide

Taylor & Francis
Taylor & Francis Group

LONDON AND NEW YORK

First published 2003
by Taylor & Francis
11 New Fetter Lane, London EC4P 4EE

Simultaneously published in the USA and Canada
by Taylor & Francis Inc,
29 West 35th Street, New York, NY 10001

Taylor & Francis is an imprint of the Taylor & Francis Group

© 2003 Taylor & Francis

Publisher's Note:
This book has been prepared from camera-ready-copy
provided by the author
Printed and bound in Great Britain by
Biddles Ltd, Guildford and King's Lynn

Every effort has been made to ensure that the advice and information
in this book is true and accurate at the time of going to press.
However, neither the publisher nor the authors can accept any legal
responsibility or liability for any errors or omissions that may be made.
In the case of drug administration, any medical procedure or
the use of technical equipment mentioned within this book,
you are strongly advised to consult the manufacturer's guidelines.

British Library Cataloguing in Publication Data
A catalogue record for this book is available
from the British Library

Library of Congress Cataloging in Publication Data
Occupational stress in the service professions / [edited by] Maureen F. Dollard,
Anthony H. Winefield, and Helen R. Winefield
p. cm.
Includes bibliographical references and index.
ISBN 0-415-26760-9
1. Job stress—Congresses. 2. Service industries workers—Psychology—Congresses.
3. Service industries workers—Health and hygiene—Congresses.
4. Service industries—Psychological aspects—Congresses.
I. Dollard, Maureen F. II. Winefield, Anthony H. (Anthony Harold), 1937–
III. Winefield, Helen R.

HF5548.85 .O256 2003
158.7′2—dc21
2002040855

ISBN 0-415-26760-9

Contents

Foreword

More Jobs, Better Jobs, and Health

Lennart Levi

For several decades, researchers in stress psychology and stress medicine have tried to create a unifying model to describe the core causes of work-related stress and a wide range of morbidity and mortality.

The two – complementary – concepts most widely applied are the demand-control-support model (Karasek and Theorell, 1990; Johnson and Hall, 1988; Johnson, 1997) and the effort–reward imbalance model (Siegrist, 1996).

Both have been tested empirically and compared with one another (i.e. 3rd International Conference on Work Environment and Cardiovascular Diseases, Düsseldorf, March 20–22, 2002; Kristensen and Siegrist, 2002).

The results so far indicate that both models capture key components of the stress-inducing characteristics of today's work environment but do so to a varying degree, depending on the profession under study and the specific conditions of work and life in each context.

Also, one person's meat is another person's poison. If your work situation exposes you to continuous over-stimulation, additional stimulation at work or outside it is unlikely to improve your overall situation. In contrast, if your work is habitually under-stimulating, such added stimulation may be highly beneficial.

Most of the current approaches to work-related stress have been rather non-specific. Many have focused on industrial jobs but less so on service professions. And they have usually not considered various professions one by one. This failure is corrected in this book on Occupational Stress in the Service Professions, edited by Maureen Dollard, Tony Winefield, and Helen Winefield. Focusing on professional groups as diverse as clergy and prostitutes, correctional officers and general practitioners, each chapter reviews the most recent literature about the service profession under study. It proceeds by identifying the specific (and non-specific) sources of work stress in each profession as well as the outcomes of such exposures, often drawing on studies conducted by the authors themselves. A theoretical critique of available research, options for therapeutic and preventive interventions and proposals for future research provide additional important components for consideration and action by relevant stakeholders.

The contributors are highly acknowledged researchers in these fields, from Australia, Canada, Germany, the Netherlands and the United Kingdom. By skilfully compiling their contributions, the editors offer the reader a fair method for evaluating a range of highly diverse occupational/organisational issues that could present psychosocial and other hazards to the workers.

Theoretically, environment and lifestyle-related disease may be prevented at any of the links in the pathogenic chain. Thus, environmental stressors might be removed, modified, or avoided by adjusting, e.g. the work environment, organization, and content. Preventive variables that interact might be increased (e.g. by improving social networks or expanding coping abilities). Emotional, behavioural, and physiological, pathogenic mechanisms might be interrupted (e.g. by blocking adrenergic beta-receptors, antismoking campaigns, psychotherapeutic counselling, tranquilizers). Precursors of disease might be treated to that they do not progress to overt disease.

Briefly, an overall programme for research and environmental and health action should aim at being:

- Systems oriented, addressing health-related interactions in the person-environment ecosystem (e.g. family, school, work, hospital, and older people's home)
- Interdisciplinary, covering and integrating medical, physiological, emotional, behavioural, social, and economic aspects
- Oriented to problem solving, including epidemiological identification of health problems and their environmental and lifestyle correlates, followed by longitudinal interdisciplinary field studies of exposures, reactions, and health outcomes, and then by subsequent experimental evaluation under real-life conditions of presumably health-promoting and disease-preventing interventions
- Health oriented (not merely disease oriented), trying to identify what constitutes and promotes good health and counteracts ill health)
- Intersectoral, promoting and evaluating environmental and health actions administered in other sectors (e.g. employment, housing, nutrition, traffic, and education)
- Participatory, interacting closely with potential caregivers, receivers, planners, and policymakers
- International, facilitating transcultural, collaborative, and complementary projects with centres in other countries.

In order to safeguard individual rights, prevent the perpetuation of harmful or useless measures, limit losses to the community's purse, and advance knowledge of the future, any of these, or other, actions must be evaluated when implemented. Such evaluation is the modern, humane substitute for nature's slow and cruel 'survival of the fittest', and is a means of enabling people to adapt with minimal trauma to a rapidly changing environment and to control some of its changes (Kagan and Levi, 1975; Levi, 1979, 1992).

Some readers are likely to regard the authors' proposals for prevention and cure as utopian. However, the awareness of the importance of the subject area is increasing very rapidly, particularly so in the European Union and its 15 Member States. Here are some examples.

The Swedish EU Presidency Conclusions of 23–24 March, 2001, read that 'regaining full employment not only involves focusing on *more* jobs, but also on *better* jobs. Increased efforts should be made to promote a good working environment for all'.

According to WHO (2001), mental health problems and stress-related disorders are 'the biggest overall cause of premature death in Europe'. Based on such considerations, the European Council of Ministers (15 November, 2001) concluded that 'stress and depression related problems... are of major importance... and significant contributors to the burden of disease and the loss of quality of life within the European Union', underlining that such problems are 'common, cause human suffering and disability, increase the risk of social exclusion, increase mortality and have negative implications for national economies'.

If this is so, this means that there is an urgent need for preventive measures across societal sectors and levels, aiming at minimising unemployment, underemployment, and overemployment, promoting 'the healthy job' concept, and humanising organisational restructuring.

The challenge to science of all this is to find out *what* to do, for *whom*, and *how*, and to bridge the science-policy gap. This volume provides important information with specific reference to the service professions.

According to the EU Framework Directive, employers have a 'duty to ensure the safety and health of workers in every aspect related to the work'. The Directive's principles of prevention include 'avoiding risks', 'combating the risks at source', and 'adapting the work to the individual'. In addition, the Directive indicates the employers' duty to develop 'a coherent overall prevention policy'. The European Commission has published its Guidance (Levi, 2000) to provide a basis for such endeavours.

Based on surveillance at individual workplaces and monitoring at national and regional levels, work-related stress (and its outcomes in terms of both cardiovascular and mental morbidity) could be prevented or counteracted by job-redesign (e.g. by empowering the employees, and avoiding both over- and under-load), by improving social support, and by providing reasonable reward for the effort invested by workers, as integral parts of overall management systems. And by adjusting occupational physical, chemical and psychosocial settings to the workers' abilities, needs and reasonable expectations – all in line with the requirements of the EU Framework Directive and Article 152 of the Treaty of Amsterdam, according to which 'a high level of human health protection shall be ensured in the definition and implementation of all Community policies and activities'.

Supporting actions should include research, but also adjustments of curricula in business schools, schools of technology, medicine and behavioural and social sciences, and in the training and retraining of labour inspectors, occupational health officers, labour union representatives, managers and supervisors. Again, the challenge to science is to provide evidence-based guidelines for all such endeavours, some of which are provided in this important volume. As such the book has wide appeal and will be of immense value to service professionals, human resource managers, policy makers, and organisational and occupational psychologists, industrial sociologists, and students of all these disciplines.

REFERENCES

Johnson, J.V. and Hall, E.M., 1988, Job Strain, workplace social support and cardiovascular disease: A cross-sectional study of a random sample of Swedish working population. *American Journal of Public Health*, **78**, pp. 1336–1342.

Johnson, J.V., 1997, Empowerment in future worklife. *Scandinavian Journal of Work Environment Health*, **23**, pp. 4–7.

Kagan, A. and Levi, L., 1975, Health and environment – psychosocial stimuli: a review. In Levi, L. (ed.), *Society, Stress and Disease. Vol. 2: Childhood and Adolescence* (London: Oxford University Press), pp. 241–260.

Karasek, R. and Theorell, T., 1990, *Healthy Work – Stress, Productivity, and the Reconstruction of Working Life* (New York: Basic Books).

Kristensen, T.S. and Siegrist, J., 2002, *Work Environment and Cardiovascular Diseases. Third International Conference*, Düsseldorf, Germany, 20–22 March 2002, Book of Abstracts.

Levi, L. (ed.), 1979, Psychosocial factors in preventive medicine. In: *Healthy People. The Surgeon General's Report on Health Promotion and Disease Prevention*. Background papers (Washington, D.C.: U.S. Department of Health, Education, and Welfare) pp. 207–252.

Levi, L., 1992, Work Stress. *European Bulletin of Environmental Health*, **1**, p. 9.

Levi, L. and Levi, I., 2000, *Guidance on Work-Related Stress. Spice of Life or Kiss of Death?* (Luxembourg: European Commission), ISBN 92–828–9806–7.

WHO, 2001, *Mental Health in Europe* (Copenhagen: World Health Organization, Regional Office for Europe).

Acknowledgements

The editors kindly thank Ms Cheryl Ayliffe for her work in formatting the book, and Ms Carolyn Boyd for formatting the index. Thanks also to Dr Rob Ranzjin, Dr Kurt Lushington, Ms Alison Wallis, Ms Sarah Cotton, and Ms Sharon Casey all from the University of South Australia, School of Psychology and Prof Dick Dienstbier, University of Nebraska, Lincoln for their help in reviewing various aspects of the book.

Contributors

Kara Anne Arnold is a PhD Candidate at the Queen's School of Business, Queen's University, Kingston, Ontario, Canada.

Julian Barling, PhD is Associate Dean, School of Business, Queen's University, Kingston, Ontario, Canada.

Ronald J. Burke, PhD is Professor of Organizational Behavior, School of Business, York University, Toronto, Ontario, Canada.

Peter Cotton, PhD is a Principal Consultant with Insight SRC and an Honorary Senior Fellow with the Department of Psychology at the University of Melbourne, Parkville, Victoria, Australia.

Sarah J. Cotton, **BA Hons** is a PhD Candidate, Work & Stress Research Group, School of Psychology, at the University of South Australia, Adelaide, South Australia, Australia.

Tom Cox, PhD is Professor of Organisational Psychology at the Institute of Work, Health & Organisations, University of Nottingham, United Kingdom.

Jan de Jonge, PhD is Associate Professor of Work and Health Psychology at the Utrecht University, Department of Social and Organizational Psychology and Research Institute for Psychology and Health, The Netherlands.

Maureen F. Dollard, PhD is Associate Professor of Organisational Psychology, and Director of the Work and Stress Research Group, School of Psychology, at the University of South Australia, Adelaide, South Australia.

Christian Dormann, **PhD** is Senior Research Fellow at the Johann Wolfgang Goethe Universität Frankfurt, Germany.

Esther R. Greenglass, PhD is Professor of Psychology at York University, Toronto, Canada.

Amanda Griffiths, PhD is Professor of Occupational Health Psychology at the Institute of Work, Health & Organisations, University of Nottingham, United Kingdom.

Susan P. Griffiths, PhD is a Lecturer at the School of Psychology, Curtin University of Technology, Perth, Western Australia, Australia.

Peter M. Hart, PhD is the Director of Insight SRC and an Honorary Senior Fellow at the Department of Psychology, University of Melbourne, Parkville, Victoria, Australia.

Pascale M. Le Blanc, PhD is Assistant Professor at the Department of Social and Organisational Psychology of Utrecht University and Deputy Director of the Research Institute for Psychology & Health, The Netherlands.

Robert L. Lonne, PhD is a lecturer in the School of Humanities and Human Services at the Queensland University of Technology, Brisbane, Queensland, Australia.

Jacques C. Metzer, PhD is Associate Professor and Head of the School of Psychology at the University of South Australia, Adelaide, South Australia, Australia.

Raymond Randall, PhD is a Lecturer in Applied Psychology at the Institute of Work, Health & Organisations, University of Nottingham, United Kingdom.

Angeli Santos is Research Associate, Institute of Work, Health & Organisations, University of Nottingham, United Kingdom.

Wilmar B. Schaufeli, PhD is Professor of Clinical and Organisational Psychology at the Department of Social and Organisational Psychology of Utrecht University and Scientific Director of the Research Institute for Psychology & Health, The Netherlands.

Paul Whetham, PhD is a lecturer at the University of South Australia, City East Campus, Adelaide, South Australia, Australia.

Anthony H. Winefield, PhD is Foundation Professor, and Director of the Work & Stress Research Group at the School of Psychology, University of South Australia, Adelaide, South Australia, Australia.

Helen R. Winefield, PhD is Associate Professor in the Departments of Psychology and Psychiatry at the University of Adelaide, Adelaide, South Australia, Australia.

CHAPTER ONE

Introduction:
Context, Theories and Intervention

Maureen F. Dollard

1.1   INTRODUCTION

At the turn of the century a confluence of economic, political and sociocultural forces are impacting profoundly on our contemporary work arrangements (Cooper *et al.*, 2001). The climate in organisations has changed dramatically in the past decade with increased demands from globalisation of the economy, and the rapid development of communication technology (Cascio, 1995; Schabracq and Cooper, 2000). There is rapid industrialisation in developing economies (Cheng, 2000).

The impact of globalisation of the economy has led to a shift from manufacturing to knowledge and service based economies. Further, the exponential rate of technological change has 'outstripped efforts to develop sociotechnical perspectives that integrate human needs and values into the management of jobs and organizations' (Cooper *et al.*, 2001).

Kendall *et al.* (2000) report that within the workplace these changes translate to *overemployment* for many workers: those in full-time jobs are experiencing increased pressure and faster pace (Bousfield, 1999), increased workload (Townley, 2000), longer shifts and longer hours (Heiler, 1998; Winefield *et al.*, 2002), and demands for high organisational performance. Overemployment has been linked to cardiovascular disease for some time (Breslow and Buell, 1960). The risk of heart attack for those working long hours (e.g. 11 hours) is 2.5 times the risk of those working an 8 hour day (Sokejima and Kagamimori, 1998).

Employment has also become more precarious as workers are employed increasingly on contract (Schalk *et al.*, 2001; Winefield *et al.*, 2002), and the permanent job itself has become more insecure, leading to predictions that by 2020 a quarter of the workforce will be in non-traditional employment arrangements (Judy and D'Amico, 1997, see Kendall *et al.*, 2000). Quinlan (2002) describes the results of recent reviews on the health effects of precarious (casual, short term, temporary, self-) employment in 11 countries, from 1986 to 2000 (Quinlan *et al.*, 2001), and on the health effects of downsizing/restructuring and job insecurity published in the international literature between 1966 and 2001 (Bohle *et al.*, 2001). Overwhelmingly the reviews found a measurable deterioration in health effects for precarious and survivor groups. The latter review found that those most affected among surviving workers were committed workers, older workers, and those subject to ongoing insecurity.

Workers are now being required to perform multiple tasks, learn new skills, and self-manage to meet competitive demands. According to Kendall *et al.* (2000) this has lead to jobs that are more fluid (Cooper *et al.*, 2001), possibly exacerbating role ambiguity and role conflict, and leading in turn to work stress and illness (Dunnette, 1998). In addition, for many workers the amount and scope of work has diminished with technological advances leading to *underemployment* (Cooper *et al.*, 2001), and this can also be risky. Research has found that those working less than 6 hours per day have 3 times the risk of heart attack than those working an 8 hour day (Sokejima and Kagamimori, 1998). Winefield *et al.* (2002) however point out that those working lower hours may have been doing so because they were already suffering from the stress of too high a workload.

Technological changes have also led to an increasing amount of poor-quality work – 'work not fit for a machine to do' – that is unsatisfying, offering low pay, low job security and unreliable hours, and often undertaken by women and cultural minorities (Winefield *et al.*, 2002).

Organisations have downsized and restructured to improve flexibility and competitiveness or as a result of economic recession (Kawakami, 2000) leading to both mental and physical ill-health (Chang, 2000). Flatter organisational structures are hazardous as workers find career options limited (Kasl, 1998).

Belkic *et al.* (2000) argue that modern work demands are squeezing out 'passive' and 'relaxed' jobs (e.g. scientists increasingly compete for funding, general practitioners participate in settings of corporate managed care), which may lead to two classes of occupations: those with high control or those with low control, but *all* with high demands. Further, disparity between managerial and unskilled jobs in actual levels of control has reportedly increased over the years with a greater relative increase for managerial staff (Lunde-Jensen *et al.*, 2000).

Reports of changes in work arrangements are widespread and many have been linked to the emergence of new costs, for the individual, the organisation and society. As noted by Levi (2002) in the Preface of this book, besides negative implications for national economies, mental health problems and stress-related disorders are the biggest overall cause of premature death in Europe according to the WHO (2001).

1.2   COST OF WORK STRESS

The cost of occupational stress is acknowledged as a problem around the world, and is a common concern in both developing and industrialised nations (Kawakami, 2000).

Some attempts have been made to quantify the impact of stress on the economy in terms of Gross Domestic Product. In Denmark work related sickness and absence is estimated to be 2.5% of GDP, in Norway 10% (Lunde-Jensen, 1994) and in the European Union, 5–10% due to work stress (Cooper *et al.*, 1996).

The impact of sickness absence in UK economy is estimated to be 12 billion pounds, 50% of which is estimated to be stress related (Cooper, 1998). In the United States, it is estimated that 54% of sickness absence is stress related (Elkin and Rosch, 1990).

The economic costs to Australia in terms of work related stress are substantial and determinable at a state level. In Australia workers are generally entitled to workers' compensation for stress when the claimant's employment significantly contributed to stress, not including situations where reasonable disciplinary action or failure to obtain a promotion, transfer or other benefit in relation to employment has occurred. The 'stress' condition is required to be 'outside the bounds of normal mental functioning' (insurer for the Commonwealth Government-Comcare), or a psychiatric condition listed in the Diagnostic and Statistical Manual of Mental Disorders – 4th Edition Revised (DSM IV-R), the American Psychiatric Association (DSM IV) or the International Classification of Diseases: Classification of Mental and Behavioural Disorders – 10th Ed (ICD-10) (i.e. in South Australia and most other states).

The cost and prevalence of such claims varies from state to state. The following details are drawn from the 'Extract from the "Comparison of Workers" Compensation Arrangements in Australian Jurisdictions" ' 2001. In New South Wales in 1999/2000 there were 1577 new claims comprising 17% of all occupational disease claims, each at an average cost of $20, 617 per claim, and the total gross cost was $33 million. The largest proportion of claims (20%) was from Health and Education where large groups of professionals coalesce. In Victoria, 5% of claims were for stress in 1997/98 (1587 new claims). Apart from circulatory disease and back injury claims, stress claims were most costly and represented the highest average payment per claim. The Victorian WorkCover Authority declared stress as a *significant* cause of 86 deaths since 1985, *including* 15 suicides.

In South Australia there were 162 claims in 1998/99 accounting for 2% of all injuries and 3.5% of all income maintenance costs. In Western Australia 601 claims were lodged in 1997/98 for work stress, 2.2% of all claims with claim cost of $23,399 twice that of other claims (an increase of 34% from 1996/97). In Queensland an increase of 19% was found in 1999/2000 and an increase of 28% in 2000/2001. The average cost of the claim was $17,249, over *twice* that of the next most expensive. A striking statistic is that the average duration of time off for psychological/psychiatric claims was 96.1 days compared to 28.9 days for other claims.

It is difficult to derive a GDP figure for Australia, as data sets between state jurisdictions are incomplete. However estimates are around $49 million in 1995/96 (National Occupation Health and Safety Commission, 1998) (excluding Victoria and Australian Capital Territory data) with an *additional* $38 million for Commonwealth workers in 1995–96 (Australian National Audit Office, 1997).

Work stress claim costs generally measure the cost of the problem to workplace insurer, but do not reflect all stress costs. For example, stress may manifest itself in other classifiable (physical health) symptoms. In US the National Institute of Occupational Health and Safety identified psychological disorders (including neuroses, personality disorder, and alcohol and drug dependency) as one of the 10 leading occupational diseases and injuries (Sauter *et al.*, 1990).

Also in Australia there is a lot of stigma associated with making stress claims, which may inhibit people from making them. There may also be organisational challenges to stress claims. Claim rates therefore do not reflect the stress risk of work environments.

Costs do not include the additional costs to organisations incurred through staff replacement and retraining, special supervision, work flow interference, unplanned absences and service complaints, and the cost of sick leave leading up to the compensation claim (Toohey, 1995). Further, they do not reflect the costs to the individual such as loss of self-esteem, loss of professional-esteem, new or exacerbated physical symptoms, loss of physical stamina, disruption to intimate life, lost hours of professional development, loss of professional sensitivity, or increased psychological distress. Work stress cost estimates therefore grossly underestimate the real cost of the problem.

In addition to psychological disorders, stress at work may lead to other costly behaviours such as smoking and aggression. O'Leary *et al.* (1996) argue that organisationally motivated aggression (revenge, retaliation), may occur when workers perceived an inequitable disbursement of rewards and other resources by the organisation. As we shall see later these conditions lead to strain according to the Effort–Reward Imbalance Model (see Section 1.4.3).

According to a report published last year in US, 'the cost of workplace violence to employers is estimated to be between \$6.4 billion and \$36 billion in lost productivity, diminished image, insurance payments and increased security' (Daw, 2001, p. 52).

Surveys confirm widespread reporting of the experience of work stress. In Europe, 28% of 15 000 workers surveyed report that stress is a work-related health problem (Paoli, 1997). In Australia, the Australia Workplace and Industrial Relations Survey (1995), report that 26% of people rate work stress as the second largest cause of work related injury and illness (behind physical strains and sprains-43%) (See Extract from the 'Comparison of Workers' Compensation Arrangements in Australian Jurisdictions' July, 2000.). In Japan 63% of workers in a nation wide survey 1997 report 'strong worry, anxiety or stress at work or in daily working life', an increase of 12% since 1982. In Japan the emergence of 'karoshi' (death from overwork) is an increasing social concern (Kawakami, 2000). In US 68% of respondents to a survey reported they had to work very fast, and 60% never had enough time to finish their work (Theorell, 1999).

Finally, income disparity resulting from 'good' and 'bad' jobs is argued to have negative health consequences for all members of society as it leads to wider social consequences and a breakdown of social cohesion and a strong community life (Wilkinson, 1996; Winefield *et al.*, 2002).

The phenomenon of work stress appears widespread and a number of new triggers are emerging in modern work arrangements. Costs, although difficult to determine appear considerable at an individual level [psychological and physical well-being of the individual], at an organisational level, and at a societal level, because of potential damage to community functioning, not merely in terms of workers' compensation statistics. Nevertheless the latter are important because they draw economic attention to the problem. In summary, widespread evidence of stress at work compels further efforts to understand, explain, predict and prevent the occurrence of occupational stress.

1.2.1 Stress, Stressors, Strain

In a very general sense work stress describes a field of study, a rubric, the 'area of practice or research focusing on social psychological characteristics of work that are detrimental to employee's health' (Beehr, 1989, p. 74). A generic definition of *job stress* given by US National Institute of Occupational Safety and Health (1999) is 'harmful physical and emotional responses that occur when the requirements of the job do not match the capabilities, resources, or needs of the worker. Job stress can lead to poor health and even injury'. Work stress, job stress and occupational stress are often used interchangeably.

Work stressors or hazards or risks are defined as environmental situations or events potentially capable of producing the state of stress (Baker, 1985; Beehr, 1989; Beehr and Bhagat, 1985; Greenhaus and Parasuraman, 1987). Stressors may be physical or psychosocial in origin and both can affect physical and psychological health, and may interact with each other (Cox *et al.*, 2000a). Physical stressors may include biological, biomechanical, chemical and radiological, or psychosocial hazards. Psychosocial hazards (stressors) are 'those aspects of work design and the organisation and management of work, and their social and environmental contexts, which have the potential for causing psychological, social or physical harm' (Cox and Griffiths, 1995). This discussion will focus on psychosocial stressors. However it should be remembered that physical hazards can lead to psychological injury (see Dollard *et al.*, 2001).

Based on a consensus of various reviews of the literature on psychosocial hazards, Cox *et al.* (2000b) have constructed a table which outlines ten different categories of job characteristics, work environments and organisational aspects which have been shown to be hazardous (stressful or harmful to health). The categories are further divided into context and content of work (Table 1.1 for a full account).

There is general agreement in the literature about what a stressor is (antecedent of stress) and what a strain is (consequence of stress). As we shall see however, stress definitions vary according to a theoretical perspective. Some theories do not specifically define the intermediary term, stress.

Strain refers to reactions to the condition of stress. These reactions may be transitory, but short-term strains are presumed to have longer term outcomes (Sauter *et al.*, 1990). Occupational strain may include psychological effects (e.g. cognitive effects, inability to concentrate, anxiety), behavioural effects (e.g. use of smoking, alcohol), and physiological effects (e.g. increased hypertension Schnall *et al.*, 2002).

Table 1.1 Stressful Characteristics of Work

Category	Risk factors/Conditions*
Job Characteristics and Nature of the Work	
Job Contents/demands	High physical, mental and/or emotional demands, lack of variety, short work cycles, fragmented or meaningless work, underutilisation, high uncertainty, continuous exposure to people through work
Workload/workplace	Work overload or underload, machine pacing time pressure, deadlines
Work schedule	Shift working, inflexible work schedules, unpredictable hours, long or unsocial hours
Job control	Low participation in decision making, lack of control over workloads
Physical environment and equipment issues	Inadequate or faulty equipment, poor environmental conditions (space, light, thermal etc.)
Social and Organisational Context of Work	
Organisational culture and function	Poor communication, low levels of support for problem-solving and personal development, lack of definition on organisational objectives
Interpersonal relationships at work	Social or physical isolation, poor relationships with superiors, interpersonal conflict, lack of social support
Role in organisation	Role ambiguity, role conflict, responsibility
Career development	Career stagnation and uncertainty, underpromotion or overpromotion, poor pay, job insecurity, low social value of work
Individual Risk Factors	
Individual differences	Coping styles, personality, hardiness
Home-work interface	Conflicting demands of work and home, low support at home, dual career problems

*Hazards have been changed to 'risk' accommodating the dynamic relationship between the person, perception of events and the event itself. Adapted from Cox *et al.* (2000b, p. 4).

Chronic stressor, stress or strain refers to an ongoing exposure, condition or reaction respectively. An example of a chronic stressor is work load, and presumably this would be linked with a chronic or long lasting reaction, as the exposure is ongoing.

Acute stressor, stress, strain refers to a short lived exposure, condition or reaction respectively. An example of an acute stressor would be a violent incident, that could lead to an acute response or a chronic response such as Post Traumatic Stress Disorder, depending on its nature.

1.2.2   Stages of the Response to Stress

Underlying the reaction to stress is a well described *physiological process*. It has been described as adaptive and enables the individual to mobilise energy resources in the body to deal with the stressor. It is only when the stressor is sustained without the opportunity for recovery that health problems occur, presumably because of continuous physiological arousal.

Selye's (1956) early exposé on the general syndrome of responses [the General Adaptation Syndrome-GAS] is particularly helpful in understanding the physiological response to stress. The alarm stage prepares the individual to 'fight or flee' the stressor. It is accompanied by enabling responses such as increased heart rate, increased blood pressure, more rapid breathing. With repeated exposure to a stressor the individual enters the stage of resistance. This is accompanied by different physiological changes and enables the individual to further withstand the stressor. Finally, according to Selye a stage of exhaustion is reached if the stressor is unrelenting, resulting in organic damage or even death. However, our modern knowledge of the impacts of stress-generated hormones and neuroendocrines on immune functioning, cardiovascular processes, etc. leads us to conclude that prolonged exposure to stressors and prolonged resistance (not necessarily 'exhaustion') can potentiate a variety of disease processes. By progressing through these stages the normal physiological response can become pathological.

It is generally agreed that there are two arousal pathways in the stress-response (Dienstbier, 1989; Mason, 1974; Schaubroeck and Ganster, 1993). The first is the *SNS–adrenal–medullary* pathway, so-called because it is one of the first processes (and sometimes the only one) to be elicited in a stressor context, especially in most work stress situations (i.e. involving effort, not merely harm or loss). Arousal involves stimulation of the adrenal medulla, by the hypothalamus acting through the sympathetic nervous system (SNS) to release peripheral catecholamines (adrenalin) and SNS synapses to release noradrenalin. The second pathway is the *pituitary–adrenal–cortical* arousal system and involves the activation of the adrenal cortex by the pituitary gland to release glucocorticoids, such as cortisol into the blood.

For a person who copes ideally with work-related stressors, baseline arousal is typically low. However, when facing a challenge requiring mental or physical coping, the SNS-adrenal medullary arousal is quick and strong, with a quick return to baseline arousal levels at the termination of the demand. In fact, the rapid return to baseline of such arousal is possible since the associated hormones and neural transmitters (i.e. principally adrenaline and noradrenaline) have half-lives of under 2 minutes. Because the energy and arousal generated by strong SNS-adrenal medullary arousal may be sufficient to meet the demand and, perhaps equally importantly, because the worker may *experience* the associated energy as sufficient to meet the demand, arousal of the pituitary–adrenal–cortical system may be minimized.

In contrast, when a worker is unable to cope easily with demands, either because of characteristics of the job (low control, continuous demand) or because of a mismatch of coping ability and job demands, a different pattern of arousal is often observed. Workers in such circumstances often show elevated baseline arousal levels of both arousal systems, with less of a demand-generated SNS–adrenal–medullary surge but slower return to baseline. Unfortunately, this pattern is often accompanied by extensive pituitary–adrenal–cortical arousal – an arousal pattern than cannot cease with the end of demand, since the half-life of cortisol (the major corticosteroid released in humans) is 90 minutes.

The less adaptive response pattern outlined immediately above is described in a variety of ways, depending upon theoretical emphasis. Karasek and Theorell (1990) describe it as a catabolic (break-down) process, while others (e.g. McCarty *et al.*, 1988) emphasise the depletion of (particularly) central nervous system neurotransmitters. Others (e.g. Schaubroeck and Ganster, 1993) focus on the negative downstream health consequences, such as cardiovascular degeneration (See Dienstbier, 1989, and Dienstbier and Pytlik Zillig, in press, for extensive discussions of these processes and of ways to enhance positive coping patterns.).

In summary it is assumed that chronic stress (i.e. without control or respite results in chronic neuroendocrine and cardiovascular over-arousal, and cardiovascular degeneration (Schaubroeck and Ganster, 1993). Presumably these physiological changes underpin both psychological and physical health problems.

1.2.3   Stress Effects (Strain)

An extensive amount of research has now concluded that prolonged exposure to work-related psychosocial hazards can have negative mental and physical health behavioural and social consequences for employees (Ganster and Schaubroeck, 1991). Effects of stress may be categorised as (physical) health effects, cognitive and psychological effects, and behavioural strain.

Health Effects

While the experience of stress may be accompanied by feelings of emotional discomfort, and may significantly affect well-being at the time, it does not necessarily lead to the development of a psychological or physiological disorder (Cox *et al.*, 2000a).

However there are a number of health effects now documented that have been linked to the experience of chronic adverse work environments, or acute traumatic events. Further, the health state itself may act as a stressor, as it may sensitise the person to other sources of stress by reducing one's ability to cope (Cox *et al.*, 2000a).

The stress response described above involves a range of systems, and it is not surprising therefore that the range of symptoms is large. Most health conditions susceptible to work stress appear to involve the cardiovascular and respiratory systems, (e.g. CHD-Belkic *et al.*, 2000, and asthma), the immune system (e.g. rheumatoid arthritis), and the gastro-intestinal system (e.g. gastric ulcers) (see Cox *et al.*, 2000a).

Psychological Effects

Psychological strain includes cognitive and psychological effects, such as:
- an inability to concentrate
- job dissatisfaction (Dollard *et al.*, 2001)

- affective disorders, including anxiety, depression (Amick *et al.*, 1998) and anger (Kendall *et al.*, 2000)
- somatic symptoms such as headaches, perspiration, and dizziness (Caplan *et al.*, 1975; Perrewe and Anthony, 1990).

Longer term psychological outcomes may include mental illness and suicide (Kendall *et al.*, 2000).

Behavioural Effects

This may be indicated by increased or excessive use of alcohol and drugs, including tobacco; or by reduced work performance, higher levels of absenteeism or sick leave, industrial accidents and staff turnover (Caplan *et al.*, 1975; Perrewe and Anthony, 1990).

Besides these outcomes, strain from the work environment may spill over into the home environment, leading to marital problems and other social issues (Sauter *et al.*, 1990).

In addition to the above, acute stress events may lead to Post Traumatic Stress Disorder.

1.2.4 Post-Traumatic Stress Disorder (PTSD)

In the case of intense, acute events, often life threatening and beyond the normal range of expectations – for example, the experience of violent incidents, witnessing a robbery, working with abused clients, or dealing with road accidents, Post Traumatic Stress Disorder PTSD may develop (see DSM-IVTR 2000). Symptoms typically include:

- intrusive recollections
- dreams
- sensitivity to stimuli associated with the initial event, and
- avoidance of activities or situations associated with the trauma.

A range of other psychological conditions may co-exist with PTSD, such as anxiety, depression, thoughts of suicide, panic disorder, anti-social personality disorder, agoraphobia, and substance abuse (see Kendall *et al.*, 2000). Chronic stressors have also been viewed as a precursor to PTSD, for example in the case of bullying (Tennant, 2001).

1.3 DEFINITIONAL CONTEXTS

Mulhall (1996) reminds us that 'the discourse on stress, or indeed anything else does not arise in a political or ideological vacuum' p. 456. As we search through the history of occupational or work stress in the scientific discourse we can see that socio-political contexts have influenced the research agenda and the way work stress is conceptualised (see Calnan *et al.*, 2000).

Indeed Calnan *et al.* (2000) observe that systematic research on job stress commenced with the pioneering work of Kahn and his colleagues (Beehr, 1989) and dominated the work stress research agenda in US during the 1960s and 1970s (Kahn *et al.*, 1964). The research approach was guided by Role Stress Theory, and emphasised personal attributes and subjective characteristics rather than the characteristics of the situation. Calnan *et al.* (2000) argue that 'this highly individualised conception of role stress effectively depoliticised the work stress discourse, facilitating its easy passage into corporate human resource management' p. 297.

At the same time in the Scandinavian countries a social democratic political approach to the issue was emerging. The climate gave rise to a radically different research approach that focused on work characteristics and occupational health, and sought work reform and industrial democracy supported by trade unions, government and employer's organisations (Calnan *et al.*, 2000).

So we see that even in Western societies different sociopolitical contexts give rise to different emphases in work stress theories. Nevertheless most work stress research and theorising has been undertaken almost exclusively in western industrialised societies, and we do not yet know the extent of its relevance and usefulness in eastern societies (Laungani, 1996; Jamal and Preena, 1998). The relevance and application of the concept however is sought in many other ways not yet theorised, in: western cultures vs eastern and emerging economies; trans-culturally and across regions; in dominant vs indigenous cultures; in rural vs metropolitan locations; in large vs small to medium enterprises; in corporations vs owner operator arrangements and so on.

Understanding of stress is further confounded by the fact that the concept of stress is not value free. Levi (1990, p. 1144) proposed that approaches to the understanding of stress may incorporate one or more of four value concerns:

- a humanistic-idealistic desire for a good society and working life
- a drive for health and well-being
- a belief in worker participation, influence, and control at the individual level
- an economic interest in competitiveness and profits of business organisations and the economic system.

Placed within this framework occupational stress becomes a social and political issue as much as a health problem (Levi, 1990).

Since the early 1960s, research in the work stress area has burgeoned, leading to different understandings about what stress is or means. As Baker (1985) argues 'since stress certainly has a multifactorial aetiology, investigators have been able to formulate, and at least partially validate, substantially different models of the causes of stress' (p. 368). We currently have a kernel of an idea of what stress is and how it impacts and what it costs. However there are many unknowns about work stress which require researchers to look closely at the local context, including the dominant and local economic, social and political ideology, in trying to understand work stress situations.

1.4 WORK STRESS THEORIES, PARADIGMS, AND FRAMEWORKS

Work stress theories are important as they attempt to describe, explain and predict stress/strain according to a coherent set of hypotheses. Cox *et al.* (2000a) argue that most current stress theorising is psychological and conceptualises work stress in terms of a negative psychological state, and the dynamic interaction between the person and the work environment. It is however difficult to provide an unambiguous taxonomy of work stress theories as there are so many (see Cooper, 1998) and while they may differ in emphasis, in many ways their content is overlapping and complementary. Taxonomies include the following dimensions: stimulus/response combinations (see Cox, 1978); sociological vs psychological; and, environmental vs individual.

Though not comprehensive Cox *et al.* (2000a), categorise them thus: (1) Interactional theories, which focus on the structural features of the persons interaction with their work environment: and (2) Transactional theories of stress which focus on the cognitive processes and emotional reactions associated with the person's interaction with their environment (Cox *et al.*, 2000a). We will now provide a systematic and comprehensible introduction to four contemporary theories of work stress. As will be seen in the forthcoming chapters these theories are employed variously by researchers as frameworks within which to conceptualise occupational stress in the human service professions. We will commence our discussion with model that is however more clearly oriented to the sociological or social-psychological (see also Dollard *et al.*, 2001).

1.4.1 Demand Control Support Model

The Job Demand-Control model argues that work stress primarily arises from the structural or organisational aspects of the work environment rather than from personal attributes or demographics of the situation (Karasek, 1979). According to Karasek *et al.* (1981, p. 695) 'strain results from the joint effects of the demands of the work situation (stressors) and environmental moderators of stress, particularly the range of decision making freedom (control) available to the worker facing those demands'.

Strain can be understood to result for those with objective high demand and objective low control. It is therefore not necessary to investigate the impact of these environmental conditions by distinguishing appraisals or various needs that may exist among workers (Karasek, 1989). The assumption is that if individual 'needs' for certain environmental conditions play a role in the stress process, they do so by operating similarly *for all people*. This conceptualisation of stress views environmental causes as the starting point, although they do not strictly preclude the importance of personal factors (Karasek and Theorell, 1990).

Objective stressor $\rightarrow$ Strain

Empirical tests of the work environment model, ideally investigate the link between objective stressors and illness (Frese and Zapf, 1988). *Stress* in these models refers to the intermediate state of arousal between the objective stressor and strain. Karasek (1989) notes that this state (stress) is rarely measured in his research.

Drawing from research in industrial sociology, animal research on 'learned helplessness' (Abrahmson *et al.*, 1978) and health psychology, the theory hypothesises that when workers are faced with high levels of demands and a lack of control over decision making and skill utilisation, that adverse health effects will result. Referring to Figure 1.1, as:

- levels of psychological work demands increase and workplace autonomy or control decreases, levels of psychological strain increase (follow Diagonal A).
- demands and control increase congruently, increases in job satisfaction, motivation, learning, efficacy, mastery, challenge, and performance will be observed (follow Diagonal B).

The high demand-low decision latitude combination is labelled job strain (high strain); the high demand-high decision latitude combination active work; the low demand-low decision latitude passive work; and the combination low demand-high decision latitude low strain work. According to the model, workers in high strain jobs (e.g. machine paced, assemblers, and service-based cooks and waiters) experience the highest levels of stress. Executives and professionals do not belong to the high strain group therefore do not experience high levels of stress despite popular suggestions (i.e. executive stress). Presumably high status workers have the opportunity to regulate high levels of demands through frequent opportunities to use control (Karasek and Theorell, 2000).

Karasek argues that high demand jobs produce a state of normal arousal (i.e. increased heart rate, increased adrenalin, increased breathing rate). This enables the body to respond to the demand. However, if there is an environmental constraint, such as low control the arousal can not be channelled into an effective coping response (e.g. participation in social activities and informal rituals). Unresolved strain may in turn accumulate and as it builds up can result in anxiety, depression, psychosomatic complaints and cardiovascular disease.

The model also takes account of the work environment imperative of productivity and postulates that increased productivity will occur when workers have jobs that combine high demands and high control (i.e. active jobs).

The model has been expanded to include an important aspect of the work environment – social support (Johnson and Hall, 1988) (the DCS model). A recent review of 81 studies of social support, in particular emotional support found that it was reliably related to beneficial effects on aspects of the cardiovascular, endocrine and immune systems (Uchino *et al.*, 1996). Potential health related behaviours did not appear to be responsible for the associations.

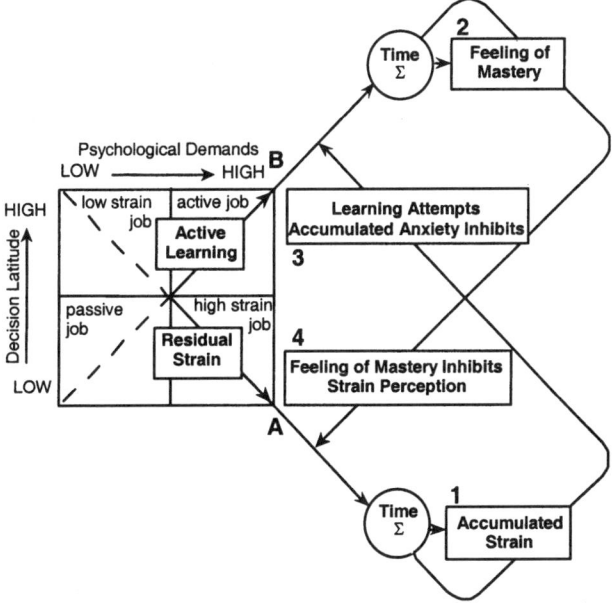

Figure 1.1 The psychological demand-decision latitude model: Dynamic associations linking environmental strain and learning to evolution of personality. Healthy Work (p. 99), by R.A. Karasek and T. Theorell (1990). Copyright 1990 by Robert Karasek. Reprinted by permission of Basic Books, a member of Perseus Books, L. L. C.

Jobs with high demands, low control, and low support from supervisors or coworkers carry the highest risk for psychological or physical disorders (high strain-isolated jobs).

Considerable empirical support for the models has been found.

- Empirical tests of the DC model have shown that large-scale multi-occupational (heterogeneous) studies have tended to provide support for *interaction* effects between demand and control predicting strain (Schnall *et al.*, 1994).
- Smaller scale studies of the DC model in homogeneous samples have found primarily main effects of demands and control (e.g. Hurrell and McLaney, 1989; Melamed *et al.*, 1991; Perrewe and Anthony, 1990; Spector, 1987a).
- Epidemiological studies provide the most convincing support for the core assumptions of the DCS model (see de Jonge and Kompier, 1997). Longitudinal studies have shown job strain to predict myocardial infarction in a study of working men over a 10 year period (Theorell *et al.*, and the SHEEP study group, 1998).

- In a review of 10 longitudinal studies of men, six showed an increase in cardiovascular disease (CVD) risk due to job strain (high demands, low control), and two showed mixed results. Of 5 cohort studies in women, four showed higher levels of elevated risk in CVD related to job strain (Belkic *et al.*, 2000).
- A study of 33 698 working women (nurses) in the United States found high strain workers showed lower vitality and mental health, higher pain, and increased risk of both physical and emotional limitations than active workers. Iso-strain work increased the risks further (Amick *et al.*, 1998).
- The Whitehall II study low support also shows a link between low levels of job control and an increased risk of coronary heart disease (e.g. plasma fibringen) (Bosma *et al.*, 1997) and social support at work, low decision latitude, high job demands, and effort–reward imbalance were associated with increased risk of psychiatric disorder over time (Stansfeld *et al.*, 1999).
- An increase in obesity is found due to job strain (Hellerstedt and Jeffrey, 1997).

Most studies of the DCS framework have examined the job strain hypothesis. The *active–passive* dimension of the model according to Theorell and Karasek (1996), has been under utilised in research and needs more attention: 'certainly patterns of active coping behaviour could affect the progression of disease development' (p. 10). Empirical support for the hypothesis has been found (e.g. Karasek, 1981; Landsbergis *et al.*, 1992).

A study of correctional officers by Dollard and Winefield (1998) showed that the level of active coping was significantly higher in active jobs than in passive jobs, consistent with the idea that workers experiencing passive jobs, with little opportunity for control, will show reduced motivation to tackle new problems. A more recent study found not only increased worker motivation but also greater health impairment in 381 insurance company workers in active jobs (Demerouti *et al.*, 2001). It was argued that the levels of demands were in fact too high, that they were not reduced by increasing control, and that neither too few or too many demands are good for employees (see Warr's Vitamin Model, 1987).

Evaluation

In comparison to all other models, empirical testing of the DCS models has dominated the occupational stress research in the past 15 years. This is probably in part due to the ease with which research of the highly specified three dimensions of the model can be implemented. On the other hand the model has been criticised for its relative simplicity and its lack of attention to psychological processes. Karasek does not deny the importance of psychological processes; rather it's a matter of emphasis. Further there is a lack of explication of emotional demands, an issue clearly of relevance to human service professions.

Although the model is essentially a sociological model, a challenge is that tests of the model are usually by self-report and in this way results represent psychological appraisals rather than the sociological or objective situation (Muntaner and O'Campo, 1993). Nevertheless there is evidence of good convergence between self-report and supervisor and subordinate-report (Cox *et al.*, 2000, p. 14), and consistency between self-report and objective ratings of the work environment (Spector, 1987b) (see Chapter 3).

Finally the DCS model is itself sometimes categorised as a psychological interactional model in the sense that it examines structural features of the person's *interaction* with the work environment (Cox *et al.*, 2000a) and also predicts coping outcomes.

The model attracts strong empirical support and has good face value in the workplace (Theorell, 1998). For example, in bureaucracies with rigid hierarchical power (control) structures, workers at the lowest levels in the organisation predictably have the highest levels of demands combined with the lowest levels of control, and the worst states of health. The model helps to develop links between healthy work (i.e. active jobs) on the one hand and productivity on the other.

It is now generally agreed that the DC model is validated when predicted additive *or* multiplicative interaction effects between the job characteristic variables and measures of strain or activity are uncovered (Theorell and Karasek, 1996). De Jonge and Kompier (1997) have observed that the interaction hypothesis is not often supported in epidemiological studies. The practical consequences of direct effects are in effect the same.

The importance of the three dimensions is reiterated in the literature and reviews of the DCS model (de Jonge and Kompier, 1997; Kristensen, 1995). In a review of work stress and employee health literature, Ganster and Schaubroeck (1991) concluded that 'the better designed studies show a significant relationship between self-reported stressors (lack of participation, lack of social support, hectic job demands, and low decision latitude) and epidemiologically meaningful outcomes' (p. 260). Further better specification of measures matching demand aspects with appropriate measures of control is argued to be important in demonstrating the buffering hypothesis (Van der Doef and Maes, 1999).

1.4.2   Burnout Theory

An interactional psychological model that is particularly relevant for people-oriented professions is the theory of burnout. Burnout theory has received widespread attention over the last 25 years. Burnout is thought to result from prolonged exposure to chronic interpersonal stressors on the job from working with troubled people. Burnout has been described as a three-dimensional syndrome characterised by emotional exhaustion, depersonalisation, and reduced personal accomplishment (Maslach and Jackson, 1981). Burnout is conceptualised as:

> an individual stress experience embedded in a context of complex social relationships, and it involves the person's conception of both self and others (Maslach, 1998, p. 87).

Although human service work is argued to impose special stressors on workers because of the client's emotional demands (Maslach, 1978, 1982) some studies have found that stressors such as clients' emotional demands, or problems associated with the professional helping role, such as failure to live up to one's own ideals as stressors, were less potent in predicting stress than those more in common with other non-helping professions (Collings and Murray, 1996; Shinn *et al.*, 1993). For example a US study of 168 protective services personnel (social workers) found that organisational variables were more strongly associated with job satisfaction and burnout than were client factors (Jayaratne *et al.*, 1995).

Overall empirical research on burnout has generally shown that job factors are more strongly related to burnout than are biographical or personal factors (Maslach and Schaufeli, 1993). Burisch (1993) argues that the burnout process begins inevitably with some frustration or loss of autonomy with which the individual failed to cope in an adequate way.

Maslach and Jackson (1982) argue that burnout results not from 'bad apples' but from the 'bad kegs' the apples are in. They note that

> although personality variables are certainly important in burnout, research has led us to the conclusion that the problem is best understood (and modified) in terms of the social and situational sources of job-related stress. The prevalence of the phenomenon and the range of seemingly disparate professionals who are affected by it suggest that the search for causes is better directed away from the unending cycle of identifying the 'bad people' and toward uncovering the operational and structural characteristics of the 'bad situations' where many good people function (p. 243).

Recent theoretical developments in burnout theory have turned to more positive conceptualisations, focusing on contrasting or opposite states of burnout-specifically engagement (Leiter and Maslach, 1998). Engagement consists of high energy (rather than exhaustion), strong involvement (rather then cynicism), and a sense of efficacy (rather than a reduced sense of accomplishment (Maslach, 1998)). The second development involves a framework for conceptualising the key causal factors in the development of burnout and this is specifically relevant here.

Given that the research shows burnout is largely a result of the organisational context, Maslach and colleagues have turned to earlier stress theories for concepts and frameworks. One concept of interest to them is the job-person fit model, not in the narrow sense of how an individual personality fits with the job, but rather how their motivations, emotions, values and job expectations fit with the job. Further, the job is viewed more broadly to encompass the organisational context.

Job-person mismatch is hypothesised to lead to burnout, the greater the mismatch the greater the burnout. The new model hypothesises that burnout is an important mediator in the earlier job-person fit models that predict that better fit produces certain outcomes such as commitment, satisfaction, performance and job tenure (Maslach, 1998). While a number of challenges remain in recasting burnout theory, Maslach and Leiter (1997) outline six areas where mismatch can occur, and the result is increased exhaustion, cynicism, and inefficacy of burnout.

The six mismatches are briefly described as follows.

- *Work overload* occurs when the job demands exceed limits.
- *Lack of control* occurs when people have little control over the work they do, either because of rigid policies and tight monitoring, or because of chaotic job conditions.
- *Insufficient reward* involves a lack of appropriate rewards for the work people do.
- *Breakdown of community* occurs when people lose a sense of positive connection with others in the workplace often due to conflict.
- *Absence of fairness* occurs when there is a lack of a system of justice and fair procedures which maintain mutual respect in the workplace.
- *Value conflict* when there is a mismatch between the requirements of the job and peoples' principles (Maslach, 1998, pp. 75–76).

Evaluation

According to Maslach (1998) the six mismatch model may be particularly useful when formulating interventions. While considerable empirical evidence exists for links between work place stressors and burnout, Soderfeldt *et al.* (1995) have argued that burnout research may be flawed as it merely reframes or renames a phenomenon that other occupational groups share. Nevertheless, burnout theory has been extended beyond human service workers by replacing the 'depersonalisation' construct with that of 'cynicism'.

More research is required on the latest formulations of the theory. Specifically the framework focuses attention on the relationship between the person and the situation, rather than on one or the other in isolation. As well it covers the central dimensions of the DCS model, and ERI theory (see below).

1.4.3 Effort–Reward Imbalance Model

The Effort–Reward Imbalance (ERI) model (Siegrist, 1996, 1998) is a more recently evolved model. This model is a transactional theory of stress in the sense that it focuses more on the interaction between environmental constraints or threats and individual coping resources. However it also derives from sociological and industrial medical frameworks, and emphasises the social framework of the job (e.g. social status of job).

According to the model (Figure 1.2), workers expend effort at work and they expect rewards as part of a socially (negotiated) organised exchange process of work. In this way ERI theory bears some resemblance to expectancy value theory of motivation. It assumes that the 'work role in adult life provides a crucial link between self-regulatory functions such as self-esteem and self-efficacy and the social opportunity structure' (p. 192). ERI theory emphasises rewards (such as money, esteem, and social control) rather than job control. When an imbalance occurs, measured empirically as the ratio of Efforts/Reward, between the efforts a worker puts in and the rewards that are received, strain results. For example, workers who have high job demands and low pay are likely to experience strain as a result of this imbalance. Similarly strain is expected when workers experience a threat to their job security (status).

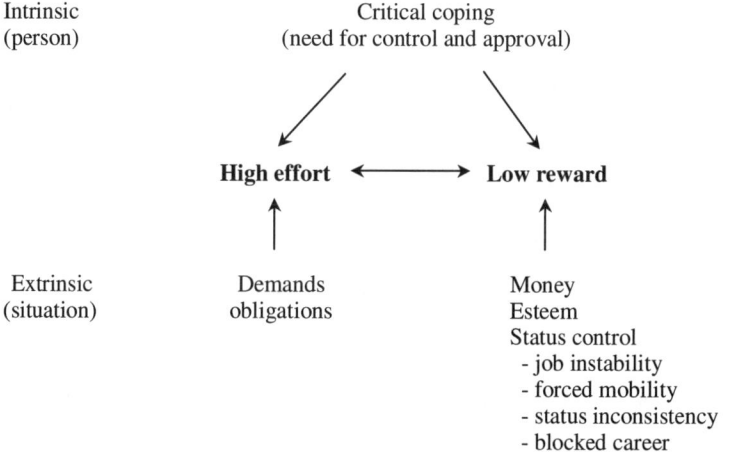

Figure 1.2 Model of effort–reward Imbalance at work.
Siegrist, J. (1998). Adverse health effects of effort–reward imbalance at work. In C.L. Cooper
(Ed.), *Theories of organizational stress* (pp. 190–205), Oxford: Oxford University Press, p. 193.
Reprinted with permission by Oxford University Press.

ERI further identifies extrinsic effort and intrinsic efforts. Extrinsic effort is conceptually similar to the job demands concept in the DCS model. On the other hand instrinsic efforts refers to a personal characteristic of coping, a pattern of excessive striving in combination with a strong desire of being approved and esteemed. This pattern is referred to as over-commitment. Conceptually it has links with the Type A behaviour pattern and negative affectivity. A worker may have a high need for control which results in over-commitment and immersion in the job and likely, a personal perception of low rewards. In empirical tests of the model, the idea is that over-commitment could moderate or mediate the imbalance between demands and rewards. This is a major departure from the DCS model which specifies no such individual variable.

Evaluation

A concern of Robert Karasek, the major proponent of the DCS model, is that we can not rule out the possibility that the development of personal attributes is independent of exposure to the work environment. The impact of this is that when ERI research tests for the effect of over-commitment (or negative affectivity on strain), true variance in strain measures due to variation in work environment measures could be removed. Karasek *et al.* (1998) argued that 'indeed, recent research (Dollard and Winefield, 1998) that explicitly test for the possibility of such over control with negative affectivity, using job experience cohorts to test whether negative affectivity is itself associated with duration of exposure to stressful job characteristics, finds that it is'.

Nevertheless the model has gained widespread attention particularly in Europe. Effort–Reward Imbalance at work has been found in studies to predict new cases of coronary heart disease, and helps to explain cardiovascular risk factors in workers. It has also been shown to be important in explaining adverse health effects such as gastrointestinal disorders, psychiatric disorders and poor subjective health (see Siegrist and Peter, 2000). Validation studies of the ERI model in other countries (e.g. Japan) appears to 'capture the Japanese workers' stress' (Kawakami and Tsutumi, 2000, p. 13).

1.4.4   Cognitive Phenomenological Theory

A transactional theory that has sometimes been used in work stress research is the widely discussed cognitive-phenomenological theory of stress (and coping). Stress is defined in this approach as a relationship between the person and the environment that is appraised as taxing or exceeding resources, and endangers well-being (Lazarus and Folkman, 1984). In this cognitive model, appraisal of stress is necessary: 'for threat to occur, an evaluation must be made of the situation to the effect that a harm is signified' (Lazarus, 1966, p. 44). Stress in these stimulus-response based models is not just the property of the body. It is the result of transactional psychobiological processes (Lazarus, 1966). The theory emphasises how individuals perceive or primarily appraise a stimulus (e.g. as a threat) 'Is this a problem?', and then secondarily appraises their coping resources (e.g. as inadequate) 'What am I able to do about it?'. Both appraisal processes, primary and secondary, are key mediators in the relation between stressor and strain (Lazarus, 1991). If a situation is perceived as stressful and important then coping is activated. It may involve attempts to modify the person-environment relationship (i.e. problem focused coping) or attempt to regulate resulting emotional distress (i.e. emotion focused coping). The situation is then re-appraised and the process repeated. If the situation is resolved, coping ceases. If it is unresolved then psychological and physiological strain persist resulting in longer term negative effects on health and well-being (Lazarus and Folkman, 1984).

Empirical tests of the model assess the relationship between cognitive appraisal and strain rather than the relation between the objective stressor and the perception or appraisal of it. Of course the model presented here is highly simplistic but captures fundamental relationships in a psychological paradigm.

Objective stressor $\rightarrow$ Cognitive appraisal $\rightarrow$ Strain

Evaluation

The essence of the cognitive model is the *meaning* given by individuals to events. This means that research on the approach requires investigations of individual transactions. While this approach provides rich insights into the cognitive processes of individuals, many limitations in its utility in work stress research can be identified (see Harris, 1991).

Nevertheless the theory has generated significant amounts of research on stress and coping (see Cox *et al.*, 2000a; Dollard, 1997; Edwards, 1998). The usefulness of this is that when workers are in employment contexts that are taxing, recommended coping approaches may help alleviate strain. This approach forms part of a range of possible intervention strategies, and will be discussed in more detail in the last section.

The theory has limitations in the work stress context, as it can not specify which aspects of the work environment would be stressful because according to the theory each individual might see the environment in a different way (Baker, 1985).

Overall Evaluation of Work Stress Theories

The four theories discussed each explain important aspects of the work stress picture and each has its own limitations. Cox *et al.* (2000a) note that transactional models are in a sense a development of the interactional models, and are *largely consistent* with them. Essentially the transactional models elucidate important cognitive and coping processes.

Kasl (1998) has recently argued that it is worth studying the relative contribution of both the DCS and ERI model to explain health and well-being, in view of their differences and complementary aspects. Theorell (1998) has conceptualised the relationship between the two models as the DCS embedded contextually within the ERI model. On the one hand ERI focuses on social structure, and while this is important for the DCS model the latter model focuses on the situational aspects of the job (Theorell, 1998). Theorell (1998) therefore argued that a good exploration of the work environment should include components of both DCS and ERI models (e.g. health promoting aspects of both models – rewards, control, and support as well as others relevant to the local context (study environment)). A complicating and unresolved point however is where to locate the over-commitment personal variable.

The utility of the models in small to medium size owner operator enterprises is not known. There is some evidence of their applicability in non-western cultures but more research is required on this (see Kawakami, 2000).

A major criticism of the work environment approach, is that it is simplistic and promulgates the notion of the individual as passive. Specifically the major challenge to both approaches comes from proponents of the psychological models who argue 'they ignore the mediation of strong cognitive as well as situational (contextual) factors in the overall stress process' (Cox *et al.*, 2000a, p. 11).

On the other hand major criticism of the psychological models comes from proponents of the work environment approaches. For example, they argue that by locating the sources of stress within the organisational structure of the work place, connections to the broader concepts of 'alienation, power, qualifications, worker's collectives, labour conflicts, management, and so forth' can be made (Kristensen, 1996, p. 254). On the other hand, when stress is understood in terms of perception and individual difference it is likely viewed as an individual problem and re-organisation of work processes may be avoided. Strategies may be directed toward adapting the worker to the existing working conditions (Baker, 1985). These opposing views highlight the potential conflict between broader notions of health and safety in the work place and the economic goals of business and industry in the investigation of work stress (Baker, 1985).

There is no comprehensive theory available to explain fully work stress. Certain *frameworks* have therefore been proposed to capture key work variables and symptoms of stress shown to be linked (see Cooper and Marshall, 1976; Cox *et al.*, 2000a). It is useful to think of work stress as on-going process with multiple and continuous feedback between a number of elements (Figure 1.3, Kagan and Levi, 1975):

- the work environment context (psychosocial stimuli (1)) as appraised by the individual
- interacting with the person, bringing certain vulnerabilities and strengths (2)
- determines the emotional, behavioural and physiological reactions i.e. strain (3)
- which under some circumstances may lead to precursors of disease (4) or disease itself (5)
- Interacting or moderating variables (e.g. social support, autonomy) may exacerbate or ameliorate the relationships.

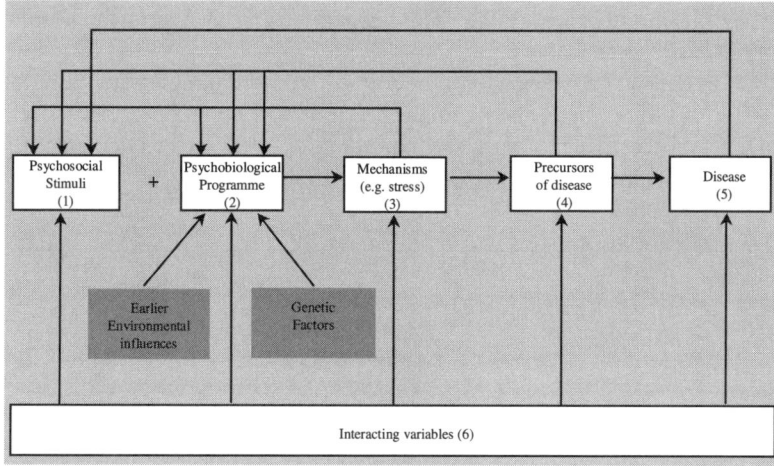

Figure 1.3 © Oxford University Press, 1975. Reprinted from Society, Stress and Disease, Volume 2: Childhood and Adolescence by Lennart Levi (1975) by permission of Oxford University Press.

Complex models may specify moderator variables, personal (self-esteem, negative affectivity, coping – see Dollard, 1997) variables and/or social (support) variables. Moderator variables are important because in the presence of high or low levels of these factors, the link between stressors and strain is strengthened or weakened. Mediator variables on the other hand intensify or weaken the link between stressors and strain more directly. For example a mediator variable may account (in part) for the observed relationship between stressor and strain (e.g. negative affectivity). Moderator and mediator effects give some insight into areas for intervention where the work environment is immutable.

In summary, there are a plethora of theories used as guiding frames for the interpretation of work stress problems (see Cooper, 1998), and we have canvassed but a few here. As work stress has multiple origins, various theories and aspects of them have found empirical support in the literature. In turn paradigms of thinking have their basis in the socio-political context of the time.

An approach we can use to bring empirical findings together from the various areas is to frame them at different epistemological stances and different ontological levels (Figure 1.4). Epistemology is the theory of knowledge embedded in a theoretical perspective and thereby in a methodology (Crotty, 1998). 'Objectivism is the epistemological view that [stressors] exist as *meaningful* entities independently of consciousness and experience, that have truth and meaning residing in them as objects ('objective' truth and meaning therefore), and that careful (scientific?) research can attain that objective truth and research' (Crotty, 1998, pp. 5–6). This is the epistemological view underpinning research of the DCS model. On the other hand the epistemological stance in research of the cognitive transactional model would necessarily be constructivism – there is no truth waiting to be discovered. The idea is that different people construct meanings in different ways, from culture to culture, and both subject and object interact to generate meaning (Crotty, 1998). This ideology underlies the phenomenological theory of (work) stress.

The major theories may also be framed at interlinking discipline levels. In broad terms the cognitive model might be viewed as a psychological model, and the DCS and the ERI model as sociological. The latter two models may still be regarded as within a social psychological paradigm (Muntaner and O'Campo, 1993) (see Figure 1.4).

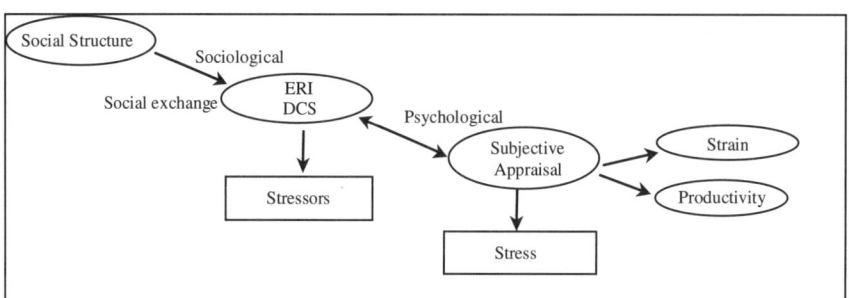

Figure 1.4 Social Psychological Paradigm.

The aim of sociological social psychology is to study the effects of social structure on human behaviour or personality. Psychological social psychology, aims to study behaviour of individuals in social context. It is argued that sociological social psychology is a discipline between two ontological levels (the sociological and the psychological), whereas psychological social psychology situates itself at a single ontological level, the psychological (e.g. working shift work causes sleep disturbance) (Muntaner and O'Campo, 1993).

It is also argued that sociological social psychology is emergentist, that what happens at a sociological level can not be explained in psychological terms (e.g. unemployment rates can not be understood through examination at the psychological level of individual workers) (Muntaner and O'Campo, 1993). Further, it is also assumed that what happens at a sociological level can determine what happens at a psychological level (e.g. increased levels of unemployment lead to more work pressure and higher levels of depression in workers, as workers engage in longer hours to keep their jobs) (Muntaner and O'Campo, 1993). In work stress theories unequal distribution of power/control (DCS) and rewards (ERI), both to a large extent determined by social structures, fail to offset demands workers face, leading to strain.

Using these arguments, theories that purport to explain and describe work stress may be visualised along a continuum where aetiology is understood to emerge upstream. Each major theory has a (contested) place on the continuum, and finds some support in the empirical literature. Essentially the continuum can be seen as the 'middle ground', an issue taken up further in Chapter 3.

1.5 INTERVENTIONS

Having described the health effects of work stress it is now important to consider the strategies used to identify, assess, and manage stress in the workplace. Strategies may be implemented at the national level, organisational level, or individual level.

1.5.1 Management Strategies

Interventions in stress management are typically classified into primary, secondary or tertiary approaches (Kendall *et al.*, 2000). Quick *et al.* (1998) have published a useful table summarising preventive strategies and surveillance indicators for organisational stress (see Figure 1.5).

Focusing on the intervention strategies, Quick *et al.* (1998) identify both organisational and individual strategies within the levels of prevention. In addition to this are national policy approaches, and group level approaches (see Guerts and Grundemann, 1999).

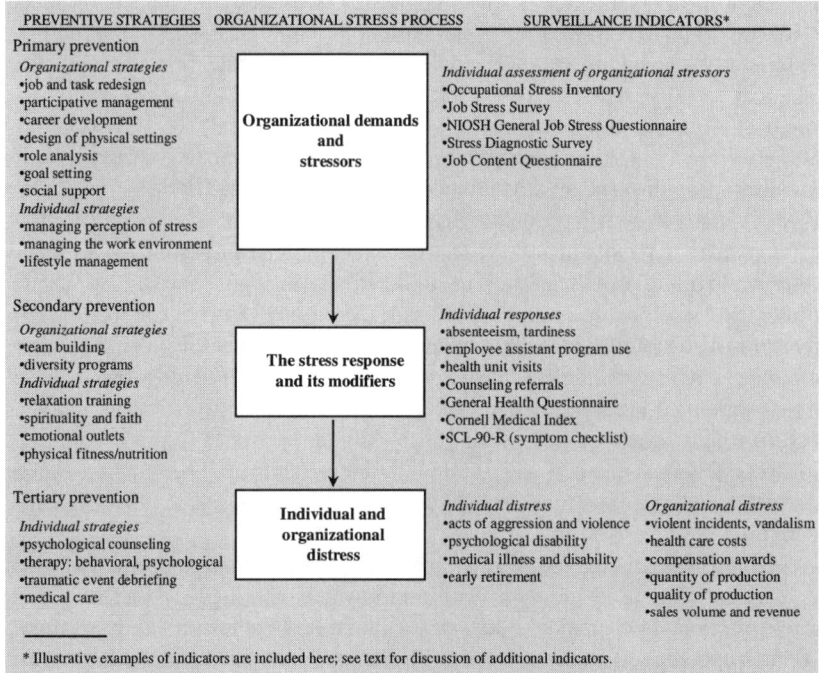

PREVENTIVE STRATEGIES   ORGANIZATIONAL STRESS PROCESS   SURVEILLANCE INDICATORS*

Primary prevention

Organizational strategies
•job and task redesign
•participative management
•career development
•design of physical settings
•role analysis
•goal setting
•social support
Individual strategies
•managing perception of stress
•managing the work environment
•lifestyle management

**Organizational demands
and
stressors**

Individual assessment of organizational stressors
•Occupational Stress Inventory
•Job Stress Survey
•NIOSH General Job Stress Questionnaire
•Stress Diagnostic Survey
•Job Content Questionnaire

Secondary prevention

Organizational strategies
•team building
•diversity programs
Individual strategies
•relaxation training
•spirituality and faith
•emotional outlets
•physical fitness/nutrition

**The stress response
and its modifiers**

Individual responses
•absenteeism, tardiness
•employee assistant program use
•health unit visits
•Counseling referrals
•General Health Questionnaire
•Cornell Medical Index
•SCL-90-R (symptom checklist)

Tertiary prevention

Individual strategies
•psychological counseling
•therapy: behavioral, psychological
•traumatic event debriefing
•medical care

**Individual and
organizational
distress**

Individual distress
•acts of aggression and violence
•psychological disability
•medical illness and disability
•early retirement

Organizational distress
•violent incidents, vandalism
•health care costs
•compensation awards
•quantity of production
•quality of production
•sales volume and revenue

* Illustrative examples of indicators are included here; see text for discussion of additional indicators.

Figure 1.5 Quick, Quick, Campbell and Nelson (1998). Preventive strategies and surveillance indicators for organisational stress. In Cooper, C., 1998 (Ed.), *Theories of organizational stress.* (Oxford: Oxford University Press) p. 259. Reprinted with permission by Oxford University Press.

1.5.2   Primary Approaches

National Policy Level

There is much to be achieved at the national policy level that can lead to organisational change. Strategies at the national level include legislation, national monitoring systems, and active transfer of knowledge.

Legislation

This is necessary, to focus attention on the issue of stress at work to establish a framework in which to address it. For example Kompier *et al.* (1994) found in their analysis of 5 European countries' regulations, policies and practices, that countries which possessed framework legislation (Sweden, The Netherlands, UK) also had well-developed work stress prevention programmes, in contrast to those who did not yet recognise work stress as an important policy issue (Germany, France). Australia is similar to the former grouping in that it has well defined legislation, but not as yet well defined prevention programmes.

National Monitoring Systems

These are important for benchmarking and drawing attention to risky occupations (Houtman and Kompier, 1995). The Netherlands is the only country with a national monitoring system (Houtman *et al.*, 1998) although various monitoring systems do exist in other countries (Kompier *et al.*, 1994).

Active Transfer of Knowledge

This may include the use of leaflets, research reports, books, conferences, training courses, video, and TV broadcasting. Kompier *et al.* (1994) found that it was the countries with framework legislation that organised the most preventive activities, conducted the most research, developed research networks (formal in Sweden and The Netherlands) and research groups, and embraced the active transfer of knowledge related to work stress.

1.5.3 Organisational Level

Sustainable Organisation Approaches

At an organisational level these strategies may be referred to as 'healthy organisation approaches' (see Kendall *et al.*, 2000) or 'organisational development' approaches. The European Commission Guidance 1998 calls for mainstreaming of stress prevention into organisational development (healthy workers in healthy organisations objectives) to underpin the imperative for more flexible firms, with high skill, high trust and high quality to improve productivity and to create conditions for competitiveness and employment (European Commission 2000).

Recent developments also suggest a whole of organisational approach to performance (bottom-line) analysis at the organisational level. Liukkonen *et al.* (1999) argue that performance analysis should also include a predictive element regarding present and future health of employees, to *reflect the organisation's collective health and capacity* (see Chapter 3 for more detail on this point). In this way companies can be influenced to examine cost–benefit analyses of work environments that lead to unhealthy behaviours as well as low productivity.

Job Redesign

Job redesign includes strategies that aim to prevent the occurrence of work stress. This may include interventions at a national policy level, or organisational level interventions (e.g. job redesign, occupational health and safety strategies, selection processes to improve fit, job enrichment, changing work schedules) or individual level where the intention is to modify the stressor (e.g. personnel policy e.g. training and development; career opportunities; selection; pre-employment medical examination).

Secondary Approaches

These focus on reducing the impact of the stress response before they become too severe. They may include training groups or individuals to be resilient to work stressors, for example through stress inoculation training, fitness programmes, and relaxation training.

Tertiary Approaches

These focus on the amelioration of an identified stress condition. Approaches may include psychotherapy, and post-traumatic assistance programmes (Kompier and Cooper, 1999), case management, injury management or disability management (Kendall *et al.*, 2000).

Stress may be managed by the individual worker in an informal sense or stress management may be part of an overall organisational policy (see van der Klink *et al.*, 2001).

1.5.4   Evaluation of Interventions

Having categorised various forms of interventions what can we say of their efficacy or effectiveness?

Organisational Interventions

Organisational Development and The Healthy Organisation
Increasingly more comprehensive approaches are being employed for stress management such as organisational development and the 'healthy organisation' approach. The approach process aims to improve the workplace by making it more humane and democratic. Cherniss (1991) points to two studies that have successfully employed OD to alleviate burnout (Golembiewski *et al.*, 1987; Hunnicutt and MacMillan, 1983). Both studies involved staff directly in identifying and solving organisational problems that contributed to stress and burnout.

Kompier *et al.* (1998) found in their analysis of 10 natural experiments in the Netherlands using a multiple case study approach that sickness absenteeism reduced and that financial benefits exceeded the costs of the intervention. Kompier and Cooper (1999) found support for the Dutch study in a further analysis of 11 natural experiments of European case studies. They suggested that a combination of organisational and individual interventions may work best especially where workers are in jobs with a low degree of latitude, and they found support for this in their analysis (see also Van der Klink *et al.*, 2001).

The lack of organisational intervention evaluations is highlighted in the Van der Klink *et al.* (2001) study which incorporated only 5 organisational interventions of 48 studies examined (cf. Geurts and Grundemann, 1999). Karasek and Theorell (1990) insist that 'these solutions seem to offer an easy alternative to complex and difficult labor/management negotiations over workplace control. To avoid the more difficult underlying issues is to deal with symptoms instead of the causes, however' (p. 7).

It is likely to be politically expedient to focus at the individual level rather than the organisational level. The impetus for maintaining the status quo is not strictly management-driven. Psychologists, medical practitioners other health professionals may unwittingly reproduce work stress situations by focusing on the individual (perhaps even economically-driven) rather than the organisation (Dollard, 1997).

Job Redesign

Several studies have attempted to redesign central DCS dimensions:
- increasing participation in decision making through increased staff meetings led to increased levels of both decision latitude and social support, and in turn to decreases in role conflict and role ambiguity and finally to reduced levels of emotional strain, job dissatisfaction, absenteeism and intention to leave (Jackson, 1983)
- devolution of control over pace of work, organisation of rest breaks, allocation of work assignments and overtime from supervisors to work teams (Wall and Clegg, 1981), and increased control over work arrival and departures (Pierce and Newstrom, 1983) resulted in consistent benefits (Murphy, 1988)
- changes in decision latitude and/or social support led to improved cardiovascular risk pattern with regard to lipids (Orth-Gomer *et al.*, 1994); decrease in sleep disturbance and gastrointestinal complaints (Wahlstedt and Edling, 1998); and a decrease in the prevalence of pain in the upper spine (see Theorell, 1998 for further detail)

Theorell (1998) notes that a few redesign studies that have focused on improved productivity, such as quality of care, have also had the consequence of improved health and decreased sick leave in employees.

Individual Approaches

Consistent support for the efficacy of cognitive behavioural approaches in the alleviation of stress symptoms have been found in the literature (Bamberg and Busch, 1996) as well as for stress inoculation training, a specific cognitive-behavioural programme (Saunders *et al.*, 1997). However their efficacy is generally measured in terms of stress state reduction rather than modifications of the work environment.

Relative Efficacy and Issues in the Evaluation of Interventions

A recent meta-analysis by Van der Klink *et al.* (2001) of 48 experimental studies (all utilising a no-treatment control group) evaluated the effectiveness of individual and organisational focused interventions on health complaints, psychologic resources and responses and quality of work life, and found a small but significant overall effect for the interventions as a whole, leading to the conclusion that *stress management interventions are effective*. The study found a moderate effect for cognitive-behavioural interventions and multi-modal interventions. A small effect size was found for relaxation and a non-significant effect was found for organisational focused interventions.

The researchers expressed surprise at the lack of support for the effectiveness of the organisational intervention, pointing to recent (uncontrolled) evaluations which did find effects for organisational interventions (Kompier *et al.*, 1998).

The question of whether organisational interventions or individual interventions are better or more effective really depends on the aims and purpose of the intervention. For example where workers are experiencing acute levels of psychological distress, relaxation approaches may be helpful.

Briner and Reynolds (1999) argue against broad brush approaches to intervention: 'employees experience different features of the organisation and the job as more or less pleasant or unpleasant, and that even when they are perceived as equally unpleasant, the effects on employee states and behaviors is likely to vary considerably from employee to employee as a consequence of individual differences and the local context in which the employee is working' (p. 660).

They argue that the choice of intervention will not be simple, and each will have costs, benefits, and limitations. As a general rule they argue that 'the approach to interventions should be evidence based (see Briner, 1997) such that the causal relationships between job conditions and negative employee states and conditions are established before interventions are designed or implemented' (Briner and Reynolds, 1999, p. 661).

In contrast to individual interventions, strictly controlled evaluations of job redesign interventions implied for example by the DCS model (i.e. increasing control or support) are difficult or impossible to arrange in this field according to Theorell (1998). He argues that published studies of job redesign interventions are typically unorthodox as: job interventions sometimes include both individual and organisational components in job redesign; study groups may not be strictly comparable; both experimental and control conditions may have been subject to redesign but to different degrees; and because the follow-up period for evaluation is insufficiently long. This makes it difficult to draw tight conclusions about the successfulness of job redesign approaches as a site of intervention for work stress.

In general there is a plea for best practice randomised, case control studies, but on the other hand there are a number of challenges and difficulties in obtaining gold standard evidence for the efficacy of interventions (see Cordery *et al.*, 1991; Kahn and Byosiere, 1992; Kompier *et al.*, 1994). Problems include:

- a lack of longitudinal studies
- a lack of control groups
- rapid organisational change due to numerous sources not necessarily the intervention
- multifactorial sources of work stress from intra-organisational and extra-organisational
- extra-organisational influences on variables such as local labour market conditions
- a reliance on self-report data
- short-term (individual) outcomes
- a lack of multidisciplinary approaches (e.g. economic aspects)

Kompier and Cooper's (1999) intervention case studies in general involved evidence obtained without a control group or randomisation but with follow-up evaluation (6 cases), a few with evidence obtained from a properly conducted study with a control group but no randomisation (2 cases), and 1 case which utilised a randomised case control group. Interestingly the latter study implemented an individual focused intervention, about which the authors comment 'it is easier (although not easy!) to develop a randomised control condition (for a training program directed at the employee) than for a program which places the emphasis on changing stressful working conditions' (p. 317). As will be seen in Chapter 3 problems in applying orthodoxy to intervention analysis has lead to more flexible action approaches to intervention and evaluation.

1.6 THE BOOK

Within society, the professions have been one of fastest growing sectors of the occupational structure. A profession is defined as a calling, vocation, or form of employment that provides a needed service to society and possesses characteristics of expertise, autonomy, long academic preparation, commitment, and responsibility (Huber, 2000, p. 34). Professionalism involves: (1) a systematic and unique body of knowledge that can be applied to a variety of problems; (2) concerns for the interest of the community and public benefit rather than self; (3) strict control of behaviour through a code of ethics learned in training and established and maintained by professional associations; and (4) a high reward structure conferred through occupational status/prestige/external recognition and high income (Barber, 1963; Haralambos and Holborn, 1991; Osigweh, 1986). The more such characteristics an occupation has and the more developed each of these, the more professionalised the occupation claims to be (Wang, 2002).

Professionalism is based on a diverse set of characteristics, skills levels, knowledge and complexity, membership in an occupational group, or perspectives, but its core values are determined by what society or others ascribe to it (Huber, 2000). According to relevant theories (Salaman, 1974; Turner and Hodge, 1970) a mature profession has a true sense of professional community, and its members are bound together by a sense of their professional identity and a set of shared common values.

These ideas of professionalism are under increasing criticism in society as questions are raised as to whether professions are self-serving rather than function in the interests of providing valuable services to society, whether they primarily service the interests of the wealthy and powerful (e.g. lawyers and accountants), whether professionism is merely a market strategy (restricting access thereby creating market demand), and whether professions are occupational groups that succeed in controlling and manipulating the labour market to maximise its rewards (Weber, 1958). [see Haralambos and Holborn, 1991 for a fuller account of this discussion].

Challenges to the maintenance of modern professions are ongoing such as employment in large organisations where deskilling of professions has occurred (e.g. nurses and teachers) (Braverman, 1974). In the public sector, where jobs are linked to the functions of government (regulating, policing, keeping welfare recipients quiet) professional social workers are experiencing role-conflict and losses of autonomy as they struggle to implement their practice (Oppenheimer, 1973, see also Chapter 11). Threats to wages and fiscal restraint within the public sector also challenges the notion of stable professional careers within the government. Severe reductions in resources to universities and at the same time demands for increased productivity has resulted in commercial imperatives that threaten academic freedom (van Emmerik and Euwema, 2001). Jobs traditionally regarded as stress-free, such as university teaching are becoming increasingly stressful as a result (A.-H. Winefield, 2000 and see Chapter 9 this volume).

The aim of this book is to explore, and describe, the current work context of a range of human service professions. The focus of the book is to contextualise the exploration of the work of the professions within the framework of work stress theory. The book does not particularly intend to argue that conditions at work for human service professions are becoming increasingly stressful, for at any time a range of stressors will impinge on the professions. Rather the idea is to provide a documentary of current tensions within each of the professions at the turn of the century. Each of the chapters will more or less: provide a description of the occupations and their value to society; identify sources of work stress; identify outcomes or effects of work stress in the profession; discuss interventions tried or known and outcomes; provide a theoretical critique of work stress research in the profession; introduce new findings (quantitative and/or qualitative); discuss and provide future research directions.

According to Hasenfeld (1983, p. 1) human service organisations refer to those whose 'principal function is to protect, maintain, or enhance the personal well-being of individuals by defining, shaping, or altering their personal attributes'.

Using this frame and taking into account the features of professionalism described above by Barber, we include in our analysis of *Occupational stress in the service professions* police, teachers, academics, general practitioners, psychologists, clergy, oncology care providers, senior nurses, social and human service workers (Chapters 4, 5, 6, 7, 8, 9, 11, 12, 13).

A divergent occupational position to that of the professional is that of the prostitute. We include the work of prostitutes despite widespread acknowledgement about being the 'oldest profession' and providing a valuable service to society, because the experience of workers in this sector juxtaposed against the 'professions' underscores the privileged position that professions (as opposed to prostitutes) tend to have: high income, social status, high power, and work content legally sanctioned (Chapter 10). Finally, in public sector human service work in particular we see the emergence of large cohorts of volunteers (unpaid workers), arguably as a cost cutting exercise. While the work of the volunteer is valuable, conflict is set up as the professional and volunteer work side by side often performing similar tasks. This threatens the occupational space of the professional, and at the same time leads to unusual stressors for the volunteer. We felt it important in our book to explore therefore occupational stress in the volunteer 'human service professional' (Chapter 14).

As outlined in Section 1.4 there are numerous theoretical frameworks about work stress and despite literary remarks that models could be combined empirically (Kasl, 1998) and theorising about the contextual relation between them has begun (Theorell, 1998), no systematic integrated theoretical model has yet been attempted. This is the project of de Jonge and Dormann in their development of the DISC (Demand-Induced Strain Compensation) Model in Chapter 2. Grounding their model specifically in the extant, emerging and unique issues in human service work, they build a model based on five principles, i.e. (1) multidimensionality of concepts, (2) triple match principle, (3) compensation principle, (4) balance principle, and (5) hierarchical principle. Specifically the model is designed to explain what aspects of jobs may activate psychological compensation processes of job-related strain or balance challenging job demands.

Next, Chapter 3 takes up the important issues of measurement and method in occupational stress research. It notes a renaissance of interest in the issues as researchers, managers and policy makers seek gold standards in the stress management evidence base.

In any research the exploration of local variables in addition to those implied by theory are fully recommended (Theorell, 1998). The book meets this challenge as in essence the chapters on the selection of human service professions canvas both local and contextual issues, as well as theoretical frameworks.

Occupational Stress in the Service Professions sets out to describe and analyse the work context of a range of important human service occupations at the turn of the 21st century. We are delighted to present the following chapters on teachers, oncology care providers, senior nurses, general practitioners, academics, prostitutes, police, psychologists, clergy, social workers and volunteers which provide important insights into the important tasks they perform, specific challenges that they face, and ideas for preventing stress and advancing healthy work.

1.7 ACKNOWLEDGEMENTS

Various aspects of this chapter were also prepared for the National Occupational Health and Safety Commission (NOHSC) Symposium on the OSH Implications of Stress, Melbourne 2001.

I would like to thank Prof. Richard Dienstbier, University of Nebraska, Lincoln for his helpful contribution to the section on stages of the stress response. Discussions with Peta Miller NOHSC and Dr Wendy Macdonald, La Trobe University led to modifications to the Cox *et al.*, 2000b table.

1.8 REFERENCES

Abrahmson, L.Y., Seligman, M.E.P. and Teasdale, J.D., 1978, Learned helplessness in humans. Critique and reformulation. *Journal of Abnormal Psychology*, **87**, pp. 49–74.

Amick, B.C., Kawachi, I., Coakley, E.H., Lerner, D., Levine, S. and Colditz, G.A., 1998, Relationship of job strain and iso-strain to health status in a cohort of women in the United States. *Scandinavian Journal of Work Environment Health*, **24**, pp. 54–61.

Australian National Audit Office, 1997, *The management of occupational stress in Commonwealth employment*. Auditor-General Audit Report No. 8.

Baker, D.B., 1985, The study of stress at work. *Annual Review of Public Health*, **6**, pp. 367–381.

Bamburg, E. and Busch, C., 1996, Employee health improvement by stress management training: A meta-analysis of (quasi-) experimental studies [in German]. *Z Arbeits Organisationpsychol*, **40**, pp. 127–137.

Barber, B., 1963, Some problems in the sociology of the professions. *Daedalus*, **92**, p. XX.

Beehr, T.A., 1989, The themes of social-psychological stress in work organisation: From roles to goals. In *Occupational stress and organizational effectiveness* (New York: Wiley), pp. 71–101.

Beehr, T.A. and Bhagat, R.S., 1985, *Human stress and cognition in organizations: An integrated perspective* (New York: Wiley).

Belkic, K., Schnall, P., Landsbergis, P. and Baker, D., 2000, The workplace and CV health: Conclusions and thoughts for a future agenda. In P.L. Schnall, K. Belkic, P. Landsbergis and D. Baker (eds), *State of the Art Reviews, Occupational Medicine, The workplace and cardiovascular disease* (Philadelphia: Hanley and Belfus, Inc.), Vol. 15, pp. 307–321.

Bohle, P., Quinlan, M. and Mayhew, C., 2001, The health effects of unemployment and job insecurity: An old debate and new evidence. *Economic and Labour Relations Review*, **12**, pp. 32–60.

Bosma, H., Marmot, M.G., Hemingway, H., Nicholson, A.G., Brunner, E. and Stansfeld, A., 1997, Low job control and risk of coronary heart disease in Whitehall II (prospective cohort) study. *British Medical Journal*, **314**, pp. 558–565.

Bousfield, G., 1999, Dutch study to explore workplace stress. *Safety and Health-The International Safety, Health and Environment Magazine*, December.

Braverman, H., 1974, *Labor and monopoly capitalism* (New York: Monthly Review Press).

Breslow, L. and Buell, P., 1960, Mortality from coronary heart disease and physical activity of work in California. *Journal of Chronic Diseases*, **22**, pp. 87–91.

Briner, R.B. and Reynolds, S., 1999, The costs, benefits, and limitations of organizational level stress interventions. *Journal of Organizational Behaviour*, **20**, pp. 647–664.

Briner, R.B., 1997, Improving stress assessment: Towards an evidence-based approach to organisational stress interventions. *Journal of Psychosomatic Research*, **43**, pp. 61–71.

Burisch, M., 1993, In search of theory: Some ruminations on the nature and etiology of burnout. In W.B. Schaufeli, C. Maslach and T. Marek (eds), *Professional burnout: Recent developments in theory and research* (Washington, DC: Taylor and Francis), pp. 75–95.

Calnan, M., Wainwright, D. and Almond, S., 2000, Job strain, Effort–Reward imbalance and mental distress: A study of occupations in general medical practice. *Work and Stress*, **14**, pp. 297–311.

Cameron, I., 1998, Retaining a medical workforce in rural Australia, *The Medical Journal of Australia*, **169**, pp. 293–294.

Caplan, R.D., Cobb, S., French, J.R.P., Harrison, R.V. and Pinneau, S.R. Jr, 1975, *Job demands and worker health* (Washington DC.: H.E.W. Publication No. NIOSH 75-160).

Cascio, W.F., 1995, Whither industrial and organizational psychology in a changing world of work? *American Psychologist*, **50**, pp. 928–939.

Chang, S.J., 2000, Trends and future directions of job stress in Korea. In Kawakami, N. (ed.), *Job stress in East Asia: Exchanging experiences among China, Japan, Korea, Taiwan and Thailand. Proceedings of the First East-Asia Job Stress Meeting, Waseda University International Conference Centre Japan*, pp. 23–26.

Cheng, Y., 2000, Review of present work conditions and research activities on job stress in Taiwan. In Kawakami, N. (ed.), *Job stress in East Asia: Exchanging experiences among China, Japan, Korea, Taiwan and Thailand. Proceedings of the First East-Asia Job Stress Meeting, Waseda University International Conference Centre Japan*, pp. 7–31.

Cherniss, C., 1991, Institutional versus organizational levels of analysis: Commentary on Leiter. Canadian Psychology, **32**, pp. 559–561.

Collings, J. and Murray, P., 1996, Predictors of stress amongst social workers: An empirical study. *British Association of Social Work*, **16**, pp. 375–387.

Cooper, C., 1998 (ed.). *Theories of organizational stress*. (Oxford: Oxford University Press).

Cooper, C., Luikkonen, P. and Cartwright, S., 1996, *Stress prevention in the workplace: Assessing the costs and benefits to organizations*. (Dublin: European Foundation for the Improvement of Living and Working Conditions).

Cooper, C.L. and Marshall, J., 1976, Occupational sources of stress: A review of the literature relating to coronary heart disease and mental ill health. *Journal of Occupational Psychology*, **49**, pp. 11–28.

Cooper, C.L., Dewe, P.J. and O'Driscoll, M.P., 2001, *Organisational stress. A review and critique of theory, research, and applications* (California: Sage).

Cordery, J.L., Mueller, W.S. and Smith, L.M., 1991, Attitudinal and behavioural effects of autonomous group working: A longitudinal field study. *Academy of Management Journal*, **34**, pp. 464–476.

Cox, T. and Griffiths, A., 1995 (in press), The assessment of psychosocial hazards at work. In M. J. Shabracq, J.A.M. Winnubst, and C. Cooper (eds), *Handbook of work health psychology*. (Chichester: Wiley and Sons).

Cox T., Griffiths, A. and Rial-Gonzalez, E., 2000a, *Research on work-related stress* (Belgium: European Agency for Safety and Health at Work).

Cox, T., 1978, *Stress* (London: Macmillan).

Cox, T., Griffiths, A., Barlowe, C., Randall, R., Thomson, L. and Rial-Gonzalez, E., 2000b, *Organisational interventions for work stress: a risk management approach* (Sheffield: HSE Books).

Crotty, M., 1998, *The Foundation of Social Research* (Allen and Unwin: Australia).

Daw, J., 2001, Road rage, air rage and now 'desk rage', *Monitor on Psychology*, **32**, pp. 52–54.

De Jonge, J. and Kompier, M.A.J., 1997, A critical examination of the Demand-Control-Support model from a work psychological perspective. *International Journal of Stress Management*, **4**, pp. 235–258.

Diagnostic and statistical manual of mental disorders: DSM-IV-TR, 4th edition, 2000 (Washington, DC: American Psychiatric Association).

Demerouti, E., Bakker, A., de Jonge, J., Janssen, P.P.M. and Schaufeli, W.B., 2001, Burnout and engagement at work as a function of demands and control. *Scandinavian Journal of Work and Environmental Health*, **27**, pp. 279–286.

Dienstbier, R.A., 1989, Arousal and physiological toughness: Implications for mental and physical health. *Psychological Review*, **96**, pp. 84–100.

Dienstbier, R.A. and Pytlik Zillig, L.M. (in press), Toughness. In C.R. Snyder and S.J. Lopez (eds), *Handbook of Positive Psychology*. (Oxford: Oxford University Press).

Dollard, M.F., 1997, *Work stress: Conceptualisations and Implications for Research Methodology and Workplace Intervention*. (PhD Thesis). Whyalla: Work and Stress Research Group, University of South Australia. ISBN 0-86803-269-7.

Dollard, M.F., 2001, Measurement and Methodological Issues in Work Stress Research. Paper presented at the Stress Related Injury Conference, Butterworths, Sydney.

Dollard, M.F. and Winefield, A.H., 1998, A test of the Demand-Control/Support model of work stress in correctional officers. *Journal of Occupational Health Psychology*, **3**, pp. 1–23.

Dollard, M.F., Winefield, H.R. and Winefield, A.H., 2001, Occupational strain and efficacy in human service workers (Dordrecht: Kluwer Academic Publishers).

Dunnette, M.D., 1998, Emerging trends and vexing issues in industrial and organizational psychology. *Applied Psychology: An International Review*, **47**, pp. 129–153.

Edwards, J.R., 1998, Cybernetic theory of organizational stess. In C. Cooper (ed.), *Theories of organizational stress* (Oxford: Oxford University Press).

Elkin, A.J. and Rosch, P.J., 1990, The person-environment fit approach to stress: Recurring problems and some suggested solutions. *Journal of Organizational Behavior*, **11**, pp. 293–307.

European Commission, *Guidance on work-related stress: Spice of life or kiss of death*, 2000, European Communities, Luxembourg.

Extract from the 'Comparison of workers' compensation arrangements in Australian jurisdictions, 2001, In *NOHSC Symposium on the OHS Implications of Stress*, pp. 139–156.

Frese, M. and Zapf, D., 1988, Methodological issues in the study of work stress: Objective vs subjective measurement of work stress and the question of longitudinal studies. In C.L. Cooper and R. Payne (eds), *Causes, coping and consequences of stress at work* (London: Wiley), pp. 375–411.

Ganster, D.C. and Schaubroeck, J., 1991, Work stress and employee health. *Journal of Management*, **17**, pp. 235–271.

Geurts, S. and Grundemann, R., 1999, Workplace stress and stress prevention in Europe. In M. Kompier and C. Cooper (eds), *Preventing stress, improving productivity: European case studies in the workplace* (London: Routledge), pp. 9–33.

Golembiewski, R.T., Hilles, R. and Daly, R., 1987, Some effects of multiple OD interventions of burnout and work site features. *Journal of Applied Behavioral Science*, **23**, pp. 295–313.

Greenhaus, J.H. and Parasuraman, S., 1987, A work and non-work interactive perspective of stress and its consequences. In J.M. Ivancevich and D.C. Ganster (eds), *Job stress: From theory to suggestion* (New York: Haworth), pp. 37–60.

Haralambos, M. and Holborn, M., 1991, *Sociology: Themes and perspectives*, 3rd edition (London: Collins Educational).

Harris, J.R., 1991, The utility of the transactional approach for occupational stress research. In P.L. Perrewe (ed.), Handbook on job stress [Special Issue]. *Journal of Social Behavior and Personality*, **6**, pp. 21–29.

Hasenfeld, Y., 1983, *Human Service Organizations* (Englewood Cliffs: Prentice Hall).

Heiler, K., 1998, *The 12 hour working day: Emerging issues.* (Working Paper No. 51). Australian Centre for Industrial Relations Research and Training.

Hellerstedt, W.L. and Jeffrey, R.W., 1997, The association of job strain and health behaviours in men and women. *International Journal of Epidemiology*, **26**, pp. 575–583.

Houtman, I.L.D., Goudswaard, A., Dhondt, S., van der Griten, M.P., Hildebrandt, V.H. and ven der Poel, E.G.T., 1998, Dutch monitor on stress and physical load: Risk factors, consequences and preventive action. *Occupational and Environmental Medicine*, **55**, pp. 73–83.

Houtman, I.L.D. and Kompier, M.A.J., 1995, Risk factors and occupational risk groups for work stress in the Netherlands. In S.L. Sauter, and L.R. Murphy (eds), *Organizational risk factor for job stress* (Washington DC: American Psychological Association), pp. 209–226.

Huber, D., 2000, *Leadership and nursing care management* (Philadelphia: W.B. Saunders Company).

Hunnicutt, A.W. and MacMillan, T.F., 1983, Beating burnout: Findings from a three year study. *Journal of Mental Health Administration*, **10**, pp. 7–9.

Hurrell, J.J. and McLaney, 1989, Control, job demands and job satisfaction. In S.L. Sauter, J.J. Hurrell and C.L. Cooper (eds), *Job control and worker health* (Chichester: Wiley), pp. 97–103.

ICD-10 International Classification of mental and behavioural disorders: Diagnostic criteria for research, 1993, (Geneva : World Health Organization).

Jackson, S.E., 1983, Participation in decision making as a strategy for reducing job-related strain. *Journal of Applied Psychology*, **68**, pp. 3–19.

Jamal, M. and Preena, S., 1998, Job stress and employee well-being among airline personnel in an Asian developing country. *International Journal of Stress Management*, **5**, pp. 121–127.

Jayaratne, S., Himle, D.P. and Chess, W. A., 1995, Job satisfaction and burnout: Is there a difference? *Journal of Applied Social Sciences*, **15**, pp. 245–262.

Johnson, J.V. and Hall, E.M., 1988, Job strain, workplace social support and cardiovascular disease: A cross-sectional study of a random sample of Swedish working population. *American Journal of Public Health*, **78**, pp. 1336–1342.

Judy, R.W. and D'Amico, C., 1997, *Workforce 2020: Work and Workers in the 21st Century* (Indianapolis: Hudson Institute).

Kagan, A.R. and Levi. L., 1975, Health and environment-psychsocial stimuli: A review. In L. Levi (ed.), *Society, stress and disease-childhood and adolescence* (Oxford: Oxford University Press), pp. 241–260.

Kahn, R.L. and Byosiere, P., 1992, Stress in organizations. In M.D. Dunnette and L.M. Hough (eds), *Handbook of Industrial and Organizational Psychology* (2nd edition) (CA: Consulting Psychologists Press), pp. 571–650.

Kahn, R., Wolfe, D.M., Quinn, R.P. and Snoek, J.D., 1964, *Organisational stress: Studies in role conflict and ambiguity* (New York: Wiley).

Karasek, R., 1989, Control in the workplace and its health-related aspects. In S.L. Sauter, J.J. Hurrell and Cooper, C.L. (eds), *Job control and worker health* (New York: Wiley) pp. 129–159.

Karasek, R.A., 1979, Job demands, job decision latitude, and mental strain: Implications for job redesign. *Administrative Science Quarterly*, **24**, pp. 285–308.

Karasek, R.A. and Theorell, T., 1990, *Healthy work: Stress, productivity and the reconstruction of working life* (New York: Basic Books).

Karasek, R.A., 1981, Job socialisation and job strain: The implication of two related psychosocial mechanisms for job strain, In B. Gardell, and G. Johnsson (eds), *Man and Working Life* (Chichester: Wiley), pp. 75–94.

Karasek, R.A., Baker, D., Marxer, F., Ahlbom, A. and Theorell, T., 1981, Job decision latitude, job demands, and cardiovascular disease: A prospective study of Swedish men. *American Journal of Public Health*, **71**, pp. 694–705.

Karasek, R.A., Brisson, C., Kawakami, N., Houtman, I., Bongers, P. and Amick, B., 1998, The job content questionnaire (JCQ). An instrument for internationally comparative assessments of psychosical job characteristics. *Journal of Occupational Health Psychology*, **3**, pp. 322–355.

Karasek, R. and Theorell, T., 2000, The demand-control-support model and CVD. In P.L. Schnall, K. Belkic, P. Landsbergis, and D. Baker (eds), *State of the Art Reviews, Occupational Medicine, The workplace and cardiovascular disease* (Philadelphia: Hanley and Belfus, Inc), Vol. 15, pp. 78–83.

Kasl, S.V., 1998, Measuring job stressors and studying the health impact of the work environment: An epidemiologic commentary. *Journal of Occupational Health Psychology*, **3**, pp. 390–401.

Kawakami, N. and Tsutsumi, A., 2000, Job stress in Japan: testing theoretical models and its findings, In Kawakami, N. (ed.), *Job stress in East Asia: Exchanging experiences among China, Japan, Korea, Taiwan and Thailand. Proceedings of the First East-Asia Job Stress Meeting, Waseda University International Conference Centre Japan*, pp. 11–14.

Kawakami, N., 2000, Preface. Job stress in East Asia: Exchanging experiences among China, Japan, Korea, Taiwan and Thailand. In *Proceedings of the First East-Asia Job Stress Meeting, Waseda University International Conference Centre Japan*, pp. 1–2.

Kendall, E., Murphy, P., O'Neill, V. and Bursnall, S., 2000, *Occupational stress: Factors that contribute to its occurrence and effective management*. Centre for Human Services, Griffith University.

Kompier, M. and Cooper, C. (eds), 1999, *Preventing stress, improving productivity: European case studies in the workplace.* (London: Routledge).

Kompier, M., De Gier, E., Smulders, P. and Draaisma, D., 1994, Regulations, policies and practices concerning work stress in five European countries. *Work and Stress*, **8**, pp. 296–318.

Kompier, M.A.J, Geurts, S.A.E., Grudeman, R.W.M., Vink, P. and Smulders, P.G.W., 1998, Cases in stress prevention: The success of a participative and stepwise approach. *Stress Medicine*, **14**, pp. 155–168.

Kristensen, T.S., 1995, The demand-control-support model: Methodological challenges for future research. *Stress Medicine*, **11**, pp. 17–26.

Kristensen, T.S., 1996, Job stress and cardiovascular disease: A theoretical critical review. *Journal of Occupational Health Psychology*, **1**, pp. 246–260.

Landsbergis, P.A., Schnall, P.L., Dietz, D., Friedman, R. and Pickering, T., 1992, The patterning of psychological attributes and distress by 'job strain' and social support in a sample of working men. *Journal of Behavioral Medicine*, **15**, pp. 379–405.

Laungani, P., 1996, Cross-cultural investigation of stress: Conceptual and methodological consideration. *International Journal of Stress Management*, **3**, pp. 25–35.

Lazarus, R.S., 1966, *Psychological stress and the coping process* (New York: McGraw-Hill).

Lazarus, R.S., 1991, Psychological stress in the workplace. In P.L. Perrewe (ed.), Handbook on job stress [Special Issue]. *Journal of Social Behavior and Personality*, **6**, pp. 7–13.

Lazarus, R.S. and Folkman, S., 1984, *Stress, appraisal, and coping* (New York: Springer).

Leiter, M.P. and Maslach, C., 1998, Burnout. In H. Friedman (ed.), *Encyclopedia of Mental Health* (San Diego, CA: Academic Press).

Levi, L., 1990, Occupational stress: Spice of life or kiss of death? *American Psychologist*, **45**, pp. 1142–1145.

Liukkonen, P., Cartwright, S. and Cooper, C., 1999, Costs and benefits of stress prevention in organizations. In M. Kompier and C. Cooper (ed.), *Preventing stress, improving productivity: European case studies in the workplace* (London: Routledge), pp. 33–52.

Lunde-Jensen, P., 1994, The costs of occupational accidents and work related accidents in Nordic countries. *Janus*, **18**, pp. 25–26.

Maslach, C., 1978, The client role in staff burnout. *Journal of Social Issues*, **34**, pp. 111–124.

Maslach, C., 1982, *Burnout: The cost of caring* (Englewood Cliffs, N.J.: Prentice Hall).

Maslach, C., 1998, A multidimensional theory of burnout In C. Cooper (ed.), *Theories of organizational stress* (Oxford: Oxford University Press).

Maslach, C. and Leiter, M.P., 1997, *The truth about burnout* (San Franscisco, CA: Jossey-Bass).

Maslach, C. and Schaufeli, W.B., 1993, Historical and conceptual development of burnout. In W.B. Schaufeli, C. Maslach, and T. Marek (eds), *Professional burnout: Recent developments in theory and research* (New York: Taylor and Francis), pp. 1–18.

Maslach., C. and Jackson, S., 1981, The measurement of experienced burnout. *Journal of Occupational Behaviour*, **2**, pp. 99–113.

Maslach, C. and Jackson, S., 1982, Burnout in health professionals: A social psychological analysis. In G. Sanders and J. Suls (eds), *Social Psychology of Health and Illness* (Hillsdale, NJ: Erlbaum), pp. 227–251.

Mason, J.W., 1974, Specificity in the organization of neuroendocrine response profiles. In P. Seeman and G. Brown (eds), *Frontiers in neurology and neuroscience research* (Toronto: University of Toronto).

McCarty, R., Horwatt, K. and Konarska, M., 1988, Chronic stress and sympathetic-adrenal medullary responsiveness. *Social Science and Medicine*, **26**, pp. 333–341.

Melamed, S., Kushnir, T. and Meir, E.I., 1991, Attenuating the impact of job demands: Additive and interactive effects of perceived control and social support. *Journal of Vocational Behavior*, **39**, pp. 40–53.

Mulhall, A., 1996, Cultural discourse amd the myth of stress in nursing. *International Journal of Nursing Studies*, **33**, pp. 455–468.

Muntaner, C. and O'Campo, P.J., 1993, A critical appraisal of the demand/control model of the psychosocial work environment: Epistemological, social, behavioral and class considerations. *Social Science Medicine*, **36**, pp. 1509–1517.

Murphy, L.R., 1988, Workplace interventions for stress reduction and prevention. In C.L. Cooper and R. Payne (eds), *Causes, coping and consequences of stress at work* (London: Wiley), pp. 301–339.

National Occupational Health and Safety Commission, 1998. *Compendium of Workers' Compensation Statistics, Australia, 1995–96*, Canberra.

National Institute of Occupational Safety and Health: *Stress at work,* 1999, Author, DHHS (NIOSH) Publication No. 99-101, Cincinnati, OH.

O'Leary-Kelly, A.M., Griffin, R.W. and Glew, D.J., 1996, Organiziation-motivated aggression: A research framework. *Academy of Management Review*, **21**, pp. 225–253.

Oppenheimer, M., 1973, The proletarianization of the professional. In P. Halmos (ed.), *Professionalization and social change* (Keele: University of Keele, Monograph 20).

Orth-Gomer, K., Eriksson, I., Moser, V., Theorell, T. and Fredlund, P., 1994, Lipid lowering through stress management. *International Journal of Behavioural Medicine*, **1**, 204–214.

Osigweh, C., 1986, Management and professionalism. *Mid Atlantic Journal of Business*, **24**, pp. 1–20.

Paoli, P., 1997, *Working conditions in Europe: The second European survey on working conditions* (Dublin: European Foundation).

Perrewe, P.L. and Anthony, W.P., 1990, Stress in a steel pipe mill: The impact of job demands, personal control, and employee age on somatic complaints. *Journal of Social Behavior and Personality*, **5**, pp. 77–90.

Pierce, J.L., and Newstrom, J.W., 1983, The design of flexible work schedules and employee responses: relationships and processes. *Journal of Occupational Behavior*, **4**, pp. 247–262.

Quick, J.D., Quick, J.C. and Nelson, D.L., 1998, The theory of preventive stress management in organizations. In C. Cooper (ed.), *Theories of organizational stress* (Oxford: Oxford University Press), pp. 246–268.

Quinlan, M., 2001/2002, Workplace health and safety effects of precarious employment. *The Global Occupational Health Network*, **2**, pp. 1–4.

Quinlan, M., Mayhew, C. and Bohle, P., 2001, The global expansion of precarious employment, work disorganisation and occupational health: A review of recent research. *International Journal of Health Services*, **31**, pp. 335–414.

Salaman, G., 1974, *Community and occupation: An exploration of work/leisure relationships* (London, UK: Cambridge University Press).

Saunders, T., Driskell, J.E., Johnston, J. and Salas, E., 1997, The effects of stress inoculation training on anxiety and performance. *Journal of Occupational Psychology*, **70**, pp. 170–186.

Sauter, S.L., Murphy, L.R. and Hurrell Jr, J.J., 1990, Prevention of work-related psychological disorders. *American Psychologist*, **45**, pp. 1146–1158.

Schabaracq, M.J. and Cooper, C.L., 2000, The changing nature of work and stress. *Journal of Managerial Psychology*, **15**, pp. 227–241.

Schalk, R., Heinen, J. and Freese, C., 2001, Do organizational changes impact the psychological contract and workplace attitudes? A study of a merger of two home care organizations in The Netherlands. In de Jonge, J., Vierick, P., Bussing, A. and Schaufeli, W.B. (eds), *Organizational psychology and health care at the start of a new millennium* (Munchen und Merig, Rainer Hampp Verlag), pp. 23–36.

Schaubroeck, J. and Ganster, D.C., 1993, Chronic demands and responsivity to challenge. *Journal of Applied Psychology*, **78**, pp. 73–85.

Schnall, P.L., Landsbergis, P.A. and Baker, D., 1994, Job strain and cardiovascular disease. *Annual Review of Public Health*, **15**, pp. 381–411.

Schnall, P.L., Belkic, K.L., Landsbergis, P.A., Schwartz, J.E., Baker, D. and Pickering, T.G., 2001, The need for work site surveillance of hypertension, Third International Conference on Work Environment and Cardiovascular Diseases, Duesseldorf.

Selye, H., 1956, *The stress of life* (New York: McGraw-Hill).

Shinn, M., Morch, H., Robinson, P.E. and Neuner, R.A., 1993, Individual, group and agency strategies for coping with job stressors in residential child care programmes. *Journal of Community and Applied Social Psychology*, **3**, pp. 313–324.

Siegrist, J., 1996, Adverse health effects of high-effort/low-reward conditions. *Journal of Occupational Health Psychology,* **1**, pp. 27–41.

Siegrist, J., 1998, Adverse health effects of Effort–Reward imbalance at work. In C. Cooper (ed.), *Theories of organizational stress* (Oxford: Oxford University Press).

Siegrist, J. and Peter, R., 2000, The Effort–Reward imbalance model. In P.L. Schnall, K. Belkic, P. Landsbergis and D. Baker (eds), *State of the Art Reviews, Occupational Medicine, The workplace and cardiovascular disease* (Philadelphia: Hanley and Belfus, Inc), Vol. 15, pp. 83–87.

Soderfeldt, M., Soderfeldt, B. and Warg, L., 1995, Burnout in social work. *Social Work*, **40**, pp. 638–646.

Sokejima, S. and Kagamimori, S., 1998, Working hours as a risk factor for acute myocardial infarction in Japan: Case-control study. *British Medical Journal*, **317**, p. 780.

Spector, P.E., 1987a, Interactive effects of perceived control and job stressors on affective reactions and health outcomes for clerical workers. *Work & Stress*, **1**, pp. 155–162.

Spector, P.E., 1987b, Method variance as an artifact in self-reported affect and perceptions as work: Myth or significant problem? *Journal of Applied Psychology*, **72**, pp. 438–443.

Stansfeld, S.A., Fuhrer, R., Shipley, M.J. and Marmot, M.G., 1999, Work characteristics predict psychiatric disorder: Prospective results from the Whitehall II study. *Occupational and Environmental Medicine*, **56**, pp. 302–307.

Tennant, C., 2001, Liability of psychiatric injury: An evidence based appraisal. Paper presented at the Stress Related Injury Conference, Butterworths, Sydney.

Theorell, T., 1998, Job characteristics in a theoretical and practical health context. In C. Cooper (ed.), *Theories of organizational stress* (Oxford: Oxford University Press), pp. 205–219.

Theorell, T. and Karasek, R.A., 1996, Current issues relating to psychosocial job strain and cardiovascular disease research. *Journal of Occupational Health Psychology*, **1**, pp. 9–26.

Theorell, T., 1999, How to deal with stress in organizations? – a health perspective on theory and practice. *Scandinavian Journal of Work Environment Health*, Special Issue, **25**, pp. 616–624.

Theorell, T., Tsutsumi, A., Hallquist, J., Reuterwall, C., Hogstedt, C., Fredlund, P., Emlund, N., Johnson, J.V. and the SHEEP study group, 1998, Decision latitude, job strain and myocardial infarction: A study of working men in Stockholm. *American Journal of Public Health*, **88**, pp. 382–388.

Toohey, J., 1995, Managing the stress phenomenon at work. In P. Cotton (ed.), *Psychological health in the workplace: Understanding and managing occupational stress* (Victoria: Australian Psychological Society), pp. 31–50.

Townley, G., 2000, Long hours culture causing economy to suffer. *Management Accounting*, **78**, pp. 3–5.

Turner, C. and Hodge, M.N., 1970, Occupations and professions. In J.A. Jackson (ed.), *Professions and professionalization* (London, UK: Cambridge University Press).

Uchino, B.N., Cacioppo, J.T. and Keicolt-Glaser, J.K., 1996, The relationship between social support and physiological processes: A review with emphasis on underlying mechanisms and implications for health. *Psychological Bulletin,* **119**, pp. 488–531.

Van Emmerik, I.J.H. and Euwema, M.C., 2001, At risk of burnout; Gender and faculty differences within academia, In de Jonge, J., Vierick, P., Bussing, A. and Schaufeli, W.B. (eds), *Organizational psychology and health care at the start of a new millennium,* (Munchen und Merig, Rainer Hampp Verlag), pp. 123–137.

Van der Doef, M. and Maes, S., 1999, The job demand-control (-support) model and psychological well-being: A review of 20 years of empirical research, *Work and Stress*, **13**, pp. 87–114.

Van der Klink, J.J.L., Blonk, R.W.B., Schene, A.H. and van Dijk, F.J.H., 2001, The benefits of interventions for work-related stress. *American Journal of Public Health*, **91**, pp. 270–275.

Wahlstedt, K.G.I. and Edling, C., 1998, Organizational changes at a postal sorting terminal-their effects upon work satisfaction, psychosomatic complaints and sick leave. *Work and Stress*, **11**, pp. 279–291.

Warr, P.B., 1987, *Work, unemployment and mental health* (Oxford: Oxford University Press).

Weber, M., 1958, *The protestant ethic and the spirit of capitalisam* (New York: Charles Scribners Sons).

Wilkinson, R.G., 1996, *Unhealthy societies: The afflictions of inequality* (London: Routledge).

Winefield, A.H., 2000, Stress in academe. In D. Kenny, J.G. Carlson, F.J. Mc Guigan and J.L. Sheppard (eds), *Stress and health: Research and clinical applications* (Sydney: Harwood), Chap 23, pp. 437–446.

Winefield, A.H., Montgomery, B., Gault, U., Muller, J., O'Gorman, J., Reser, J. and Roland, D., 2002, *The psychology of work and unemployment in Australia today: An Australian Psychological Society Discussion Paper.* Australian Psychologist, **37**, pp 1–9.

Wall, T.D. and Clegg, C.W., 1981, A longitudinal study of group work redesign. *Journal of Occupational Behavior*, **2**, pp. 31–49.

Wang, X., 2002, Developing a true sense of professional community: An important matter for PM professionalism. *Project Management Journal*, **33**, pp. 5–11.

WHO, 2001, Mental health in Europe (Copenhagen: World Health Organization, Regional Office for Europe).

CHAPTER TWO

The DISC Model: Demand-Induced Strain Compensation Mechanisms in Job Stress

Jan de Jonge and Christian Dormann

2.1 INTRODUCTION

Present-day work is related to particular behavioural (e.g. posture in nursing jobs), cognitive (e.g. complex problem solving in counselling jobs), and emotionally demanding tasks (e.g. aggressive customers in retail jobs) of the employee (Hockey, 2000). Such demanding aspects may lead to (short and long term) psychological and physical dysfunctioning, such as burnout, depression, cardiovascular diseases, and musculoskeletal problems (e.g. Hoogendoorn *et al.*, 2000; Schaufeli and Enzmann, 1998; Schnall *et al.*, 2000). Sickness absenteeism and work disablement may well be the consequences (Allegro and Veerman, 1998; Cooper, 1994).

Globally, job stress is still a major and rising concern in most countries. For example, a recent European survey among 16,000 workers showed that about 30% reported work activities as the main cause of their health problems (Merllié and Paoli, 2001; Paoli, 1997). However, job stress not only represents a societal problem, but also a scientific one; an important question is what researchers have learned in the past decades. For instance, what progress has been made regarding the identification of *causal* relationships between work and health/well being, about *interactive* relations between work-related factors, and about their *specific* relationship with particular health and well being outcomes? (see also Warr, 1994; Zapf *et al.*, 1996). In other words, it is very important for occupational health psychology to answer such questions because the answers jointly determine how work should be (re) designed in a way that is beneficial to both companies and employees. We do know a lot about the relationship between general job stressors and general stress reactions, but specific relations between specific stressors and strains are not yet well elaborated (cf. Cooper *et al.*, 2001; Houkes, 2002). The present chapter tries to gain more insight into the *specific* relationships between work-related psychosocial risks and employee health and well being through the development of a new theoretical framework.

Furthermore, one of today's key developments is a growing service sector, characterised by, among other things, working with clients and information processing. In addition, service workers are facing different and also specific stressors compared to other kinds of workers (such as client demands, aggression, and confrontation with death and dying). Research has also shown that the pace of work in this sector has become increasingly induced by clients. This in turn, may lead to more stress-related problems (Merllié and Paoli, 2001). On the one hand, service theories (Hasenfeldt, 1983, 1992; Zeithaml and Bitner, 2000) try to capture the specific characteristics of work in service organisations. It is argued, for example, that working with and for people ('clients') is more complex than working with inanimate objects. However, these theories do not tell us much about the (adverse) health and well being effects of service jobs (Söderfeldt *et al.*, 1996). On the other hand, many theoretical models within the domain of occupational health psychology are not specifically geared to service work (de Jonge *et al.*, 1999b; Söderfeldt, 1997). Therefore, a second major aim in developing a new model is the applications of such a model to service work in gaining a greater insight into these particular work situations.

The outline of this chapter is as follows. We will define service work and briefly outline the distinct attributes of services in the second section. In the third section we will discuss a general perspective on job stress and associated terms such as job demands and job resources. The fourth section is concerned with two theoretical frameworks which have gained a paradigmatic function in the domain of occupational health psychology: the model of Karasek and his team on the one hand, and the model of Siegrist and colleagues on the other. The strength and weaknesses of these models will be discussed. We will outline the similarities (and differences) between the models and argue that it is remarkable that these frameworks have not been integrated so far. The present chapter ends trying to fill this gap by introducing an integrated model, called the Demand-Induced Strain Compensation (DISC) Model of job stress, applied to service work.

2.2 WHAT ARE SERVICES?

The following definition of services may be useful: 'Services include all economic activities whose output is not a physical product or construction, is generally consumed at the time it is produced, and provides added value in forms (such as convenience, amusement, timeliness, comfort, or health) that are essentially intangible concerns of its first purchaser' (Quinn *et al.*, 1987). Most researchers agree that this definition describes most of the essential features of services (e.g. Parasuraman *et al.*, 1988): (a) intangibility; (b) simultaneous production and consumption; (c) heterogeneous and individualised; (d) perishable; and (e) direct contact between provider and consumer. Each of these terms stands for several aspects (Zeithaml and Bitner, 2000). For instance, intangibility refers to the fact that services cannot be inventoried, patented, and they cannot as easily be displayed as products. Simultaneous production and consumption refers also to the fact that customers co-produce the service, that customers and employees mutually affect each other and, thus, that mass production is not really possible. By heterogeneous services it is meant that the outcomes of services (i.e. customers' experiences) depend on the actions taken by the employee and on many other factors organisations cannot fully control. Perishability refers to the impossibility to synchronise supply and demand in services and that services cannot be resold or returned ('Wouldn't it be nice if a bad haircut could be returned or resold to another consumer?', Zeithaml and Bitner, 2000, p. 14). And, finally, services require direct contact between employees and customers, though this directness may vary from strong to weak.

2.2.1 What are service jobs?

From a psychological point of view, even a comprehensive list of businesses does not help much in understanding what makes service jobs unique to employees when compared to non-service jobs. Rather, it is more promising to look at different tasks carried out by service employees.

One important distinction between service tasks is whether they are direct person-related jobs or indirect person-related jobs (Mills, 1986). In direct person-related jobs the primary task is – in one way or another – to 'modify' the clients physically or psychologically, for example, as in the case of counsellors, nurses, social workers, and teachers. Hasenfeldt (1983, 1992) has called these kinds of jobs 'human service jobs'. According to him, service jobs are characterised by two main properties. First, human service professionals work directly with people whose well being should be protected, maintained, or enhanced by defining, shaping or altering their personal attributes. People are, in a sense, their raw material. Second, human service professionals are mandated to protect and to promote the welfare of their clients. Other service jobs are only indirect person-related and they represent most of the service jobs, for example, retail employees (e.g. Nerdinger, 1994). Nevertheless, even employees in indirect person-related *jobs* have to perform direct person-related *tasks*. Consequently, service jobs should be distinguished on the basis of the amount of interaction work done. This may be face-to-face or voice-to-voice such as in call centers (Dormann *et al.*, in press). When dealing with the issue of stress in the service sector, we, thus, have to focus on the tasks service employees carry out. Focussing on broader concepts such as service industries or service jobs does not provide a useful framework in which to investigate differences between services such as preparing dishes in a kitchen or providing people with food by working a farm.

2.2.2 What are particular demands arising in service jobs?

The experiences customers want to have, impose a variety of cognitive, emotional, and behavioural (physical) demands on service employees. These demands are imposed on the service employee through the way customers behave and how they express their desires and wishes. For example, cognitive demands include solving customer's problems fast and without errors. A principal cognitive problem exists when the goals to be reached are not clear. In stress research this is referred to as role ambiguity. What is unique for services is that service employees may not only be exposed to role ambiguity due to unclear expectations of their supervisor, but also to a second source of role ambiguity, which is related to their customers (e.g. Dormann and Zapf, 2001). We will elaborate more on these and further aspects in the following sections. In addition, although several demands imposed on the employee by their organisation may not be unique to service tasks *per se* (e.g. time pressure), a unique aspect in services is that these demands may be opposite to the demands stemming from the customers (e.g. empathy, which requires time). Consequently, service employees are often confronted with conflicting demands, which may lead to conflicts in their professional roles.

2.2.3   Further characteristics of service jobs: job strains and job resources

Service employees may also respond to their jobs in a different manner than employees in other jobs. Generally, these responses are called stress reactions or strains (Koslowski, 1998). For instance, professional burnout can be considered an important job-related strain in this area (Schaufeli *et al.*, 1993). A prominent facet of the burnout concept is called depersonalisation (Maslach, 1998). This means that a service employee's attitude towards his or her customers is similar to an attitude one has towards inanimate objects (e.g. when a surgeon removes the appendix from patient #7 rather than from Ms Treloar). By definition, depersonalisation of customers cannot occur if employees do not work with customers.

We will argue later that service jobs can not only be characterised by particular demands and particular responses to the tasks, but also by certain job-related resources. We define job-related resources as situational characteristics that facilitate job-related goal attainment. Like job demands, a variety of cognitive, emotional, and behavioural (physical) job resources can be distinguished in service work, such as timing control, emotional control, and instrumental support from colleagues, respectively. For example, timing control includes the freedom to do a complex task when one feels able to concentrate on it, emotion control includes freedom to display emotions in a manner that is consistent with one's personality, and instrumental support includes direct help when one feels unable to accomplish a task alone.

What makes service jobs distinct is that task goals may vary from customer to customer. Customers' individual aims and wishes may demand certain kinds of cognitive, emotional and behavioural regulatory processes that are not often found in other jobs. Thus, a much greater variety and latitude of job resources are required to cope with the demands than compared to other kinds of jobs. Such additional, different, job-related resources may be, for example, emotional control (Zapf *et al.*, 2001) or outcome control (Söderfeldt *et al.*, 2001).

2.3   WHAT IS JOB STRESS?

It is virtually impossible to write about job stress without observing that there is little agreement as to how 'stress' should be defined (Hart and Cooper, 2001; Kahn and Byosiere, 1992; Koslowski, 1998). Several authors consider it impossible or pointless to define job stress and hold the view that the concept is no more than a general term denoting a certain field of research (Buunk *et al.*, 1998). However, following the perspective of Lazarus (e.g. Lazarus, 1993), we would like to suggest that the confusion about the definition of job stress can be avoided by taking the view that, in general, job stress primarily refers to the occurrence of *negative emotions* which are evoked by demanding situations in the workplace (Gaillard and Wientjes, 1994).

Examples of such negative emotions are fear, anger, guilt, shame, grief, envy and jealousy (Buunk *et al.*, 1998). Whereas emotions are defined as relatively short-lived affective reactions that have a specific object as a reference (e.g. fearing aggressive clients, anger about computer break down etc.) moods are more enduring states (e.g. from hours sometimes up to days) without such referents (Ekman and Davidson, 1994; Fridja, 1993). Emotions and moods have a physiological (e.g. blood pressure), behavioural (e.g. facial expression) and a feeling component. The latter is also referred to as affect (e.g. Larsen, 2000).

Negative emotions should therefore be regarded as the crucial dependent variables in the job stress process. However, the call for consideration of emotions as reactions to the work itself and its environmental conditions (Gaillard and Wientjes, 1994; Lazarus, 1993) has only resulted in a very few studies (e.g. Brief *et al.*, 1998; Fisher, 2000; Temme and Zapf, 1997; Weiss *et al.*, 1999). On the other hand, research on moods as dependent variables has been pervasive in job stress research from its early beginning. What is known, for example, is that negative moods lead to more negative emotions, and several processes whereby this happens have also been identified (Larsen, 2000). However, far more interesting for researchers concerned with job stress is to know whether emotions lead to more enduring mood states. This is very important for stress researchers because most knowledge on stress reactions is based on studies using moods or mood-laden concepts rather than emotions as dependent variables. Although it may be the case that a high density of negative emotions over time may result in a negative mood, which persists much longer than the last emotional reaction, the empirical knowledge justifying this assumption is not very sound (Ekman, 1994). For example, it is well known that positive emotions and negative emotions do not simultaneously occur and are thus, almost perfectly negatively correlated (e.g. Diener and Emmons, 1984; Fisher, 2000). Positive and negative moods, however, are usually uncorrelated (e.g. Diener and Emmons, 1984; Diener *et al.*, 1995; Watson and Tellegen, 1985). Thus, there remains much research to be done.

Although studies and theorizing about emotions at work have just begun to accumulate, mood-related concepts have been worked out more clearly. Warr (1987, 1994), for instance, conceptualised mental health and well being at work primarily in terms of affective states that are encountered on the job, like anxiety, depression and displeasure. Furthermore, negative moods are often accompanied by physiological changes, such as increased secretion of particular hormones and increased heart rate. Employee health and well being can generally be described as optimal physical, psychological and social functioning of the (service) worker. We will expand on this later in this chapter when we describe the DISC Model.

Broadly speaking, employee health and well being may be related to two core aspects of the job: job demands and job resources (Frese and Zapf, 1994; see also Hobfoll, 1989, 1998, 2001). First, the term 'job demands' is often used in job stress research, referring particularly to aspects of the work itself (Koslowski,

1998). According to Jones and Fletcher (1996), job demands refer to the degree to which the working environment contains stimuli that require some effort. In other words, job demands are the things that have to be done, within a particular time, and so on. Job demands lead to negative consequences if they require additional effort beyond the usual way of achieving the work goals (see also Demerouti *et al.*, 2001a). Semmer (1996) has referred to these effects as obstacles to task fulfilment. The additional effort required may result in additional cognitive, emotional, and/or behavioural activity. Generally, if job demands require additional effort to accomplish one's work goals, they are likely to elicit negative emotional responses and negative moods like anxiety or depression. On the other hand, however, job demands can be good in the right circumstances: they can be challenging, provide stimulation and growth and stimulate the utilisation of a worker's skills and abilities (Jones and Fletcher, 1996; Warr, 1987). We will discuss what these right circumstances could be in Section 2.5. In one sense, all job demands include some type of additional effort placed on the employee.

Second, 'job resources' refer to different aspects of the work itself, which can be used by the worker:

1. to reduce job demands directly or indirectly (i.e. either avoiding or combating demands) and the associated physiological, psychological and social costs (Antonovsky, 1987; Demerouti *et al.*, 2001a; Karasek, 1979);
2. to achieve personal and/or work goals (Demerouti *et al.*, 2001a; Diener and Fujita, 1995; Hackman and Oldham, 1980);
3. stimulate personal growth and skill development (Demerouti *et al.*, 2001a; Karasek, 1979).

According to Hobfoll (1989, 2001), job resources are necessary for understanding job stress. Resources can be broadly conceptualised as a kind of energetic reservoir that is tapped when the individual has to respond to environmental stimuli (like job demands). However, although the resources 'power up' these responses, individuals – when confronted with job stress – strive to minimise net loss of resources. In addition, when workers are not confronted with job stressors, they strive to develop resource surpluses in order to offset the possibility of future loss ('energy accumulating behaviour'; Hobfoll, 1989).

There are two important issues with regard to job resources that are worth mentioning. It is well known that *availability* of resources and *the use* of resources should be distinguished, but differential effects associated with this distinction have rarely been investigated. For example, it may have detrimental effects on health if individuals are aware that job autonomy is unavailable to them, even though there may be no need to make any decision or change the way things are going (i.e. a 'golden cage'). On the other hand, the mere availability of resources may not prevent individuals from developing health problems. For instance, in the case of high time pressure, one may have to use one's decision autonomy to postpone certain tasks. Without doing so, the demands may lead to overtaxation and, by this, to health problems. As Frese and Zapf (1994) argued, however, using resources should be considered as an *extra effort* necessary for task accomplishment, which, by definition, represents a demand (stressor in their terminology) that causes strain. In instances where resources are used, a second issue is worth consideration, which has not yet received much attention: resources may be more or less depleted. We believe that *transience* (i.e. the degree to which a job resource is depleted through usage) is a further characteristic which is useful to distinguish different kinds of job resources. For instance, having knowledgeable colleagues who are always willing to share their knowledge means that informational support represent a *perpetual* (rather than transient) source of informational support. To provide another example, the reader may think of an individual using his or her decision autonomy to postpone certain tasks. This resource is likely to be transient because it is unrealistic to assume that job autonomy goes far enough to postpone *all* tasks. Hobfoll (1989, 2001) argued that depletion of resources is stressful in itself. On the one hand, thus, job resources may help to compensate for job demands, implying that an interaction between demands and resources causes strain. On the other hand, using these resources and depleting them represents a new demand. In the latter case, the positive (i.e. compensation) and negative effects (i.e. representing an additional demand) of using job resources may sometimes counteract, so that an interaction effect is unlikely to be observed.

Furthermore, job resources could have different psychological functions according to their primary source (see also de Jonge *et al.*, 2000a; Herzberg, 1966). This primary source could be split up into two factors, i.e. (1) job context resources such as salary, and (2) job content resources like job autonomy. In general, we believe that job context resources are particularly likely to relate to negative aspects of health and well being (e.g. strain such as negative emotions and/or mood, irritation, and depression), whereas job content resources are particularly likely to relate to positive aspects of health and well being (e.g. positive emotions and/or mood, active learning, motivation and growth).

To recapitulate, job demands refer to those aspects of the job which require sustained behavioural, physical, cognitive and/or emotional effort. They can be positive in the right circumstances (e.g. increase activation and utilise abilities), but can also elicit negative emotional responses and consequently drain energy. Job resources refer to those aspects of the job which can: (1) reduce job demands and related efforts and/or costs (direct or indirect); (2) help to achieve personal and/or work goals; and (3) stimulate personal growth and development (Demerouti *et al.*, 2001a). People not only strive to retain resources and minimise losses when confronted with stress, but also try to save them in times of low job stress (Hobfoll, 1989, 2001). Finally, the presence or absence of particular job resources due to their primary source (i.e. context or content) may explain particular health outcomes as well.

2.4   Job stress and employee health/Well being: two theoretical frameworks

In the last two decades, two theoretical frameworks have been quite successful in generating and guiding job stress research: the Demand-Control-Support Model originally proposed by Karasek (1979) and supplemented by Johnson (1986), and the Effort–Reward Imbalance Model of Siegrist *et al.* (1986). Particularly the Demand-Control-Support Model has served as a theoretical basis for much of the empirical research on job stress, whereas the ERI Model still has to prove itself as a major model. Nevertheless, the latter model offers new explanations and predictions, and thus provides new theoretical insights and heuristic value.

The original Demand-Control Model (DCM) (e.g. Karasek, 1979, 1998; Karasek and Theorell, 1990), which did not include the aspect of social support, is a dynamic, situation-centered model. This model considers two psychosocial job characteristics to be important determinants of the (adverse) health of workers: (1) 'psychological job demands', and (2) 'job decision latitude'. Karasek defines psychological job demands as psychological stressors present in the work environment (e.g. high pressure of time, high working pace, difficult and mentally exacting work). The term 'job decision latitude' comprises two theoretically distinct constructs: (1) the worker's authority to make decisions on the job ('decision authority', or 'autonomy'), and (2) the breadth of skills used by the worker on the job ('skill discretion'; Karasek and Theorell, 1990). The model postulates that the interaction of two key job characteristics, psychological job demands and job decision latitude, is important in the prediction of employee health (including work motivation and learning behaviour), in addition to the unique effects of both concepts. Accordingly, the DCM has two main hypotheses. The first hypothesis of the model assumes that adverse health (like exhaustion and fatigue) will occur when job demands are excessively high and job decision latitude is low. The second hypothesis predicts that motivation, active learning behaviour and personal growth will occur when job demands are high (but not overwhelming) and as well, as job decision latitude is high. Although the active job is demanding, it also is combining freedom to use skills with a lot of autonomy. The role of workplace social support in the model was theoretically elaborated by Johnson and Hall (1988). The extended Demand-Control-Support Model predicts that the most unfavourable health effects are expected for a combination of high demands, low decision latitude, and low social support.

There is much empirical support for the DCM showing that the combination of (high) job demands and (low) job decision latitude is an important predictor for poor psychological health (including burnout) and illness. Epidemiological studies examining cardiovascular diseases offer the most support for the model (Schnall *et al.*, 1994). However true interactions (demands * control, or demands * control * support) have not been found consistently. Furthermore, the active learning hypothesis has rarely been examined; that is, the majority of empirical studies investigated health-related dependent variables and did not consider variables such as motivation, learning behaviour, or challenge (de Jonge and Kompier, 1997; Karasek, 1998). Beehr and colleagues (2001) have argued that testing a true, multiplicative, interaction between demand and control is necessary to validate the Demand-Control theory. They further state '… if main effects are all that constitute the theory, then demands and lack of control are simply a set of independent stressors with no necessary relationship to each other' (p. 117). In addition, in testing a three-way active learning hypothesis (i.e. demand * control * support), Schaubroeck and Fink (1998) showed that the combination of high demands with high control plus low support, or with low control plus high support, causes health problems because one key ingredient for successful coping is lacking. In such cases high demands are likely to produce failure to cope effectively. Since failures occur despite the presence of one (and only one) key ingredient of potential successful coping, unfortunately, failures to cope successfully are usually attributed internally (Schaubroeck and Fink, 1998). Finally, the model has not been adapted to (human) service jobs thus far, although several studies exist which reported tests of the DCM in this area with consistent results (for an overview, see de Jonge, 1995; Söderfeldt, 1997).

Despite these conclusions, the model does predict both health and motivational outcomes, and several reviews emphasise that the model is appropriate for further theoretical and empirical examination (de Jonge and Kompier, 1997; Theorell and Karasek, 1996; van der Doef and Maes, 1998, 1999). However, what makes the DCM a unique and interesting theoretical model is the interactive effect on job-related strain particularly (Beehr *et al.*, 2001).

An alternative theoretical model, the Effort–Reward Imbalance (ERI) Model (Siegrist, 1996, 1998; Siegrist *et al.*, 1986) puts its emphasis on the reward rather than the control structure of work (Marmot *et al.*, 1999). The focus of the ERI Model is on the centrality of paid employment in adult life. In this model, two key characteristics are important, that is, (1) efforts at work, and (2) occupational rewards. Work-related efforts refer to the individual's effort to cope with potentially negative external stimuli like workload and physical exertion (i.e. additional effort). The formulation of effort in the model considers also an intrinsic, personal characteristic called 'overcommitment', which corresponds to the individual's own drive to fulfil his or her expectations in combination with a strong desire of being approved and esteemed. Occupational rewards include concerns about salary and fringe benefits, as well as job security, promotion prospects, self-esteem, respect, and support. The ERI Model claims that lack of reciprocity between 'costs' and 'gains' (i.e. high effort/low reward conditions) causes a state of emotional distress which can lead to cardiovascular risks and other strain reactions (like poor subjective health and sickness absence). Having a demanding, but unstable job, performing at a high level without being offered any promotion aspects, are examples of stressful imbalance. Finally, it is claimed that the specific combination of extrinsic and intrinsic efforts provides a more sensitive indicator of job stress than the DCM, which only considers situation-centred characteristics (Calnan *et al.*, 2000). For instance, people characterised by the intrinsic effort 'overcommitment' tend to exaggerate their extrinsic efforts and to underestimate their rewards, leading to a more stressful imbalance and worser health in the long run (de Jonge *et al.*, 2000b).

The number of published empirical studies with the ERI Model are growing rapidly, and the combination of high effort and low reward at work was found to be a risk factor for cardiovascular health, subjective health (including burnout), mild psychiatric disorders, and reported symptoms (e.g. Bosma *et al.*, 1998; de Jonge *et al.*, 2000b; Peter *et al.*, 1998a, 1998b; Stansfeld *et al.*, 1998). There is also some evidence for the moderating effect of overcommitment (i.e. overcommited employees report stronger adverse health effects due to an effort-reward imbalance; e.g. de Jonge *et al.*, 2000b; Siegrist and Peter, 1994), though there are also studies that did not find moderating effects of overcommitment (e.g. de Jonge and Hamers, 2000) or even report opposite results (e.g. van Vegchel *et al.*, 2001a). Finally, similar to the DCM, the model has not been adapted to (human) service jobs thus far, although several studies have tested the model in human service work (e.g. Bakker *et al.*, 2000; de Jonge and Hamers, 2000; van Vegchel *et al.*, 2001a).

2.5   AN INTEGRATED VIEW: THE DEMAND-INDUCED STRAIN COMPENSATION (DISC) MODEL OF JOB STRESS

From the theoretical frameworks discussed above and from many similar ones in the field of job stress, it is difficult to decide which framework is relevant in a particular setting (e.g. human service work). As each framework has its adherents and opponents, and many studies have both supported and contradicted hypotheses, the choice is even more difficult (Koslowski, 1998). Over the past ten years, several researchers have critically examined these two theoretical frameworks (e.g. de Jonge and Kompier, 1997; Le Blanc *et al.*, 2000; Koslowski, 1998; Kristensen, 1995; Schwartz *et al.*, 1996; Terry and Jimmieson, 1999; van der Doef and Maes, 1998, 1999). Criticisms stem from both a theoretical and a methodological point of view. The most important caveats will be presented here, which have led to an initial development of a new integrated view merely based upon the conceptual and methodological limitations of both models. To put it differently, the shortcomings of current theoretical frameworks have been translated into new ideas, concepts and assumptions, trying to combine the different perspectives that have been offered so far. Moreover, the integrated model pays more attention to the particular job demands, job resources, and health outcomes arising in service jobs.

Generally speaking, both the DCM and the ERI Model offer a basic guide to predict job-related strain. They are both balance, or better, compensation models on a basic level, but they are global and general in stating their disbalance and/or balance principle. More specifically, no particular (psychological) mechanism of how (specific) job demands should be compensated by (specific) job resources has been described or explained. In addition, no particular mechanism of how a well-balanced mixture of particular job demands and matching job resources leads to active learning has been explained in detail. Finally, the frameworks are currently not adapted to organisations like (human) services because the peculiarities of client work are not considered (Söderfeldt, 1997). This is surprising as a growing number of people are working in service jobs (with clients or patients; Dollard *et al.*, 2001), which seem to be increasingly demanding (Merllié and Paoli, 2001).

A first comment refers to the concept and measurement of the key constructs of both the DCM and the ERI Model. On one hand, there are authors who have adopted an integrated view in which there are very close ties between theory and the measurement of the concepts of the DCM (e.g. Beehr *et al.*, 2001); testing the theory should only be valid if the original scales are used. On the other hand, we and various authors are of the opinion that the conceptualisation of key constructs in both models is theoretically not very clear, and that the corresponding scales do not specifically reflect the supposed concepts. It is argued that although the theory is convincing, the scales are too global in measurement, and that different job characteristics are lumped together in the respective scales (e.g. Cooper *et al.*, 2001; de Jonge and Kompier, 1997; van Vegchel *et al.*, 2001a; Wall *et al.*, 1996). Thus, their use and specific value in today's organisations (and particularly in service jobs) is at least doubtful. For instance, Wall *et al.* (1996) reiterated a common objection of the scales developed by Karasek (1985) because his resource-scale is a compound of job decision latitude (resembling autonomy) and skill discretion (resembling learning opportunities). A compounded scale like Karasek's one does not provide a precise measure of job autonomy. This makes it unlikely that the proposed interaction effects (demands * autonomy) can be established in statistical analysis (de Jonge *et al.*, 2000c). Wall *et al.* (1996) showed clearly that support for the DCM is greater when a focused measure of job control is used (see also van der Doef and Maes, 1999).

A second point of attention is that issues of optimal match between job characteristics (e.g. why decision latitude should buffer psychological job demands better than, for instance, social support) are not elaborated in either model. For example, it has been suggested that job autonomy should be further differentiated into several aspects that are relatively independent (e.g. Breaugh, 1985; de Jonge, 1995; Frese, 1989; Sargent and Terry, 1998). Among others, these include control over scheduling (e.g. 'I can decide for myself when to do what kind of work') and control over work methods (e.g. 'I can decide the way I accomplish my tasks'), and control over interactions with clients (e.g. 'I can decide when I stop an interaction with a client'). These different aspects of control are thought to moderate different kinds of demands such as time pressure, obstacles at work, and emotionally demanding clients, respectively. Results from Sargent and Terry (1998) revealed some support for this assumption, suggesting that job demands were moderated by task control but not by other (more peripheral) aspects of control. Cooper *et al.* (2001) concluded in their Demand-Control review that 'Evidence to date shows some support for the Karasek model, particularly when salient job demands and areas of control are clearly defined and are *matched* with each other.' (p. 140; italics ours).

The above-mentioned two criticisms apply in principle to the scales used to test the ERI model and the buffering hypothesis of social support. For instance, the effort-scale of the ERI model (Siegrist and Peter, 1997) consists of a variety of items with a different content, such as physical load, time pressure, responsibility, and working overtime. This makes it unclear which 'effort' is important for the effort–reward imbalance in the prediction of adverse health outcomes (cf. van Vegchel *et al.*, 2001a). In addition, some authors have omitted the item related to physical demands from surveys of white-collar workers (e.g. Calnan *et al.*, 2000).

Third, though the focus of the two models is different (content versus context resources), similarities in underlying psychological processes exist, but have not been emphasised. The job resources in both models can be viewed as factors fostering self-regulatory processes (Carver and Scheier, 1998). Whereas rewards in the ERI Model are particularly thought to support emotional self-regulation via increased self-esteem or self-efficacy (Siegrist, 1996), the job resources in the DCM are more concerned with the cognitive regulation of the task. Social support may have both functions plus an additional one because it is often viewed as a three-dimensional concept; it may comprise instrumental support (fostering behavioural regulation), informational support (fostering cognitive regulation), or emotional support (fostering emotional regulation; House, 1981). Thus, the DCM and the ERI Model both allow basically for emotional self-regulatory processes.

A fourth issue is the indefinite number of outcome variables used in testing both models. Although both the DCM and the ERI Model were initially designed to predict physiological outcomes to a large extent, a rationale for this restriction was not clearly provided and subsequent analysis have employed a great variety of health-related variables. We believe that a theoretically sound explanation why certain health/well being outcomes should be predicted by either model has not been provided thus far. However, the prediction of cardiovascular diseases in the ERI Model can be considered a notable exception in this respect (Siegrist, 1996). Finally, the interplay between positive and negative outcomes (and underlying mechanisms) is not well elaborated. While some level of job demands seems to be necessary to achieve new learning and growth, a too high level is obviously harmful. This led us to the notion of an optimal level of job demands as proposed by Selye (1936) and Warr (1987).

2.5.1   The DISC Model: Five principles

The basic aim of our new model, called Demand-Induced Strain Compensation (DISC) Model, is to integrate the models mentioned before by unification of common principles and separation of different foci. In addition, the applicability to human service jobs will be discussed. In general, the DISC Model rests on regulatory processes, which explain how humans compensate for poor psychological states in terms of strains (e.g. Vancouver, 2000). In addition, the DISC Model will also predict constructive reinforcement of behaviour patterns ('active learning'). Finally, the DISC Model is particularly designed to explain what aspects of jobs may activate psychological compensation processes of strains, or balance challenging job demands. Thus far, five preliminary principles of the DISC Model are:

1. **Multidimensionality of concepts:** Job demands, job resources, and job-related strains (and concepts concerned with positive well being) are not unidimensional concepts. At a very basic level they may comprise of behavioural, cognitive, and/or emotional components.

2. **Triple match principle:** Job demands and job resources on the one hand and (adverse) health/well being outcomes on the other will be as closer related as qualitatively similar processes connect them. In other words, (1) job demands should match job resources, (2) job demands should match (adverse) health/well being outcomes, and (3) job resources should match (adverse) health/well being outcomes (see Figure 2.1). Specifically, referring to the first principle, cognitive demands and cognitive resources are more likely to affect cognitive forms of strains, emotional demands and emotional resources are more likely to affect emotional forms of strains, and, finally, behavioural demands and behavioural resources are more likely to affect behavioural (physical) forms of strains.

3. **Compensation principle:** Not only will qualitatively similar processes and the involved concepts positively relate to each other, but depending on their quantitative aspects (e.g. hedonic tone, activation level, attention), the negative effects of job demands can be compensated by activation of antagonistic processes promoted by matching job resources. Poor health and poor well being may diminish once a compensating matching job resource is available.

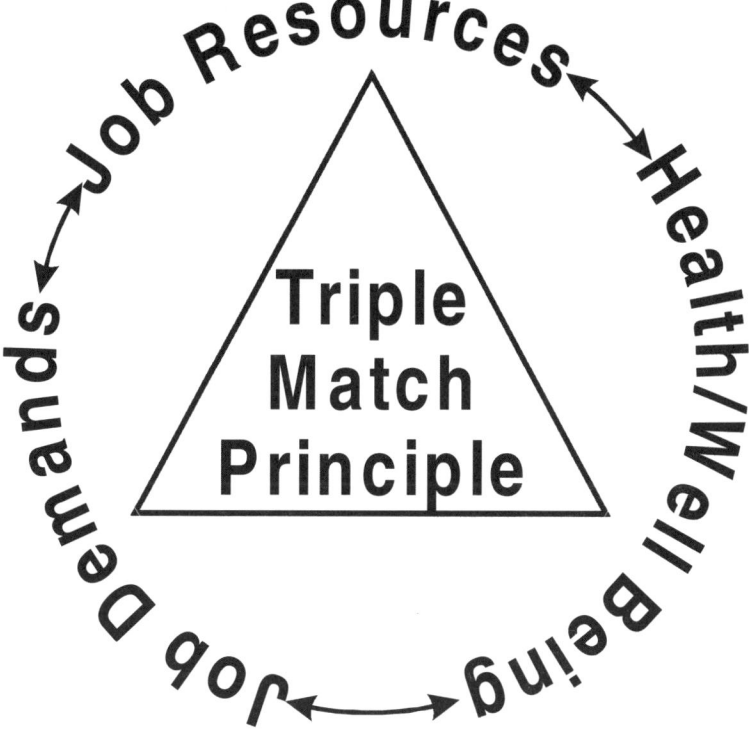

Figure 2.1 The Triple Match Principle (TMP) of the DISC Model.

4. **Balance principle:** The DISC Model suggests that only through discerning usage of limited job resources may active learning and growth occur. Resources usage is elicited by job demands. Excessive usage of resources may lead to their depletion, which causes strain and thereby hampers learning and growth. Therefore, there should be a well-balanced mixture of particular high (but not overwhelming) job demands and their corresponding antagonists (i.e. matching job resources) to engage active learning and growth.

5. **Hierarchical principle:** Besides the strong relations that are proposed to exist between concepts founded on qualitatively similar processes, emotions play a central role in the processing. Behavioural, cognitive, and emotional demands may all elicit emotional responses, and virtually all empirically investigated strains include an emotional component.

In the following, we elaborate more on these five principles. The first principle is concerned with the **multidimensionality** of job demands, job resources, and job-related strains. Although this seems to be a trivial notion, we believe that previous studies and models ignoring the multidimensional nature provided results that are too complex to highlight the main principles underlying work stress. For example, a simple measure such as time pressure may have a cognitive component (e.g. because it is almost always likely to be accompanied by concentration demands), a behavioural component (e.g. fast and risky movements), and an emotional component (e.g. fear of being pressed, anger about the supervisor who always pushes). Similarly, job resources may also consist of all three dimensions. For instance, social support is well known to have a cognitive-informational component (e.g. colleagues providing necessary information to answer a patient's request), a behavioural component (e.g. direct helping when the patient is heavy and has to be lifted up), and an emotional component (e.g. colleagues providing sympathy/re-assurance after a verbal attack of a patient; Dormann and Zapf, 1999; House, 1981). Finally, job-related strain variables might be expressed in different ways and can be roughly categorised in at least three dimensions: (1) cognitive, (2) behavioural (physical), and (3) emotional (Koslowski, 1998; Le Blanc *et al.*, 2000). This suggests that the strain dimensions should correspond to the demands and resources dimensions to a large extent. However, strain variables are also complex in themselves. Depression, for example, also comprises cognitive components (e.g. difficulties in making decisions), behavioural components (e.g. problems in standing up in the morning), and emotional components (e.g. sadness; American Psychiatric Association, 1994). We admit that it is difficult to disentangle these three components in research practice, but the *primary* content of such a measure (i.e. cognitive, behavioural, or emotional) may be more easily identified.

The second principle is what we call the **Triple Match Principle** (TMP; see also Cohen and Wills, 1985; Frese, 1999; Sargent and Terry, 1998). The TMP proposes that strong relations among the concepts should be observed if they are based on qualitatively identical processes. In fact, it is far from trivial to argue that, for example, behavioural demands (e.g. physical workload in terms of kilograms to be lifted per hour) should lead to emotional strain (e.g. emotional exhaustion). Nevertheless, the literature is full of studies aiming to demonstrate cross-domain relations, and, with few exceptions, stress researchers have not yet proposed specific cause-effect relations (e.g. Broadbent, 1985; Dormann and Zapf, 2002; Frese, 1977). For example, emotional exhaustion has only very rarely been related to emotional demands, but heading the list of empirical research on emotional exhaustion are cognitive and behavioural demands such as role conflict and time pressure (Schaufeli and Enzman, 1998). A consistent theoretical explanation for looking at cross-domain relations, however, has not yet been provided. We do not deny that such relations exist, and below we will provide a psychological explanation as to why they occur. Nevertheless, we believe that stress researchers have to take into account that their concepts may involve qualitatively distinct processes. Even if cross-domain relations exist, we believe a good theory should explicitly explain why cross-domain effects are worth investigating. Such a theory is likely to be complex (as the respective part of our theory shows up to be, too), and the simplicity principle suggests that less complex theories should be preferred. In this regard, the DISC Model is parsimonious because it proposes that the strongest effects, on average, should occur within the same domain. Another important aspect that the TMP suggests is, that not only should demands and resources match strains, but that resources should also match demands (Frese, 1999).

The TMP outlined above grounds the basis of the **compensation principle** (i.e. our third principle). We believe that job demands and job resources from qualitatively identical domains have the highest probability to counteract each other. In a sense, one can think of demands and resources stemming from the same domain as activating qualitatively similar processes with reversed quantitative sign. For instance, suppose an employee is confronted with a verbally aggressive patient, which elicits negative emotions. The employee's negative emotions may be reduced or even turn into positive ones if a colleague outside the room is emotionally supportive (e.g. telling a joke). This is so because positive and negative emotions cannot exist at the same time (e.g. Fisher, 2000). Similarly, if an employee is confronted with a too complex task overtaxing his or her working memory, a colleague providing informational support may cause insight, which relieves working memory. When one is assisted by colleagues in carrying a heavy patient, the negative impacts on the musculoskeletal apparatus may be less. Generally, it should be noted that compensation does not always mean putting the job resources into action. Availability of job resources may sometimes be enough (Frese, 1989; Miller, 1979). For instance, knowing that colleagues will provide support in cases one is not able to meet the job demands could be enough to compensate the demands. As explained previously, using resources that are depleted through usage is stressful and may cause health problems. This is not to say that it would be better not to use the resources, but it is important to note that job resources are no panacea for all kinds of demanding situations at work. Using transient resources may lead to a net reduction of negative effects of demands. This effect would be even larger, if the unavoidable negative effects of transient resource usage did not counterbalance it.

Thus, one may conclude that for the sake of preventing health problems, resources should be established, which are perpetual (because a loss would be stressful) and buffer the negative effects of demands by their mere availability (because using them implies additional effort, which is stressful by definition). For instance, monetary rewards as suggested by Siegrist (1996) may have such effects. When demands such as time pressure at work become too high because of layoffs, a simple brief thought about one's own job security may lead to a re-interpretation of the situation, and, perhaps, it may help to feel more relaxed and thereby reduces strain. Control or autonomy may sometimes have a similar function. The experiments of Glass *et al.* (1969) showed that control, which is available but not used (and not depleted), is able to buffer the negative impact of stressful working conditions. Of course, using transient resources may also lead to stress-buffering, but the effects are probably difficult to establish in statistical analyses.

The former arguments have to be qualified in order to lead to the fourth common principle, that is, the **balance principle**. Previous models like the DCM suggested that active learning and growth will occur in the presence of high job demands coupled with high resources. We do not believe that this is entirely true without further qualification because of several theoretical and empirical arguments. The first point is to make clear what the balance principle stands for, and the second point is its distinction from compensation. For instance, suppose that a nurse is confronted with an emotionally demanding (e.g. impertinent) patient. Impertinent patients are likely to cause strain, which can be compensated by matching job resources. As argued before, job resources have to be distinguished regarding their degree of actual usage and depletion. An example for a matching job resource that is neither really used nor depleted may be the availability of emotional support from colleagues (e.g. knowing that colleagues always like to joke about this patient or mimic his or her behaviour). This resource may buffer (i.e. compensate for) the negative effects of the demanding patient. We define this as a *unidirectional sequence*: job resources compensate job demands, but the resources have no additional effect. In particular, the resource is unlikely to activate growth and active learning. Another example would be the safety-signal function of control (Miller, 1979), which suggests that the mere availability of control informs a person that he or she could intervene if demands become too high. However, if the nurse is allowed to make decisions, but she never makes any, learning will not occur. We argued before that perpetual job resources that reduce job-related strain by their mere availability are best, but they are only best when they are aimed to reduce adverse health and well being effects of job demands. Unfortunately, perpetual job resources that reduce strain by their mere availability may have negative side effects. We believe that active learning as suggested by Karasek (1998) may not occur in such instances. Rather, active learning and growth are dependent on the use of transient resources. That is, learning and strain prevention are partly incompatible.

According to the DISC Model, growth and active learning occur only through actual usage of limited job resources. Without using resources, action represents nothing more than automatic re-action. It is unlikely that new behavioural patterns occur. Similarly, Demand-Control theory predicts that learning occurs in a challenging situation and when the employee with job decision latitude makes a choice about how to respond to this job demand (Karasek, 1998). Moreover, we believe that learning is more likely to occur if the job resource is transient. What would students' learning look like if there were an unlimited number of trials to pass an examination? Who does not believe that the motivation to learn would be much stronger if the number of trials was limited? In general, the motivation to learn and grow is low, when perpetual resources are available that make it too easy to cope with existing demands. Transient job resources are more valuable than perpetual ones in this respect. When Hobfoll (1989) argued that people try to minimise a net loss of resource, we believe he had transient resources in mind.

Active learning provides one of the few means to minimise the loss of transient job resources. For instance, we previously argued that the autonomy to decide when one accomplishes a particular task represents a transient job resource because it is unlikely that one has the autonomy to postpone *all* tasks. Postponing implies resource loss, which may be more stressful than the time pressure present. Thus, individuals do not want to start postponing immediately when facing time pressure. But, how, then, can the time pressure be reduced? Individuals have to learn. This may include, for example, learning to perform faster, learning to perform with fewer errors, learning to use better cognitive strategies, learning to apply less exhausting strategies thereby reducing the need for recovery, and so on. We define this as the *bidirectional sequence* of used and transient matching job resources: They may prevent job-related strain, and, more important, they promote learning and growth.

The new behavioural response, if effective and usable, will be incorporated in the employee's repertoire of activities ('potential solutions'). In accordance with Demand-Control theory, we further argue that the potential activity level will increase in the future due to an increasing number of solutions to deal with challenging job demands (Karasek, 1998).

From this point, two questions are important in guiding the principles of the DISC Model. The first question is whether job decision latitude is the only job resource that is capable of matching challenging job demands. Given the first principle of the DISC Model (i.e. multidimensionality of job resources), we believe that the energy aroused by particular, challenging, job demands is translated into direct action *only* by the application of a particular, matching, job resource. In some cases this could be job control, but in other cases this could be, for example, emotional rewards and/or instrumental support (pending the type of job demands). An ideal as well as optimal situation would be complete harmony of behaviours, emotions, and cognitions caused by matches of particular job demands and corresponding transient job resources, which comes close to what has been called 'flow' (Csikszentmihalyi, 1997) and 'engagement' (Schaufeli and Bakker, 2001). Flow could be described as a state of optimal experience that is mainly characterised by focused attention, complete control, loss of self-consciousness, distortion of time, and intrinsic enjoyment. In a milder form this resembles the concept of engagement, which could be defined as a positive affective-cognitive state of feeling well that is mainly characterised by vigour, dedication, and absorption (see also Demerouti *et al.*, 2001b).

Our second question concerns the level of job demands. We believe that if job demands are too high (even if they are initially challenging), they may cause damage even if (matching) job resources are available, and may be activated for either compensation or balancing. This is in line with the ideas of Warr's Vitamin Model (1987, 1994). In addition, the higher a particular demand is, the more likely it is that other job demands cannot be compensated or balanced anymore. This may not apply if there is only one extremely high job demand because it may be well compensated for if the matching job resource is extremely high as well. However, the greater the number of extremely high job demands, the greater the likelihood that poor health will develop. In such cases, the demands cannot be compensated for any longer because the necessary job resources are more and more depleted. Furthermore, these job resources are no longer available for learning activities, which require at least a minimum amount of these resources as well. Thus, balancing is also difficult in the case of too high demands. Finally, it could be that appropriate actions to compensate or balance could not be taken because of too high a level of negative affective states (e.g. anxiety).

The final basic assumption is concerned with the central role of emotions in the stress process as suggested in the **hierarchical principle** (see also Spector, 1998). Although emotions have largely been left out from empirical research on job stress (which, at best, looked at moods as dependent variables) several researchers are starting to believe that stress is all about emotions (e.g. Gaillard and Wientjes, 1994; Lazarus, 1993). Indeed, we believe that stress without emotions is like a balloon without covering – nothing to be really looked at. As we mentioned before, we believe that many demands, resources, and strains possess an emotional component or elicit emotional processes. In terms of strain, emotions are likely to be always present. Even if seemingly emotion-free concepts such as physical symptoms are used as dependent variables (e.g. breathlessness), they are probably often accompanied by emotional reactions such as fear (e.g. of a heart attack), anger (e.g. about a supervisor), shame (e.g. because of reduced physical efficiency). It is also important to note that emotions are a central aspect of human service work (Zapf *et al.*, 1999). It is, therefore, interesting to note that cognitive (e.g. time pressure, role conflict, method control) demands and resources have dominated stress research on service employees; Studies concerned with emotional demands and resources, the emotions elicited by such demands and resources, and how emotions may be linked to long-term strain reactions have really just begun (e.g. Zapf *et al.*, 1999, 2001).

Emotional demands typically involve enhancement (upregulation) or inhibition (downregulation) of emotions, either explicitly or implicitly. An example of explicit upregulation of emotions is the following. Organisations often require their employees to express positive emotions towards their customers, which can be accomplished by so-called deep acting (Hochschild, 1983). When engaged in deep-acting (opposed to surface acting), employees use techniques such as imagination to elicit the emotions they are supposed to express towards their customers. For example, imagining an uncongenial customer as naked in the mall may create a genuine smile, as required. Perhaps more often than explicit upregulation of emotions are instances in which employees are implicitly tempted to downregulate their emotions. When negative emotions such as fear arise, employees usually have to reduce their fear actively in order to perform well. For example, a nurse who is anxious about getting infected by the disease of a patient has to downregulate his or her fear in order to approach the patient and do the required job. Emotional demands are particularly serious if downregulation and upregulation demands are present simultaneously. Typically, the nurse has not only to downregulate negative emotions such as fear, but also to upregulate positive emotions in order to express them towards the patients. This simultaneity of opposing upregulating and downregulating demands was termed emotional dissonance by Hochschild (1983), and it has been shown to be one of the most serious antecedents of health-related symptoms in a variety of studies (for an example, see Zapf *et al.*, 2001). Several techniques to regulate one's emotions exist, and a comprehensive overview is presented by Larsen (2000). For emotion regulation, such as controlling the timing of interaction with customers, matching emotional job resources are required (i.e. interaction control; Zapf *et al.*, 1999, 2001).

Figure 2.2 displays the processes suggested by the DISC Model (potential main effects of resources are not shown for simplicity). In general, goal-directed action at work starts with a behavioural demand. Typically each task requires at least some behavioural action to be completed. Even cognitive tasks such as solving an equation or emotional tasks such as smiling at awkward customers finally involve behavioural actions. This is dealt with in action-regulation theory (cf. Frese and Zapf, 1994). In a broad sense, thus, behavioural demands are the starting point of action at work. Sometimes the behavioural demand almost directly translates into overt action in highly automatic fashion, e.g. during typewriting. However, often the behavioural demands reach consciousness and are then represented as cognitive demands (e.g. to figure out what a customer deserves). In principle, researchers agree that cognitions are a precursor of emotions (e.g. Clore, 1994; Fridja, 1994; Izard, 1994; Lazarus, 1994). Thus, cognitive demands may be further complemented by emotional demands (e.g.

showing a smile to an awkward customer). As suggested by the matching principle, the primary outcomes of each demand occur in the same qualitative domain. That is, behavioural demands lead to behavioural reactions, cognitive demands lead to cognitive reactions, and emotional demands lead to emotional reactions. It is very well documented that once emotional reactions are elicited, they affect subsequent cognitions. For example, emotions affect memory (e.g. Bower, 1981) and thinking (e.g. Bless *et al.*, 1996; Clore *et al.*, 1994), which then mediates the effect of emotions on behaviour. Furthermore, reactions in all domains may contribute to strain or health because strain and health are usually multidimensional concepts (e.g. depression).

Finally, let us recapitulate the two mechanisms of compensation and balance. Compensation is unidirectional: taxing job demands ('stressors') need matching job resources with a minimum loss of energy. We may call this 'survival' (cf. Hobfoll, 2001). On the other hand, balance is bidirectional: the interplay between job demands on the one hand and matching, transient, job resources on the other, is important, and might be called 'investment'. It is some sort of energy depletion with advantages in the long run. A human service worker, for instance, needs both (challenging) demands and usable, matching, job resources to learn and to grow, and thus to feel well.

The tentative model as shown in Figure 2.2 suggests a clear ordering of interaction effects. The strongest interaction should occur for the following model (Equation 2.1):

Emotional Reactions = Emotional Demands (ED) + Emotional Resources (ER) + ED * ER$\qquad$(2.1)

The reason is that the any other process shown in Figure 2.2 does not disturb the interplay between these three concepts. This is a very important prediction for service jobs because emotional demands are likely to be of particular importance here, but empirical studies on any of the three emotional concepts is almost lacking (for an overview see Zapf *et al.*, 1999, 2001). The second and third rank regarding the strengths of the interaction should, *ceteris paribus*, occur for Equations 2.2 and 2.3:

Cognitive Reactions = Cognitive Demands (CD) + Cognitive Resources (CR) + CD * CR$\qquad$(2.2)

Behavioural Reactions = Behavioural Demands (BD) + Behavioural Resources (BR) + BD * BR$\qquad$(2.3)

It is difficult to predict the ranking of buffering effects when cross-domain interactions are considered because the strength of the relations between demands in different domains and reactions in different domains are usually not known. However, as a rule of thumb, one can argue that the fewer paths from the overall model in Figure 2.2 are involved, the stronger that buffering and main effects are likely to be.

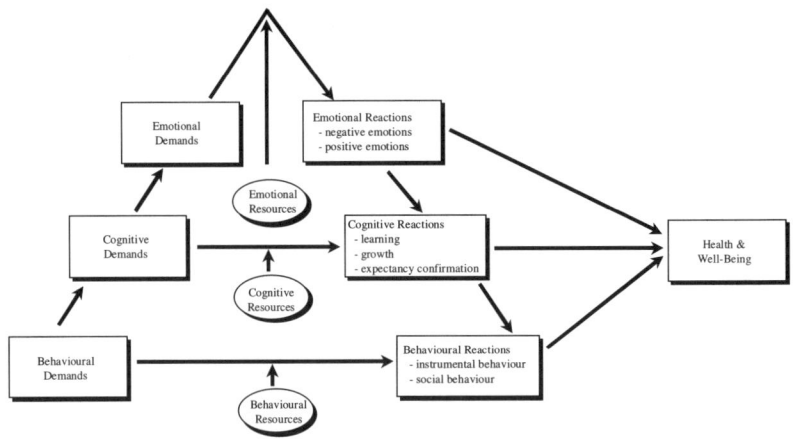

Figure 2.2 Behavioural, cognitive, and emotional processes within the DISC Model.

If more broad measures of health are considered, it is also difficult to predict the ordering of buffering (and main) effects because it depends on the behavioural, cognitive, and emotional content of such measures. However, the DISC Model suggests that most measures of subjective health and well being are more strongly affected by emotions than by cognitions and by behaviour because more pathways exist along which emotions may exert their effects (see Figure 2.2). Thus, in general, when global measures of health and well being are investigated, the DISC Model predicts that emotional demands and resources are more important than cognitive ones, which, in turn, are more important than behavioural ones.

To visualise our current line of reasoning: variables, which are supposed to interact, can be roughly arranged into a cake-like partitioned disc, with counteracting job demands and job resources in opposition (see Figures 2.3a and 2.3b).

2.5.2 Empirical evidence

In broad terms, there is evidence for the basic assumptions of the DISC Model. This is of course not surprising, as many of the unexplained results of the models mentioned in part stimulated the genesis of the DISC Model. For instance, it might be that interactions *not* found in empirical model tests were due to a mismatch of job demands and job resources (de Jonge *et al.*, 2000c; van der Doef and Maes, 1999). If the job resources under investigation were transient ones, which were depleted when used for compensation, finding interaction effects was also unlikely. In such instances, the use of transient resources represents an additional demand on its own, which hides the interaction effects. Conversely, empirical support for assumptions of the DISC Model suffers from those weaknesses that are inherent in

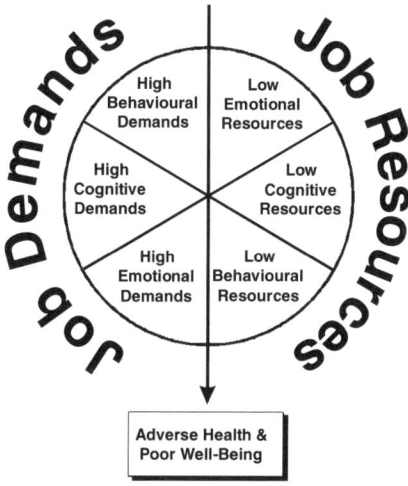

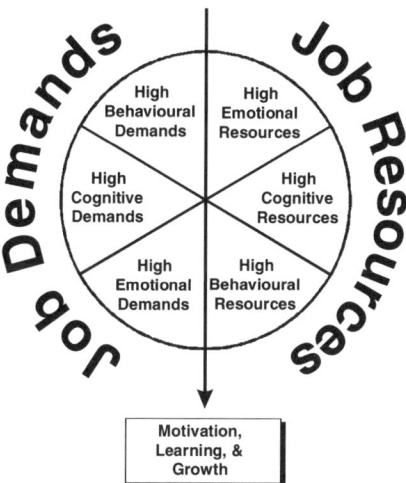

Figures 2.3a and 2.3b The Demand-Induced Strain Compensation Model of job stress: The principles of disbalance (a) and balance (b), respectively.

conventional data-gathering procedures, and which limit the support for any model in the field of job stress. For instance, most studies are cross-sectional, and depend upon perceptions of job characteristics. In addition, pure empirical DISC studies have not been carried out thus far, with one notable exception (i.e. van Vegchel *et al.*, 2001b). With this in mind, we will finally present a few empirical studies as illustrations for several basic assumptions of the DISC Model.

The first longitudinal study that tested several assumptions of the DISC Model was conducted by van Vegchel and colleagues (2001b). In a two-wave panel sample of 2,255 Swedish human service workers (one-year time interval), they found some evidence for the Triple Match Principle (TMP) of the DISC Model. First, all interactive, matching, effects in this particular group of workers were found with emotional demands (instead of cognitive demands). Second, workplace social support (mainly emotion-laden) was the most important job resource in this respect, which compensated the (negative) impact of emotional demands. Finally, outcome variables of these matching effects were primarily affective or emotional in nature (i.e. job satisfaction, emotional exhaustion, and depersonalisation). Another remarkable study has been performed by Peeters and Le Blanc (2001). In their study of 816 oncology care providers (see Chapter 5), they found evidence for our TMP (i.e. the compensation principle in particular) with respect to the burnout-component depersonalisation. Social support from colleagues was able to buffer the effect of emotional demands (i.e. confrontation with death and dying) on depersonalisation. A third illustration is a study by Ybema and Smulders (2001). In a representative sample of the Dutch working population (n = 4,334), results yielded a significant interaction effect between emotional demands (i.e. hiding of emotions) and social support from the supervisor with regard to emotional exhaustion. Supervisory support was able to buffer the adverse effect of emotional demands on emotional exhaustion, which is completely in line with the TMP (i.e. compensation principle) of the DISC Model.

Post hoc evidence for the balance principle of the DISC Model is harder to find. Only a few studies could be mentioned in this respect (for an overview, see Demerouti *et al.*, 2001b). For instance, in a sample of 895 health care professionals, de Jonge *et al.* (1999b) found some evidence for the TMP in this respect. More specifically, they found a balanced mixture of high cognitive job demands and high cognitive control in the prediction of a cognitive-laden outcome, that is, work motivation.

Mismatch of job demands and job resources can even show opposite effects in terms of adverse health and well being (i.e. job resources are not compensating, but increase the adverse effects of job demands on employee health). This might be due to cross-domain interactions that are not in line with the assumptions of the DISC Model. For instance, in a sample of 5,730 Swedish human service professionals, Söderfeldt *et al.* (2001) found such an opposite interaction effect between emotional demands and outcome control regarding psychosomatic health complaints. This can be compared to findings of de Jonge *et al.* (1999a) in a sample of 212 health care professionals, finding an identical opposite interaction effect as well. High job control appeared to have detrimental effects in both studies. These results might indeed imply that there was a poor match between the particular job demands and the particular job resource, leading to odd results. Because the resource did not match the demands, but individuals might use the non-matching transient resource, an additional increase in strain might occur because the loss of this resource was threatening. The use of other concepts of job control (like emotional control) might have resulted in a better match with emotional demands, and in a more accurate prediction of (emotional) health/well being outcomes. Another illustration of this phenomenon with a different health outcome is a study by Hollmann and colleagues (2001). In a sample of 431 human service workers from geriatric nursing homes, their results showed that job control (i.e. control over work method and working time) did *not* buffer the effects of physical demands with respect to musculoskeletal complaints. However, contrary to this, high control enhanced the effects of physical demands regarding this kind of health problem. A final study with a clear mismatch between demands and resources has been reported by Sargent and Terry (1998). In a two-wave panel study of 87 administrative staff of a university, they found an inconsistent interaction effect between underutilisation of skills and decision control (i.e. a non-matching resource) in the prediction of Time 2 job satisfaction. The negative relationship between underutilisation of skills and job satisfaction was more marked for employees who reported a high rather than a low degree of involvement in work or organisational decisions.

2.6 EPILOGUE

Clearly, job stress is an important problem in most industrialised countries nowadays. This is particularly true in service work, as working with demanding clients seems to be one of the key job demands in service work. It is worth mentioning that the number of jobs in manufacturing industries is declining relative to the number in the service industries (Zeithaml and Bitner, 2000). In addition, job stress is not merely a societal problem, but also a scientific one. From a scientific point of view, it may seem somewhat disappointing that the field of job stress and health/well being is full of inconsistent and contradictory theoretical and empirical findings. The present chapter therefore tries to overcome some of these problems by introducing a new theoretical framework and underlying psychological mechanisms.

Job stress was defined as the occurrence of negative emotions that are evoked by demanding situations in the workplace. After clear conceptualisations of job demands and job resources, we launched an integrated theoretical view on job stress by unification of common principles and separation of different foci. This view, visualised and outlined by the Demand-Induced Strain Compensation (DISC) Model, was particularly designed to explain what aspects of jobs might activate psychological compensation processes of job-related strain, or balance challenging job demands.

An important starting point in this integrated view was to look at qualitatively different concepts and processes, namely behaviour, cognition, and emotion. In addition to this multidimensional principle, four guiding principles suggested were the compensation principle, the triple match principle, the balance principle, and the hierarchical principle. In general, the DISC Model may explain why current job stress models are *not* able to show clear and consistent interactions between job demands and job resources, in the prediction of job-related strain.

Further, the DISC Model tries to explain why compensation and balancing are separable mechanisms in processes of stress and learning, respectively. Finally, there is no doubt that measures of job demands and job resources should be more specifically and unidimensionally designed to fit the different behavioural, cognitive and emotional concepts and underlying processes. For instance, this implies that an ideal social support measure should be split up into behavioural, cognitive and emotional dimensions.

Though further empirical evidence regarding the DISC principles is badly needed (longitudinal studies in particular), we believe the DISC Model is able to shed new light on *specific* relationships between work-related psychosocial risks and employee health and/or well being. It also allows for a more rigorous interpretation of particular (causal) mechanisms between job demands/resources on the one hand, and employee health/well being on the other (including learning and growth). The first results look promising, and – given the DISC principles will stand further empirical testing – it should provide corresponding, tailor-made, operational instructions for real practice in, e.g. service work. In this case, job stress interventions should be targeted primarily at specific job demands and counteracting, matching, job resources to reduce particular adverse health and well being problems. Similarly, learning and personal growth of a service worker can be enhanced through a well-balanced mixture of specific high job demands and corresponding, matching, job resources.

Given the changes in the nature of work as outlined in the introduction of this chapter, the DISC Model seems to have a promising future. One major task and challenge in this context is the translation of the DISC theory into effective interventions in service work by means of case studies or quasi-experiments.

2.7 ACKNOWLEDGEMENTS

The authors would like to thank Wilmar Schaufeli, Natasja van Vegchel, Marie Söderfeldt, Arnold Bakker, Claudia Gross, Pascale Le Blanc, Sarah Cotton, Phil Heiligers, Gabriele Häslich, and the two anonymous reviewers for their helpful comments.

2.8 REFERENCES

Allegro, J.T. and Veerman, T.J., 1998, Sickness absence. In *Handbook of Work and Organizational Psychology,* Vol. 2 (Work Psychology), 2nd edn, edited by P.J.D., Drenth, H.K. Thierry, and Ch.J. de Wolff (Hove: Psychology Press), pp. 121–144.

American Psychiatric Association, 1994, *Diagnostic and Statistical Manual of Mental Disorders*, 4th edn. (Washington, DC: American Psychiatric Association).

Antonovski, A., 1987, The salutogenic perspective: Toward a new view of health and illness. *Advances*, **4**, pp. 47–55.

Bakker, A.B., Killmer, C.H., Siegrist, J. and Schaufeli, W.B., 2000, Effort–reward imbalance and burnout among nurses. *Journal of Advanced Nursing*, **31**, pp. 884–891.

Beehr, T.A., Glaser, K.M., Canali, K.G. and Wallwey, D.A., 2001, Back to basics: Re-examination of Demand-Control theory of occupational stress. *Work & Stress*, **15**, pp. 115–130.

Bless, H., Clore, G.L., Schwarz, N., Golisano, V., Rabe, C. and Wölk, M., 1996, Mood and the use of scripts: Does a happy mood really lead to mindlessness? *Journal of Personality and Social Psychology*, **71**, pp. 565–579.

Bosma, H., Peter, R. Siegrist, J. and Marmot, M., 1998, Two alternative job stress models and the risk of coronary heart disease. *American Journal of Public Health*, **88**, pp. 68–74.

Bower, G.H., 1981, Mood and memory. *American Psychologist*, **36**, pp. 129–148.

Breaugh, J.A., 1985, The measurement of work autonomy. *Human Relations*, **38**, pp. 551–570.

Brief, A.P., Butcher, A.H. and Roberson, L., 1998, Cookies, disposition, and job attitudes: The effects of positive mood-inducing events and negative affectivity on job satisfaction in a field experiment. *Organizational Behavior and Human Decision Processes*, **62**, pp. 55–62.

Broadbent, D.E., 1985, The clinical impact of job design. *British Journal of Clinical Psychology*, **24**, pp. 33–44.

Buunk, B.P., de Jonge, J., Ybema, J.F. and de Wolff, Ch.J., 1998, Psychosocial aspects of occupational stress. In *Handbook of Work and Organizational Psychology*, Vol. 2 (Work Psychology), 2nd edn, edited by P.J.D. Drenth, H.K. Thierry, and Ch.J. de Wolff (Hove: Psychology Press), pp. 145–182.

Calnan, M., Wainwright, D. and Almond, S., 2000, Job strain, effort–reward imbalance and mental distress: a study of occupations in general medical practice. *Work & Stress*, **14**, pp. 297–311.

Carver, C.S. and Scheier, M.F., 1998, *On the Self-Regulation of Behavior* (Cambridge: Cambridge University Press).

Clore, G.L., 1994, Why emotions require cognition. In *The Nature of Emotion. Fundamental Questions*, edited by Ekman, P. and Davidson, R.J. (New York: Oxford University Press), pp. 181–191.

Clore, G.L., Schwarz, N. and Conway, M., 1994, Cognitive causes and consequences of emotion. In *Handbook of Social Cognition,* 2nd edn, edited by Wyer, R.S. and Srull, T.K. (Hillsdale, NJ: Erlbaum), pp. 323–417.

Cohen, S. and Wills, T.A., 1985, Stress, social support, and the moderating hypothesis. *Psychological Bulletin*, **98**, pp. 310–357.

Cooper, C.L., 1994, The costs of healthy work organizations. In *Creating healthy work organizations*, edited by Cooper, C.L. and Williams, S. (Chichester: John Wiley), pp. 1–6.

Cooper, C.L., Dewe, P.J. and O'Driscoll, M.P., 2001, *Organizational Stress: A review and critique of theory, research, and applications* (Thousand Oaks, CA: Sage Publications).

Csikszentmihalyi, M., 1997, *Finding Flow. The Psychology of Engagement with Everyday Life* (New York: Basic Books).

de Jonge, J., 1995, *Job Autonomy, Well-Being, and Health: A Study Among Dutch Health Care Workers* (Maastricht, The Netherlands: Datawyse).

de Jonge, J. and Kompier, M.A.J., 1997, A critical examination of the Demand-Control-Support Model from a work psychological perspective. *International Journal of Stress Management*, **4**, pp. 235–258.

de Jonge, J., Mulder, M.J.G.P. and Nijhuis, F.J.N., 1999a, The incorporation of different demand concepts in the Job Demand-Control Model: Effects on health care professionals. *Social Science and Medicine*, **48**, pp. 1149–1160.

de Jonge, J., van Breukelen, G.J.P., Landeweerd, J.A. and Nijhuis, F.J.N., 1999b, Comparing group and individual level assessments of job characteristics in testing the Job Demand-Control Model: A multilevel approach. *Human Relations*, **52**, pp. 95–122.

de Jonge, J., Bosma, H., Peter, R. and Siegrist, J., 2000a, Twee werkstress-modellen en psychische gezondheid: het Job Demand-Control Model en het Effort-Reward Imbalance Model. [Two job stress models and mental health: the Job Demand-Control Model and the Effort–Reward Imbalance Model.] *Gedrag & Gezondheid*, **28(3)**, pp. 106–122.

de Jonge, J., Bosma, H., Peter, R. and Siegrist, J., 2000b, Job strain, effort–reward imbalance and employee well-being: A large-scale cross-sectional study. *Social Science and Medicine*, **50**, pp. 1317–1327.

de Jonge, J., Dollard, M.F., Dormann, C., Le Blanc, P.M. and Houtman, I.L.D., 2000c, The demand-control model: Specific demands, specific control, and well-defined groups. *International Journal of Stress Management* **7**, pp. 269–287.

de Jonge, J. and Hamers, J.P.H., 2000, Inspanningen en beloningen in het werk van verplegenden en verzorgenden: Een kwestie van balans of disbalans? [Efforts and rewards in the work of health care workers: A matter of balance or imbalance?]. *Verpleegkunde*, **15**, pp. 64–73.

Demerouti, E., Bakker, A.B., Nachreiner, F. and Schaufeli, W.B., 2001a, The Job Demands-Resources Model of burnout. *Journal of Applied Psychology*, **86**, pp. 499–512.

Demerouti, E., Bakker, A.B., de Jonge, J., Janssen, P.P.M. and Schaufeli, W.B., 2001b, Burnout and engagement at work as a function of demands and control. *Scandinavian Journal of Work, Environment, and Health*, **27**, pp. 279–286.

Diener, E. and Emmons, R.A., 1984, The independence of positive and negative affect. *Journal of Personality and Social Psychology*, **47**, pp. 1105–1117.

Diener, E. and Fujita, F., 1995, Resources, personal strivings, and subjective well-being: A nomothetic and idiographic approach. *Journal of Personality & Social Psychology*, **68**, pp. 926–935.

Diener, E., Smith, H. and Fujita, F., 1995, The personality structure of affect. *Journal of Personality & Social Psychology*, **69**, pp. 130–141.

Dollard, M.F., Winefield, H.R. and Winefield, A.H., 2001, *Occupational Strain and Efficacy in Human Service Workers: When the Rescuer Becomes the Victim* (Dordrecht: Kluwer Academic Publishers).

Dormann, C. and Zapf, D., 1999, Social support, social stressors at work and depression: Testing for main and moderating effects with structural equations in a 3-wave longitudinal study. *Journal of Applied Psychology*, **84**, pp. 874–884.

Dormann, C. and Zapf, D., 2001, *Customer-related Stressors and Burnout: Development of the AADA-scales.* Paper submitted for publication.

Dormann, C. and Zapf, D., 2002, Social stressors at work, irritation, and depression: Accounting for unmeasured third variables in a multi-wave study. *Journal of Organizational and Occupational Psychology*, **75**, pp. 33–58.

Dormann, C., Zapf, D. and Isic, A. (in press) Emotionale Arbeitsanforderungen und ihre Konsequenzen bei Call Center-Arbeitsplätzen [Emotional job demands and their consequences in call center jobs]. *Zeitschrift für Arbeits- und Organisationspsychologie*.

Ekman, P., 1994, Moods, emotions, and traits. In *The Nature of Emotion. Fundamental Questions*, edited by Ekman, P. and Davidson, R.J. (New York: Oxford University Press), pp. 56–58.

Ekman, P. and Davidson, R.J., 1994, *The Nature of Emotion* (New York: Oxford University Press).

Fisher, C.D., 2000, Mood and emotions while working: Missing pieces of job satisfaction? *Organizational Behavior and Human Decision Processes*, **21**, pp. 185–202.

Frese, M., 1977, *Psychische Störungen bei Arbeitern: Zum Einfluß gesellschaftlicher Stellung und Arbeitsplatzmerkmalen* [Psychological Disorders in Workers: The Influence of Socio-Economic Status and Characteristics of the Workplace] (Salzburg, Austria: Müller).

Frese, M., 1989, Theoretical models of control and health. In *Job Control and Worker Health*, edited by Sauter, S.L., Hurrell, J.J. and Cooper, C.L. (Chichester: Wiley), pp. 107–128.

Frese, M., 1999, Social support as a moderator of the relationship between work stressors and psychological dysfunctioning: A longitudinal study with objective measures. *Journal of Occupational Health Psychology*, **4**, pp. 179–192.

Frese, M. and Zapf, D., 1994, Action as the core of work psychology; A German approach. In *Handbook of Industrial and Organizational Psychology*, Vol. 4, edited by Triandis, H.C., Dunnette, M.D. and Hough L.M. (Palo Alto: Consulting Psychologists Press), pp. 271–340.

Fridja, N.H., 1993, Moods, emotion episodes, and emotions. In *Handbook of Emotions*, edited by Lewis, M. and Haviland, I.M. (New York: Guilford), pp. 381–403.

Fridja, N.H., 1994, Emotions require cognitions, even if simple ones. In *The Nature of Emotion. Fundamental Questions*, edited by Ekman, P. and Davidson, R.J. (New York: Oxford University Press), pp. 197–202.

Gaillard, A.W.K. and Wientjes, C.J.E., 1994, Mental load and work stress as two types of energy mobilization. *Work & Stress*, **8**, pp. 141–152.

Glass, D.C., Singer, J.E. and Friedman, L.N., 1969, Psychic cost of adaptation to an environmental stressor. *Journal of Personality and Social Psychology*, **12**, pp. 200–210.

Hackman, J.R. and Oldham, G.R., 1980, *Work Redesign* (Reading, Mass: Addison-Wesley).

Hart, P.M. and Cooper, C.L., 2001, Occupational stress: Toward a more integrated framework. In *Handbook of Industrial, Work and Organizational Psychology*, Vol. 2, edited by Anderson, N., Ones, D.S., Kepir Sinangil, H. and Viswesvaran, C. (London: Sage Publications), pp. 93–114.

Hasenfeld, Y., 1983, *Human Service Organizations* (Englewood Cliffs: Prentice Hall).

Hasenfeld, Y., 1992, The nature of human service organizations. In *Human Services as Complex Organizations*, edited by Hasenfeld, Y. (Newbury Park: Sage), pp. 3–23.

Herzberg, F., 1966, *Work and the Nature of Man* (Cleveland: World Publishing Company).

Hobfoll, S.E., 1989, Conservation of resources: A new attempt at conceptualizing stress. *American Psychologist*, **44**, pp. 513–524.

Hobfoll, S.E., 1998, *Stress, Culture, and Community: The Psychology and Philosophy of Stress* (New York: Plenum Press).

Hobfoll, S.E., 2001, The influence of culture, community, and the nested-self in the stress process: Advancing Conservation of Resources Theory, *Applied Psychology: An International Review*, **50**, pp. 337–370.

Hochschild, A.R., 1983, *The Managed Heart* (Berkeley: University of California Press).

Hockey, G.R.J., 2000, Work environments and performance. In *An Introduction to Work and Organizational Psychology: A European Perspective*, edited by Chmiel, N. (Oxford: Blackwell Publishers), pp. 206–230.

Hollmann, S., Heuer, H. and Schmidt, K.H., 2001, Control at work: A generalized resource factor for the prevention of musculoskeletal symptoms? *Work & Stress*, **15**, pp. 29–39.

Hoogendoorn, W.E., van Poppel, M.N.M., Bongers, P.M., Koes, B.W. and Bouter, L.M., 2000, Systematic review of psychosocial factors at work and private life as risk factors for back pain. *Spine*, **25**, pp. 2114–2125.

Houkes, I., 2002, *Work and individual determinants of intrinsic work motivation, emotional exhaustion and turnover intention* (Maastricht, The Netherlands: Datawyse).

House, J.S., 1981, *Work Stress and Social Support* (London: Addison-Wesley).

Izard, C.E., 1994, Cognition is one of four types of emotion-activating systems. In *The Nature of Emotion. Fundamental Questions*, edited by Ekman, P. and Davidson, R.J. (New York: Oxford University Press), pp. 203–207.

Johnson, J.V., 1986, *The Impact of the Workplace Social Support, Job Demands, and Work Control under Cardiovascular Disease in Sweden.* (Doctoral dissertation, John Hopkins University) (Stockholm: University of Stockholm).

Johnson, J.V. and Hall, E.M., 1988, Job strain, work place social support, and cardiovascular disease: A cross-sectional study of a random sample of the Swedish working population. *American Journal of Public Health*, **78**, pp. 1336–1342.

Jones, F. and Fletcher, B., 1996, Job control and health. In *Handbook of Work and Health Psychology*, edited by Schabracq, M.J., Winnubst, J.A.M. and Cooper, C.L. (Chichester: Wiley), pp. 33–50.

Kahn, R.L. and Byosiere, P., 1992, Stress in organizations. In *Handbook of Industrial and Organizational Psychology*, Vol. 3, 2nd edn, edited by Dunnette, M.D. and Hough, L.M. (Palo Alto, CA: Consulting Psychologists Press), pp. 571–650.

Karasek, R.A., 1979, Job demands, job decision latitude, and mental strain: Implications for job redesign. *Administrative Science Quarterly*, **24**, pp. 285–307.

Karasek, R.A., 1985, *Job Content Instrument: Questionnaire and User's Guide, Revision 1.1* (Los Angeles: University of Southern California).

Karasek, R.A., 1998, Demand/Control Model: A social, emotional, and physiological approach to stress risk and active behaviour development. In *Encyclopaedia of Occupational Health and Safety*, edited by Stellman, J.M. (Geneva: ILO), pp. 34.6–34.14.

Karasek, R.A. and Theorell T., 1990, *Healthy Work* (New York: Basic Books).

Koslowski, M., 1998, *Modeling the Stress-Strain Relationship in Work Settings* (London: Routledge).

Kristensen, T., 1995, The demand-control-support model: Methodological challenges for future research. *Stress Medicine*, **11**, pp. 17–26.

Larsen, R.J., 2000, Toward a science of mood regulation. *Psychological Inquiry*, **11**, pp. 129–141.

Lazarus, R.S., 1993, From psychological stress to the emotions: A history of changing outlooks. In *Annual Review of Psychology, 1993* (Palo Alto: Annual Reviews), pp. 1–21.

Lazarus, R.S., 1994, Appraisal: The long and the short of it. In *The Nature of Emotion. Fundamental Questions*, edited by Ekman, P. and Davidson, R.J. (New York: Oxford University Press), pp. 208–215.

Le Blanc, P.M., de Jonge, J. and Schaufeli, W.B., 2000, Job stress and health. In *An Introduction to Work and Organizational Psychology: A European Perspective*, edited by Chmiel, N. (Oxford: Blackwell Publishers), pp. 148–177.

Marmot, M., Siegrist, J., Theorell, T. and Feeney, A., 1999, Health and the psychosocial environment at work. In *Social Determinants of Health*, edited by Marmot, M. and Wilkinson, R.G. (Oxford: Oxford University Press), pp. 105–131.

Maslach, C., 1998, A multidimensional theory of burnout. In *Theories of Organizational Stress*, edited by Cooper, C.L. (Oxford: Oxford University Press), pp. 68–85.

Merllié, D. and Paoli, P., 2001, *Ten Years of Working Conditions in the European Union* (Dublin: European Foundation for the Improvement of Living and Working Conditions).

Miller, S.M., 1979, Controllability and human stress: Method, evidence and theory. *Behavior Research & Therapy*, **17**, pp. 287–304.

Mills, P.K., 1986, *Managing Service Industries: Organizational Practices in Postindustrial Economy* (Cambridge, Mass.: Ballinger).

Nerdinger, F., 1994, *Zur Psychologie der Dienstleistung: Theoretische und empirische Studien zu einem wirtschaftspsychologischen Forschungsgebiet* [On the Psychology of Services: Theoretical and Empirical Studies of a Economic-Psychological Research Area.], (Stuttgart, Germany: Schäffer-Poeschel).

Paoli, P., 1997, *Second European Survey on Working Conditions 1996* (Dublin: European Foundation for the Improvement of Living and Working Conditions).

Parasuraman, A., Zeithaml, V.A. and Berry, L.L., 1988, SERVQUAL: A multiple-item Scale for measuring consumer perceptions of service quality. *Journal of Retailing*, **64**, pp. 12–40.

Peeters, M.C.W. and Le Blanc, P.M., 2001, Towards a match between job demands and sources of social support: A study among oncology care providers. *European Journal of Work and Organizational Psychology*, **10**, pp. 53–72.

Peter, R., Alfredsson, L., Hammar, N., Siegrist, J., Theorell, T. and Westerholm, P., 1998a, High effort, low reward, and cardiovascular risk factors in employed Swedish men and women: Baseline results from the WOLF study. *Journal of Epidemiological and Community Health*, **52**, pp. 540–547.

Peter, R., Geißler, H. and Siegrist, J., 1998b, Associations of effort–reward imbalance at work and reported symptoms in different groups of male and female public transport workers. *Stress Medicine*, **14**, pp. 175–182.

Quinn, J.B., Baruch, J.J. and Cushman Paquette, P., 1987, Technology in services, *Scientific American*, **257**, pp. 50–58.

Sargent, L.D. and Terry, D.J., 1998, The effects of work control and job demands on employee adjustment and work performance. *Journal of Occupational and Organizational Psychology*, **71**, pp. 219–236.

Schaubroeck, J. and Fink, L.S., 1998, Facilitating and inhibiting effects of job control and social support on stress outcomes and role behavior: A contingency model. *Journal of Organizational Behavior*, **19**, pp. 167–195.

Schaufeli, W.B. and Enzmann, D., 1998, *The Burnout Companion to Study and Practice: A Critical Analysis* (London: Taylor & Francis).

Schaufeli, W.B. and Bakker, A., 2001, Werk en welbevinden. Naar een positieve benadering in de arbeids- en gezondheidspsychologie [Work and well-being: Towards a positive Occcupational Health Psychology]. *Gedrag & Organisatie*, **14**, pp. 229–253.

Schaufeli, W.B., Maslach, C. and Marek, T., 1993, *Professional Burnout: Recent Developments in Theory and Research* (New York: Taylor & Francis).

Schnall, P.L., Landsbergis, P.A. and Baker, D., 1994, Job strain and cardiovascular disease. *Annual Review of Public Health*, **15**, pp. 381–411.

Schnall, P., Belki, K., Landsbergis, P. and Baker, D., 2000, Why the workplace and cardiovascular disease? *Occupational Medicine*, **15**, pp. 1–6.

Schwartz, J.E., Pickering, T.G. and Landsbergis, P.A., 1996, Work-related stress and blood pressure: Current theoretical models and considerations from a behavioral medicine perspective. *Journal of Occupational Health Psychology*, **1**, pp. 287–310.

Selye, H., 1936, A syndrome produced by diverse noxious agents. *Nature*, **138**, p. 32.

Semmer, N., 1996, Individual differences, work stress and health. In *Handbook of Work and Health Psychology*, edited by Schabracq, M.J., Winnubst, J.A.M. and Cooper, C.L. (Chichester: Wiley), pp. 51–86.

Siegrist, J., 1996, Adverse health effects of high-effort/low-reward conditions. *Journal of Occupational Health Psychology*, **1**, pp. 27–41.

Siegrist, J., 1998, Adverse health effects of effort–reward imbalance at work: Theory, empirical support, and implications for prevention. In *Theories of Organizational Stress*, edited by Cooper, C.L. (Oxford: Oxford University Press), pp. 190–204.

Siegrist, J. and Peter, R., 1994, Job stressors and coping characteristics in work-related disease – issues of validity. *Work & Stress*, **8**, pp. 130–140.

Siegrist, J. and Peter, R., 1997, *Measuring Effort–Reward Imbalance at Work: Guidelines* (Düsseldorf: Institut für Medizinische Soziologie).

Siegrist, J., Siegrist, K. and Weber, I., 1986, Sociological concepts in the etiology of chronic disease: The case of ischemic heart disease. *Social Science and Medicine*, **22**, pp. 247–253.

Söderfeldt, M., 1997, *Burnout? (Doctoral dissertation)*, (Lund, Sweden: Lund University).

Söderfeldt, B., Söderfeldt, M., Muntaner, C., O'Campo, P., Warg, L. and Ohlson, C., 1996, Psychosocial work environment in human service organizations: A conceptual analysis and development of the Demand-Control Model. *Social Science and Medicine*, **42**, pp. 1217–1226.

Söderfeldt, M., Axtelius, B. and Bejerot, E., 2001, Demands and control in human service work: An interaction analysis. In *Organizational Psychology and Health Care at the Start of the new Millennium*, edited by De Jonge, J., Vlerick, P., Büssing, A. and Schaufeli, W.B. (München, Germany: Rainer Hampp Verlag), pp. 55–67.

Spector, P.E., 1998, A control theory of the job stress process. In *Theories of Organizational Stress*, edited by Cooper, C.L. (Oxford: Oxford University Press), pp. 153–169.

Stansfeld, S.A., Bosma, H., Hemmingway, H. and Marmot, M.G., 1998, Psychosocial work characteristics and social support as predictors of SF-36 health functioning: The Whitehall II Study. *Psychosomatic Medicine*, **60**, pp. 247–255.

Temme, G. and Zapf, D., 1997, *Arbeitsemotionen und Arbeitszufriedenheit : Untersuchungen über ihre Zusammenhänge aus Sicht der Betroffenen* [Job-Related Emotions and Job Satisfaction: Studies on Their Relation From the Employees' Perspective]. Abschlussbericht and die Deutsche Forschungsgemeinschaft. Universität Konstanz, Germany: Sozialwissenschaftliche Fakultät.

Terry, D.J. and Jimmieson, N.L., 1999, Work control and employee well-being: A decade review. In *International Review of Industrial and Organizational Psychology*, Vol. 14, edited by Cooper, C.L. and Robertson, I.T. (Chichester, UK:. Wiley), pp. 89–148.

Theorell, T. and Karasek, R.A., 1996, Current issues relating to psychosocial job strain and cardiovascular disease research. *Journal of Occupational Health Psychology*, **1**, pp. 9–26.

van der Doef, M. and Maes, S., 1998, The job demand-control(-support) model and physical health outcomes: A review of the strain and buffer hypotheses. *Psychology and Health*, **13**, pp. 909–936.

van der Doef, M. and Maes, S., 1999, The job demand-control (-support) model and psychological well-being: A review of 20 years of empirical research. *Work & Stress*, **13**, pp. 87–114.

van Vegchel, N., de Jonge, J., Meijer, T. and Hamers, J.P.H., 2001a, Different effort constructs and effort–reward imbalance: Effects on employee well-being in ancillary health care workers. *Journal of Advanced Nursing*, **34**, pp. 128–136.

van Vegchel, N., Söderfeldt, M., de Jonge, J., Schaufeli, W., Hamers, J. and Dormann, C., 2001b, *Job Stress and Well-Being in Human Service Work: A Longitudinal Study* (Paper presented at the VIIth ENOP Conference, Stockholm, Sweden).

Vancouver, J.B., 2000, Self-regulation in organizational settings: A tale of two paradigms. In *Handbook of Self-regulation*, edited by Boekaerts, M., Pintrich, P.R. and Zeidner, M. (San Diego, CA: Academic Press), pp. 303–341.

Wall, T.D., Jackson, P.R., Mullarkey, S. and Parker, S.K., 1996, The demands-control model of job strain: A more specific test. *Journal of Occupational and Organizational Psychology*, **69**, pp. 153–166.

Warr, P.B., 1987, *Work: Unemployment and Mental Health* (Oxford: Oxford University Press).

Warr, P.B., 1994, A conceptual framework for the study of work and mental health. *Work & Stress*, **8**, pp. 84–97.

Watson, D. and Tellegen, A., 1985, Toward a consensual structure of mood. *Psychological Bulletin*, **98**, pp. 219–235.

Weiss, H.M., Nicholas, J.P. and Daus, C.S., 1999, An examination of the joint effects of affective experiences and job beliefs on job satisfaction and variations in affective experiences over time. *Organizational Behavior and Human Decision Processes*, **78**, pp. 1–24.

Ybema, J.F. and Smulders, P., 2001, *Adverse Effects of Emotional Work: Does Social Support Help?* (Book of abstracts, EAOHP-conference, Barcelona).

Zapf, D., Dormann, C. and Frese, M., 1996, Longitudinal studies in organizational stress research: A review of the literature with reference to methodological issues. *Journal of Occupational Health Psychology*, **1**, pp. 145–169.

Zapf, D., Vogt, C., Seifert, C., Mertini, H. and Isic, A., 1999, Emotion work as a source of stress. The concept and development of an instrument. *European Journal of Work and Organizational Psychology*, **8**, pp. 371–400.

Zapf, D., Seifert, C., Schmutte, B., Mertini, H. and Holz, M., 2001, Emotion work and job stressors and their effects on burnout. *Psychology and Health*, **16**, pp. 527–545.

Zeithaml, V.A and Bitner, M.J., 2000, *Service marketing. Integrating customer focus across the firm* (Boston, MA: McGraw-Hill).

CHAPTER THREE

Measurement and Methodological Issues in Work Stress Research

Maureen F. Dollard and Jan de Jonge

3.1 INTRODUCTION

A renaissance of interest in measurement and methodological issues in work stress research is emerging as researchers, managers and policy makers seek gold standards in its evidence base for policy and practice. Increasingly cost–benefit analyses of work stress/interventions are sought and there is growing recognition that health and well-being measures are a quintessential component of organisational performance analysis. The aim of this paper is therefore to review both measurement and research design issues. As will be seen in the following discussion, the issues are inextricably linked. Moreover, each of the major models of work stress requires different measurement procedures. The issues raised are relevant to both theory – and problem – driven research. The implications of these issues are widespread, for consumers of research, those who attempt to measure risk in the work environment, and for those who attempt to implement change, develop policy, or manage injury.

3.2 CONCEPTUAL FRAMEWORKS

Work stress research in general attempts to draw links between taxing aspects of the work environment (demands, stressors), perceptions and appraisals of these, and manifestations of strain including physiological, psychological, and behavioural changes that may result (Baker, 1985; Greenhaus and Parasuraman, 1987). Despite 'work stress' being used in an enormous number of research articles, researchers have still not reached agreement on the meaning of this concept (Hart and Cooper, 2001). Moreover, different conceptualisations of work stress give rise to important and complex measurement issues in research. To illustrate this point we will focus here on two dominant paradigms of thinking in the field. The first is the psychological paradigm, where *stress* is described as a relationship between the person and the environment that is appraised as taxing or exceeding resources, and endangers well-being (Lazarus and Folkman, 1984). The second paradigm consists of so-called situation-centred theoretical models, where *stress* is particularly linked to factors or events outside a particular worker (like work characteristics; see e.g. Warr, 1987).

In psychological models the most important ingredient is variously described as *cognitive appraisal* of the work environment (Lazarus and Folkman, 1984), *perceived* stress (Greenhaus and Parasuraman, 1987) or *subjective* person-environment fit (French, Caplan and Harrison, 1984). Empirical tests of the models assess the relationship between cognitive appraisal and strain (line b) rather than the relation between the objective stressor and the perception or appraisal of it (line a) (see French *et al.*, 1984, p. 2). Of course the model presented here is highly simplistic but it captures fundamental relationships in a psychological paradigm.

Objective stressor	$\rightarrow$	Cognitive appraisal	$\rightarrow$	Strain
	a	perceived stress	b	

On the other hand work environment models emphasise the objective work environment as the starting point, and causal in the development of strain. According to Karasek *et al.* (1981, p. 695) 'strain results from the joint effects of the demands of the work situation (stressors) and environmental moderators of stress, particularly the range of decision making freedom (control) available to the worker facing those demands'. Work stress in this approach, is therefore linked to the structure of work and the production process (Baker, 1985).

Strain can be understood to result for those with objective high demand and objective low control. It is therefore not necessary to investigate the impact of these environmental conditions by distinguishing appraisals or various needs that may exist among workers (Karasek, 1989). The assumption is that if individual 'needs' for certain environmental conditions play a role in the stress process, they do so by operating similarly for *most of* the people. This conceptualisation of stress views environmental causes as the starting point, although it does not strictly preclude the importance of personal factors (Karasek and Theorell, 1990; Warr, 1979). In this respect, such situation-centred models inevitably contain some person-centred asssumptions for reasons of 'fine tuning' (Le Blanc *et al.*, 2000).

$$\text{Objective stressor} \quad \rightarrow \quad \text{Strain}$$

Empirical tests of the work environment model ideally investigate the link between objective stressors and illness (Frese and Zapf, 1988). *Stress* in these models refers to the intermediate state of arousal between the objective stressor and strain. Karasek (1989) notes that this state (stress) is rarely measured in this kind of research.

3.3 CONCEPTIONS AND MEASUREMENT

3.3.1 Self report vs more objective measures

As foreshadowed, measurement depends to a large degree on the conceptual framework in which the research takes place. In cognitive-phenomenological frameworks it makes theoretical sense to use a phenomenological approach such as self-report to measure, or infer the construct. Self report can provide a rich source of patterns of thoughts and emotion and according to Lazarus and Folkman (1984) 'self-reports allow us to learn more about coping and adaptational outcomes, than any other source, despite the difficulties in validation' (p. 322).

Lazarus and Folkman (1984) argue that cognitive appraisal is an outcome of the interaction between the objective environment and personality factors. When the environment is unambiguous, for most people they argue, perception and appraisal follow the objective environment quite well.

Assessing the appraisal process could be achieved directly by asking such questions as: 'To what extent is X a problem for you?' (Newton, 1989). One advantage of cognitive theories is in the operationalisation of stress which requires a single measurement from the subject rather than commensurate measures (e.g. P-E fit theories).

There are many well known problems with self report including problems of memory, language ambiguity, social desirability and demand characteristics. The questionnaire method has particular problems including central tendency bias, response acquiescence, and consistency artifact that can lead to a variety of systematic response distortions (de Jonge *et al.*, 1999; Porac, 1990).

Nevertheless, within psychological frameworks that emphasise qualitative processes, self-report is defensible on theoretical grounds. Subjective–objective measures are theoretically plausible in cognitive phenomenological frameworks (Glick *et al.*, 1986). At the same time objective measurements of the work environment in psychological models are generally unacceptable (Lazarus and Folkman, 1984; Perrewe and Zellars, 1999).

On the other hand work-environment paradigms that emphasise the objective environment have potential problems with self-report, not only on methodological grounds but also on theoretical grounds (i.e. self-report measurement of an objective phenomenon). But self-report is the most common measure of objective work conditions even in tests of work environment models. The predominance of self-report is probably a result of convenience and potential to tap core job characteristic concepts (Karasek and Theorell, 1990).

While most researchers would view 'subjective' as originating within or dependent on the mind of the individual rather than an external object, not all insist that objectivity exists totally independent of the mind of the individual. Frese and Zapf (1988) define subjective as 'tied to one's cognitive and emotional processing (e.g. perceptions and appraisals)' whereas objective 'means that a particular individual's cognitive and emotional processing does not influence the reporting of social and physical facts' (pp. 378–379).

Frese (1999), Frese and Zapf (1988, p. 382) argue that 'some researchers seem to suppose that all subjects' responses lead to a subjective bias because of cognitive and emotional processes. We disagree because it depends on the wording of the item whether there is high cognitive and emotional processing'.

They argue that self-report can vary in level of objectivity depending on the level of cognitive and emotional processing required by question instructions. For example 'How much do you feel stressed by the burden of this job?' is high on cognitive and emotional functioning in comparison to 'How many overtime hours do you work?' which is much less so. Psychological models that assess subjective appraisal (stress) theoretically are tied to the high cognitive/emotional processing assessments.

Conversely, empirical tests of work environment models, that link stressor to strain, would use assessments low in cognitive/emotional processing. The measure would ask 'does this work condition exist?'. This is in keeping with the Karasek methodology that rarely measures the intermediate state of stress directly in job research (Karasek, 1989). A number of work environment measures seem to fit the criterion of low level of emotional and cognitive processing such as the Work Environment Scale (WES), (Moos, 1986), Hackman and Oldham's Job Diagnostic Survey (JDS) (1975), and the Job Content Questionnaire (Karasek, 1995) (see also Quick *et al.*, 1998, and Figure 1.5).

Use of the self-report method in the work environment model is based on the assumption that self-report can provide an adequate assessment of *objective* conditions. There is some experimental evidence that confirms the objective validity of verbally reported task measurements (Griffen *et al.*, 1987). Fried and Ferris (1987) concluded after a review and meta-analysis of a number of studies, that the problems potentially associated with self-rated data of job characteristics are less serious than initially believed: 'the present review seems to increase the confidence in the substance of this data' (p. 299) that objective and perceived job characteristics are related. Fried and Ferris (1987) found that in general there is a 'trend toward similarity in the relationships of incumbents' job ratings and others' job ratings to potential criterion variables' (p. 289).

Alternative approaches, like group assessments, also claim to tap (a part of) the context in which individual workers operate. By group assessments we mean that the scores of job incumbents with the same job and working in nearly identical workplaces aggregated into one general score (cf. de Jonge *et al.*, 1999). These measures refer to that part of a particular job characteristic that different workers doing nearly the same job have in common (see also Semmer *et al.*, 1996). In other words, group assessments are based on the concept of the ideal typical worker (i.e. an average worker with sufficient skills to perform his or her tasks). According to Frese and Zapf (1988, 1994), group assessments are more objective measures in the sense that the influence of idiosyncratic – individual – perceptions and possibly illusory answers are reduced. In addition to this, the expertise of workers is taken into account and problems of brief periods of observation are avoided. Finally, group assessments seem to be less subject to methodological problems such as attenuation, because unsystematic error variance is reduced. Thus, group data are likely to be more reliable than individual assessments. The claim that group assessment is more objective is corroborated by several empirical findings. For instance, a meta-analysis of 16 convergence studies conducted by Spector (1992) showed that aggregate-level correlations between work characteristics and outcomes were similar to individual level correlations. Moreover, it appeared that the convergent validity at aggregated level was rather large, and even larger than at individual level.

To study the relationship between work conditions at multiple levels (i.e. group and individual), while still trying to predict individual variance in strain measures, one could use a recently developed technique, called 'multilevel analysis' (e.g. Snijders and Bosker, 1999). In multilevel research, the data structure in the population is hierarchical and contextual, and the data are viewed as a multi-stage sample from this hierarchical population. For example the individual worker can be viewed as nested in a context of a job, an organistion, an industry and a nation, and all of these could have an (hierarchical) effect. Multilevel analysis has several advantages in comparison with conventional statistical techniques. First, data from more than one hierarchical level can be included in the same analysis. Second, the statistical dependence between individuals of the same unit (or units of the same institution) is taken into account through the random variation across departments and organisations. Finally, the multilevel model separates sampling error due to variation between departments from variation within departments. In recent years multilevel analysis has been used more and more to study particular stressor–strain relationships at different organisational levels such as groups and individuals, with quite promising results (e.g. Söderfeldt *et al.*, 1997; de Jonge *et al.*, 1999; VanYperen and Snijders, 2000; Elovainio *et al.*, 2000). For instance, de Jonge *et al.* (1999) were able to explain additional variance at the group level that could not be explained at the individual level. Their results indicate that group level job characteristics are important in explaining health outcomes, lending support to a more objective approach of measuring job characteristics by means of questionnaires using aggregation techniques.

Although a relationship exists between objective work conditions and job perceptions, the strength of the correlation varies. For example, Kirmeyer and Dougherty (1988) reported a correlation of $r = .35$ between objective workload and perceived workload, whereas Griffin (1983) reported correlations from $r = .65$ to $r = .75$ between the subjective perceptions of job components (autonomy, task variety) and their objective manipulation. There is also good convergence between self-report and supervisor and subordinate-report (Bosma and Marmot, 1997; Jex and Spector, 1996; Spector *et al.*, 1988).

Fried and Ferris (1987) note that 'because not all of the reliable variance in job perceptions is explained by objective job conditions, however, other factors (e.g. social cues, method variance, etc.) must be acknowledged as potential sources of variation' (p. 309). Even so they conclude that perception of job characteristics is *largely* unaffected by irrelevant cues (Fried and Ferris, 1987).

In general, most social psychological stress research proceeds with the assumption that the job conditions lead to affective states, via job perceptions. A plausible alternative explanation for self-report cross-sectional correlational data is that the affective state of the individual making the assessment, influences the perception of the work environment. If this is the case then the perception can not be regarded as an independent variable against which to examine the affective state of the individual (Parkes, 1982). This is particularly problematic in stress research where the hypothesised independent variable can conceivably be either the cause of strain or the effect of strain.

James and Tetrick (1986) used a two-stage least squares analysis and explored the possibility of cognition preceding affect and/or affect preceding cognition and found support for a post-cognitive, non-recursive model to represent the relationship between job conditions (as measured by supervisor ratings), job perception (measured via employee questionnaires) and job satisfaction (assessed by job satisfaction questionnaires). Essentially this means that job satisfaction occurs after job perceptions in the causal order and job perceptions and job satisfaction are then reciprocally related. Job characteristics appeared to be a stronger cause of job satisfaction than vice versa.

They did not find support for the precognitive position that job perceptions serve to explain or to justify affective reactions to objective job characteristics as suggested by Zajonc (1984): 'I am satisfied; therefore the job must be challenging' (p. 77). This provides good preliminary evidence that the perception of the objective situation is largely unaffected by affective processes (helpful to the self-report work environment model), and provides empirical support for the expected causal ordering among cognitions of jobs and affective responses to jobs. This is also consistent with the psychological model that emotional responses in humans are elicited by cognitive processes, and that 'cognitive appraisal is a necessary as well as sufficient condition of emotion' (Lazarus and Folkman, 1984, p. 275).

Lazarus and Folkman (1984) recommend persisting with self-report data to derive empirically based principles and check these out with other methods, such as experimentation, and the use of more objective measures such as behavioural indicators, and physiological measures. While they do not argue against multimethod measurement, they provide a strong argument for the retention of the self-report method despite its scientific deficits.

Latack (1986) argued that multitrait-multimethod evidence is a challenge in stress coping research, with possibilities for behavioural observation by others as a method of assessing some coping strategies. However, she noted some limitations. The first related to the inappropriateness of objective assessment of intrapsychic modes of coping. Further, she questioned the appropriateness of sources of objective assessment. For instance she wondered whether evidence from supervisors concerning subordinates' coping behaviour might come from more appropriate sources (e.g. spouses, relatives or close friends).

The use of observers in research has been used as a methodological strategy for obtaining objective measurement of independent and/or dependent variables (Kiggundu, 1980; Kirmeyer and Dougherty, 1988). But the strategy itself is not beyond criticism with claims that observers also show distortion and bias. This problem is amply highlighted by Frese and Zapf (1988), Glick *et al.* (1986) and Karasek and Theorell (1990).

Problems include: (1) limited time observation; (2) unobservability of mental processes; (3) effects of observation on work behaviour; (4) halo and stereotyping effects; (5) representativeness of workplaces to be observed (Frese and Zapf, 1988; de Jonge *et al.*, 1999); and (6) human inferences of non incumbents, e.g. inferring affective states from job characteristics (Glick *et al.*, 1986). Further, observers' own cognitive and emotional processing can produce noise, which in turn is likely to lead to conservative correlations between independent (source) measures and psychological dysfunctioning (Frese and Zapf, 1988).

3.3.2 Measures of strain

Another problem for stress research is the extent to which questionnaire measurement can be used to assess strain, in particular affective states. Research has tended to focus on the use of self-report measures of affective states such as psychological distress (e.g. GHQ), depression, anxiety, emotional exhaustion, and job satisfaction. As noted in Section 3.3.1, Frese and Zapf (1988) argued that the level of affective and cognitive processing involved in self-report could be influenced by questionnaire instructions. For example, 'How stressed are you about the amount of work you need to do?' would arouse more affective processing than 'How many tasks do you perform in a day?'

Porac (1990) further suggested that questionnaire measures of affective states require complex cognitive processes, involving complex verbal memories and self-reflection, and that such 'reflective experience' (thinking about) is the product of second-order affective processes rather than the first-order 'lived experience' (reactive). Porac's commentary on 'the job satisfaction questionnaire as a cognitive event' cogently argues for the recognition of first- and second-order affective processes and disputes that research using questionnaires can measure first-order affective states. Any test for 'pre-cognition' responses to job satisfaction questionnaires is a contradiction in terms, according to Porac (1990).

Porac (1990) argued that whilst the construct validity problem existed with all questionnaire studies it did not make them useless but rather, they required validation against direct nonverbal measures of first-order appraisals (Porac, 1990). Such first-order assessment might entail measuring the arousal component of the first-order affective reaction with the galvanic skin response, or measuring the memory structures activated during an affective state via reaction times.

The objective measurement of outcome measures in both the psychological and work environment models would be theoretically consistent. Lazarus and Folkman (1984) explain that there are problems with indexing emotions objectively using physiological indications, such as hormonal or physical responses, or facial responses to certain emotions (Ekman and Friesen, 1975): 'we are in a sense entrapped by the need to verify one unknown, the experiences of emotion, by reference to other unknowns such as the meaning of the person's actions in a particular environmental context' (p. 323).

Other less problematic objective measures recommended for work stress research include organisational data such as actual days lost to sickness, accidents, health care utilisation costs, actual illness and physiological measures, including, diastolic and systolic blood pressure, heart rate, skin temperature, and stress hormones, i.e. adrenalin and cortisol (Schaubroeck and Ganster, 1993).

3.3.3 Summary

In summary, it may not be appropriate in psychological theoretical paradigms to measure objective job characteristics because the researcher is interested in the perception or appraisal of the environment, and in such cases self-report may be a valid and useful source of information.

When objective information is considered ideal, as in work environment models it is important to ascertain the approximation of self-report to objective assessments. Despite the shortcomings of the self-report method, empirical evidence has provided support for the accuracy of self-report measures (whether or not aggregated) of the objective environment. While some concerns have been raised about the influence of irrelevant cues on perception, these do not appear to rule out the potential influence of objective characteristics of the job.

In the assessment of strain, calls for more objective assessments have been made. The measurement of affective states by questionnaire is a cognitive event, an event likely to be in common with perception of the work conditions and likely to lead to a common method effect. For good practice in both occupational psychology and psychometrics, especially if decisions are challenged in law, measures need to be:

- Supported by evidence of reliability and validity
- Appropriate and fair in the situation (Cox *et al.*, 2000)

3.4   THE INFLUENCE OF NEGATIVE AFFECTIVITY ON THE WORK-STRESS PROCESS

There are important methodological reasons for assessing negative affectivity (NA) in work stress research. Most work stress research has relied on self-report assessments of work related influences on individual stress outcomes. In addition to the sources of the common method effect already mentioned, it is also proposed that NA may also underlie observed relationships between self-reported stressors and self-reported strain (Watson and Clark, 1984). First, it has been asserted that NA spuriously inflates the relationships between the variables because of a tendency for high NA individuals to perceive higher levels of job stress because they focus on the negative aspects of other people and the world in general. Further, as NA is the disposition to experience negative affect across time and regardless of the situation, high NA subjects would be likely to report higher levels of distress than low NA individuals (Watson and Clark, 1984). On this basis, Brief *et al.* (1988) have raised concerns that results from self-report work stress research which have not assessed dispositional influences, may be spurious.

Second, it has been asserted that NA might contaminate the measures of stressors and strain, in that it might share variance with the measures (Watson and Pennebaker, 1989). An individual high in NA will experience high levels of subjective distress that may be tapped by subjective measures of stress and well-being (Watson and Pennebaker, 1989). For example, Watson *et al.* (1987) using 150 healthy adults to complete tests of NA, stress reactions, and reports of physical symptoms, found support for the hypothesis that NA might be the construct underlying perceptions of stress, symptoms, and negative moods, because of the high intercorrelation among the measures (mean intercorrelation among the measures = .50). Watson *et al.* (1987) claim that on current tests of the hypothesis, 'we continue to find that all of these measures reflect the common influence of NA' (p. 145).

A recent test of both these suppositions found evidence that NA inflated the relationship between work stressors and strain, but using improved statistical methodology (confirmatory factor analysis) failed to establish that NA measured a factor in common with measures of subjective strain (Schaubroeck *et al.*, 1992). The results did however corroborate the first supposition and similar to other studies (Brief *et al.*, 1988; Morrison *et al.*, 1992; Parkes, 1990; Terry *et al.*, 1993) found that NA acted to inflate the relationship between work-stress measures and subjective strain symptoms. However, even when these studies controlled for the effects of NA, work stressors remained significant predictors of strain.

Although evidence has been uncovered regarding inflated relationships between work stress measures and subjective measures of strain, some researchers have been unable to determine the extent of the effect as confounding or as one of substantive influence. This is partly because only self-report measures of strain were used (Parkes, 1990). However, a recent study that did use objective physiological assessments of strain was unable to draw definitive conclusions, because the work stressors themselves failed to correlate with the physiological indicators (Schaubroeck *et al.*, 1992). More research is required on this issue.

Despite evidence that NA affects perceptions of health there is little evidence that NA is predictive of long-term health status, or objective indices of health status, such as number of visits to the doctor during the past year, or number of days missed due to illness (Watson *et al.*, 1987; Watson and Pennebaker, 1989). NA had no relationship with six objective measures of strain used by Schaubroeck *et al.* (1993), but was correlated with all subjective measures of strain. Costa and McCrae (1985) however found that neuroticism was a very strong predictor of chest pains (i.e. angina pectoris) but it was unrelated to objective indicators of coronary heart disease (CHD), including death from CHD or myocardial infarction which involved actual damage to the heart. Neuroticism has been shown to increase the risk of diseases such as gastric ulcer (Friedman and Booth-Kewley, 1987). Self-reported ill health (probably influenced by NA) also has been shown to be a risk factor for ischaemic heart disease mortality (Orth-Gomer *et al.*, 1986).

Because of its general lack of correlation with objective indices of health and performance it may be that NA acts rather as a nuisance factor 'a major source of variation in measures of stress and symptoms that is ultimately unrelated to long-term objective health status' (Watson *et al.*, 1987, p. 148). Furthermore there appears to be little evidence according to Watson *et al.* (1987) to suggest that high NA, although personally distressing and affecting perceptions of stress and job satisfaction, impacts negatively on performance and should therefore be of concern to employers. However there does appear to be a number of discrepant findings, to the extent that Watson and Pennebaker (1989) and Costa and McCrae (1987) in their comprehensive reviews were not prepared to rule out the role of NA in stress etiology (Schaubroeck *et al.*, 1993).

Measurement of NA therefore appears to be a critical methodological requirement in empirical tests of the work stress process (Brief *et al.*, 1988). Correlations between negative affectivity and strain could be in part due to the trait measure capturing some aspects of a state measure. For example, workers experiencing high levels of state anxiety over a period of time may respond to questions about 'how they generally feel' as though it were a state measure because they have been exposed to chronic stressors over time. If this were the case then negative affect would also be correlated with stressor measures.

A recommended solution proposed by (Brief *et al.*, 1988; Payne, 1998; Watson and Pennebaker, 1989) and endorsed by Ganster and Schaubroack (1991) is to control for person-based measures of negative affectivity. Recent strong criticisms of this have been proposed in Karasek *et al.* (1998). He argues that the cure could be worse than the problem, and could easily be overdone leading to Type II statistical errors. True variance in strain measures due to variation in strain measures could be removed with negative affectivity. Karasek *et al.* (1998) acknowledged that 'indeed, recent research (Dollard and Winefield, 1998) that explicitly test for the possibility of such over control with negative affectivity, using job experience cohorts to test whether negative affectivity is itself associated with duration of exposure to stressful job characteristics, finds that it is' (p. 350).

Spector *et al.* (2000) proposed a range of mechanisms by which NA could affect the stressor–strain relationship. They conclude that assuming that NA has only a biasing effect, partialling it from stressor–strain relationships, is the wrong approach. Rather they conclude that better quality of data is the way forward. Despite some evidence to the contrary (Watson *et al.*, 1987) a number of studies have found that NA is not linked with job satisfaction (Chen and Spector, 1991; Janman *et al.*, 1988; Schaubroeck *et al.*, 1992). The use of job satisfaction as an index of strain in work stress research would be a methodological improvement in the sense that observed relationships would not be due to NA. Other methodological improvements would be the use of more objective measures of stressors (a problem for psychological models) and strains.

3.5 THE RELATIONSHIP BETWEEN INDEPENDENT AND DEPENDENT MEASURES

3.5.1 Trivial correlations and common method effects

Most stress research attempts to examine the relationship between stressors, or stress appraisals, and outcome measures (strain). This presents another potential problem for self-report stress research within psychological paradigms, and to a lesser extent work environment paradigms: that of trivial correlations between subjective stress measures and ill-health (Kasl, 1978). This could occur because of the overlapping content that may exist in the measures, or because of variance that is due to the common method used to measure both the independent and dependent variable. Relationships could be due to a similar cognitive event (Porac, 1990) or underlying dispositions.

Methodologically, solutions to similar method variance and trivial correlations are: (1) to use multiple methods in assessment; (2) to use objective measures of the independent and/or dependent measures; (3) to use psychometrically validated measures (as well as classifying independent variables quite different from those for the outcome variables); and (4) to measure the indicators with different response formats (de Jonge, 1995; Parker and Wall, 1998; Spector, 1987b).

The use of subjective judgements of stressors may lead to an over-estimation of the correlation between stressors and dysfunctioning that may be due to common method variance; there is also evidence that objective (observers) judgements of stressors may lead to an underestimation of the 'true' correlation (Frese and Zapf, 1988). When researchers have used both objective and subjective measures, the latter have been found to be better predictors of outcomes (Kiggundu, 1980; Kirmeyer and Dougherty, 1988; Parkes, 1982). Because of the range of methodological problems it is difficult to know if these findings are methodologically based or whether subjective assessments are really more important in the work stress process.

Many researchers talk about the common method variance problem as a limitation of their research design when it typically features the self-report method. With a few exceptions (see review by Fried and Ferris, 1987; Glick *et al.*, 1986), researchers rarely try to assess the extent of the problem. Inferences are usually made by drawing evidence from studies which have used multi-methods (e.g. interviews, card sorts, and observations), to establish the accuracy of the self-report method (Morrison *et al.*, 1992). The argument is that if accuracy is established, variance due to method effects will be trivial.

Taken together the results of Fried and Ferris (1987) and Glick *et al.* (1986) suggest that although empirical relationships exist between self-report job characteristics and outcome, they are only in part due to common method effects. Researchers are advised to be concerned about methodological problems but several authors instil confidence that problems with self-rated data of job characteristics (not job stress) are *less serious* than initially believed (e.g. Fried and Ferris, 1987, Semmer *et al.*, 1996; Spector, 1987, 1992; Wall and Martin, 1987).

3.5.2 Theoretical perspectives on common method effects

Common method effects are generally interpreted in concrete terms such as single method, similar formats or similar situation effects or in abstract terms such as affirmative response, halo and central tendency effects (Frese and Zapf, 1988). Some view them as problems to be eradicated, whilst others view them as interesting phenomena to be explained in their own right. For example response effects may be considered as a particular cognitive dynamic occurring at the time of questionnaire completion, not as biases but inherent elements in personal description of feelings about work (Porac, 1990). Substantive processes like these could account for consistency effects and common method variance in stress research.

The issue of whether method versus substance underpins observed relationships in stress research has matured beyond a simple concern with methodological artefacts into a strong theoretical debate (Glick *et al.*, 1986). Different theoretical positions have emerged to predict different levels of common method effects.

For example, the job characteristics approach emphasises objective job characteristics as major antecedents of relevant perceptions and responses. This approach predicts only minor method effects even when the data are primarily self-reported (Glick *et al.*, 1986). However, it should be noted that non-affective items should be used in this case (i.e. items which contain a minimum of cognitive processing, which are precisely defined and are as neutral as possible) (Frese and Zapf, 1988). Even if different methods are used, substantive relationships should still be found.

On the other hand, the social information processing approach (Salancik and Pfeffer, 1978) which has its roots in social comparison theory (Festinger, 1954), social learning theory (Bandura, 1969), and cognitive consistency theory

(Abelson and Rosenberg, 1958), emphasises social and cognitive processes, and assumes that these processes will result in strong method effects when data are primarily self-reported. It hypothesises that: (1) perceived job characteristics and affective responses are related to each other primarily because they share antecedent cognitive and social processes; (2) these effects may be reflected in common method variance (Roberts and Glick, 1981); and (3) objective job characteristics are only marginal determinants of both perceived job characteristics and employee responses' (Glick *et al.*, 1986, p. 443). One of the central tenets of the theory is that job characteristics are defined through the set of informational cues received about the job from others. The theory questions whether job characteristics are fixed and objective and suggests instead that they are socially constructed (Weick, 1979).

Some studies have demonstrated the impact of social influences on task perceptions. For example, a simulation study showed that 'griping' co-workers contributed significantly to perceptions of whether a clerical task was enriched or not (White and Mitchell, 1979). Other experimental studies have also demonstrated the impact of social and informational cues on perceptions of task characteristics (see O'Reilly and Caldwell, 1979). A study of correctional officers tended to overestimate their peers' alienation and to assume that the majority of officers were more custody oriented (less job-enrichment oriented) than they were (Toch and Klofas, 1982).

Both the job characteristics approach and the social information processing approach recognise that job characteristics *and* individual cognition affect employees' perceptions of their jobs and their response to job characteristics. Both also propose a direct link between perceived job characteristics and outcome. The theories however vary in *emphasis*.

In addition to those identified in social information theory, other psychological processes such as attributional processes, may also be reflected in common method effects. According to attribution theory, attributions of causations for one's own behaviour are generally attributed to situational or external influences, whereas the behaviour of others is generally attributed to dispositional or internal influences. The tendency to exaggerate the dispositional causes of behaviour for others and underestimate its situational causes is referred to as the fundamental attributional error (Jones, 1979). The phenomenon has been shown experimentally, to provide an alternative explanation to organisational characteristics determining performance. For example Staw (1975) showed that knowledge of group performance was used as a cue to attribute characteristics to self, group members and organisations, rather than organisational characteristics.

In stress research, some workers who experience stress may view causation differently from others who experience stress, and from those who do not experience stress. Along these lines Kasl (1986) has argued that the existence of a mental or physical disorder could influence the perception of the work environment, either as a direct result of the disorder, or indirectly through the psychological process of attribution. This poses the problem of causal reversal (especially for cross-sectional research) and raises questions about whether the attribution effect constitutes a plausible alternative interpretation of correlations between organisational characteristics and strain. The attribution hypothesis posits that health status is an independent variable and that correlations between strain and self-report data may be described by the following causal sequence:

Strain $\rightarrow$ Attribution of causations $\rightarrow$ Self-report of job concerns

In sum, it is expected that method variance would be greater in psychological models than in work environment models because of the psychological processes that influence measurement of job stress in the former, and the use of more objective methods of measurement of job stressors in the latter.

3.5.3 Summary

To summarise this section, although self report methods have several limitations they are recommended on theoretical grounds in empirical tests of psychological models. Although the problem of common method variance is 'ubiquitous and vexing' it needs to be faced and the benefits of self report need to be considered in relation to other sources of data (Lazarus and Folkman, 1984). The relationship between subjective stress and strain found in psychological models will be based on both substance and method.

In tests of work environment models, self report can also be employed (Karasek, 1979) using measures with questions framed more 'objectively' (Frese and Zapf, 1988). Theoretically the relationships between stressors and strain in cross-sectional models will be based on substance and method, with method accounting for less variance than in the psychological model. Although the problem of overlap in content between independent and dependent measures is more likely to be an issue for empirical tests of psychological models, it could also occur in work environment models but it can be overcome by more objective assessment of measures (Karasek and Theorell, 1990).

3.6 RESEARCH DESIGNS

A major limitation with much of the work stress research is that mainly cross-sectional designs have been used to test the relationship between the work environment and strain. These involve taking observations at only one point in time, and they can only establish correlates between the variables, not probable causes of strain. Further, cross-sectional designs can not elucidate the transactional processes that may occur in the experience of work stress (Kasl, 1986). Longitudinal designs enable the empirical validation of causal inferences, and therefore reduce (but not solve) the causal issue. Testing at multiple points in time can elucidate the transactional, development process. For these reasons longitudinal designs and are strongly recommended in work stress research (Edwards, 1992; Frese and Zapf, 1988).

There is no question that the preferred research design in assessing the onset of occupational strain, is a prospective research design where initial levels of strain, biological risk factors, and other potential confounding variables are included in the initial data collection, along with the work stress predictors (see Kasl, 1986). According to Frese and Zapf (1988) we can not demonstrate causality but we can demonstrate plausible causal relationships between two variables if there is an association between two variables, if there is evidence about the direction of causality and if other explanations can be ruled out. Utilisation of a full panel design, with an adequately planned time lag, taking stabilities of variables into account, and using covariance structural modelling has been recommended to demonstrate plausible causation (Frese and Zapf, 1988; de Jonge *et al.*, 2001). However the design is costly, time consuming and inevitably a considerable loss of data will occur over time. Sometimes, for political or applied reasons the design may not be appropriate. Nevertheless the use of the prospective design offers a range of possibilities in occupational stress research.

First, the design can be used to evaluate the impact of environmental change that may come as a result of entering the work environment (Cherry, 1978), changing work contexts (Parkes, 1982), planned organisational change (Ashford, 1988; Wall and Clegg, 1981) or intervention (van der Klink *et al.*, 2001). The gold standard research design for intervention studies is the randomised case-control study (NHMRC, 1999).

Second, the design may be used to evaluate the impact of accidental (uncontrolled) change such as job loss or retrenchment, or industrial accident (Linn *et al.*, 1985).

Third, the design may be used to examine the evolution of the coping process in response to stress (Aldwin and Revenson, 1987), or outcomes of injury management.

Fourth, the design may be used to study the longer term impact of stress on strain symptomatology at a later point in time (Bromet *et al.*, 1988; Miles, 1975; Parkes, 1991). Most stress theories are based on the premise that exposure to chronic stress will be manifest in increasing levels of strain symptomatology, and that the effect of stressors accumulate over time. One way of assessing this assumption is with the use of a prospective design. In ten such studies of men six showed an increase in cardiovascular disease (CVD) risk due to job strain (high demands, low control), and two showed mixed results. Of 5 cohort studies in women, four showed higher levels of elevated risk in CVD related to job strain (Belkic *et al.*, 2000).

Fifth, prospective models can provide information about the stability of measures used. When dependent measures are selected they need to be potentially reactive to the work environment. Some measures would be expected to remain fairly stable, such as personal disposition variables, whereas other variables, such as state measures would be expected to change according to external circumstances.

Sixth, there is good reason to expect reverse causation in stress research as suggested earlier. This may occur due to drift (a person with high emotional exhaustion may receive less support over time as they may not be able to sustain a relationship with their supervisor, e.g. Daniels and Guppy, 1997) or changes in perception due to changes in emotional state. In a review of 16 longitudinal studies of work stress which tested for reverse causation, 6 were found to show evidence (Zapf *et al.*, 1996). Longitudinal designs with full panel data, can enable the examination of reciprocal relations by the utilisation of multivariate statistical techniques such as structural equation modelling. de Jonge *et al.* (2001) using such techniques found Time 1 job characteristics influenced Time 2 job satisfaction and found only weak evidence for a lagged reverse relationship between emotional exhaustion (Time 2) and job demands (Time 1).

With any research design there are associated limitations. For the longitudinal design major challenges are how to deal with attrition, and unanticipated changes in the organisation (e.g. turnover of key proponents of the research, restructuring). Further the design is most costly in terms of time and resources.

A practical issue with longitudinal designs concerns the timing of tests: when to test for impact? Various time frames have been used: two months (Parkes, 1991) four months (Miles, 1975), six months (Ashford, 1988; Parkes, 1991), and one year (Bromet *et al.*, 1988). Others have recommended 10 years of shiftwork for the development of psychosomatic problems, whereas stressors related to mental load may have an impact on dysfunctioning within some months (Frese and Zapf, 1988). According to Frese and Zapf (1988) the type of stressor and its intensity are important in determining time course. For example, the outcomes of psychosomatic complaints, depression and anxiety, should develop quicker than psychosomatic illness. Some time ago Miles (1975) noted that there was no articulated theory specifying time lags for causal models and this observation is still relevant today.

3.7 GROUNDED, COMPREHENSIVE AND PARTICIPATIVE APPROACHES

3.7.1 Participative approaches

Given the problems in starting research from a single theoretical view point, given the multitude of theories available (see also Chapter 1) a growing body of research is beginning at the site of the struggle, using participatory action research approaches and evolving local theory (Dollard *et al.*, 1997; Landsbergis and Vivona-Vaughan, 1995). We do not suggest that approaches should be atheoretical but rather scope should be provided for the development of local theory – a wide understanding of work stress in a local context. This involves workers participating: in the definition of issues or problems (if they exist); in the development of methodology and data collection to inform the problem; in making sense of the data; in defining interventions; in helping to implement them; and in the evaluation of the results (Wadsworth, 1998). Indeed, Chapter 5 in this volume describes a PAR approach to implementing burnout prevention training for oncology care providers.

Whereas 'stress surveys' tend to measure hazards or outcomes, 'risk assessment' approaches to measurement are being used increasingly in the European Union and are supported by legislative frameworks. Risk assessment 'intends to establish an association between hazards and health outcomes, and to evaluate the risk to health from exposure to a hazard' (Cox *et al.*, 2000, p. 8). A step-wise approach to risk assessment is recommended as follows: (1) preparation and introduction of the project; (2) problem identification and risk-assessment;

(3) choice of measures and planning of interventions (control strategies); (4) implementation of interventions; (5) evaluation of interventions and re-assessment of risks (similar to those of the control cycle Cox and Cox, 1993); and (6) review of information needs and training needs of employees exposed to hazards (Cox and Griffiths, 1995; Janssen *et al.*, 1996). This particular approach is elegantly illustrated in Chapter 6, which describes such an approach with senior nurses.

The philosophy of how risk assessment might be implemented follows lessons learned from the evaluation of 11 European organisational case studies of work stress interventions by Kompier and Cooper (1999). They recommend that successful organisational interventions: (1) need to be stepwise and systematic; (2) require an adequate diagnosis or risk analysis; (3) combine both work-directed and person-directed measures; (4) use a participatory approach (worker involvement); and (5) have top management support. Obvious overlaps can be seen between the 'risk assessment' approach and the PAR methodology outlined above, and these recommendations. In other words it is evident that principles in intervention methodology are converging transnationally in contemporary approaches to applied work stress research (particularly involving intervention).

For organisations to be 'learning' environments this process needs to be ongoing, and iterative so that continuous improvements can be made in productivity and in ways conducive to good health for workers. Key stakeholders may be closely involved at every stage.

3.7.2   Triangulation

Triangulation in research is a technique growing in popularity and has the potential to overcome the limitations of mono-method, or mono-theoretical research. Triangulation utilises multiple theoretical frameworks and sources of data, to strengthen the trustworthiness of the research findings.

The goal of triangulation is to strengthen data analysis through confirmation and completeness (Breitmayer *et al.*, 1993). *Data Triangulation* utilises multiple sources for example: (a) data obtained from the focus groups; (b) data obtained using the self-report instruments; and (c) objective data obtained from organisations regarding performance (Hamilton and Bechtel, 1996). *Methodological Triangulation* involves the simultaneous use of two or more methodologies to study a single phenomenon. This may involve the use of positivist methodology seeking theory testing quantitative data (e.g. questionnaires), and the use of grounded methodology seeking theory generation and data of a qualitative nature (focus groups) (Hamilton and Bechtel, 1996). *Theory triangulation* uses two or more conceptual frameworks to explore a single phenomenon under investigation (Hamilton and Bechtel, 1996). For example Theorell (1998) argued for the simultaneous use of two work stress theories in relevant research: the Demand-Control-Support model (Karasek and Theorell, 1990), and the Effort–Reward Imbalance Model (Siegrist, 1996). *Analysis Triangulation* uses different methods of analysis such as multivariate statistics, as well as qualitative techniques such as content analysis (Minichiello *et al.*, 1995). Analysis triangulation assists in the development of a richer perspective (Hamilton and Bechtel, 1996).

A timely 20 year review of qualitative research in organisational psychology has been recently published by Lee *et al.* (1999). Their aim is to bolster interest in the use of qualitative research in organisational psychology. Lee (1999) argues that qualitative research is well suited for the purposes of description, interpretation, and explanation, whereas quantitative research is well suited for prevalence, generalisability, and calibration. Further qualitative research effectively address 'how and what' psychological and sociological processes evolve over time, whereas quantitative research effectively addresses issues of 'how much' (p. 167). They claim that contemporary researchers are seeking additional tools in their analysis of organisational research questions less amenable to insight or solution based on traditional quantitative techniques. They provide exemplary case studies of research including ethnography (where the researcher spends a considerable amount of time interacting within an organisation or work setting), and in-depth interviews (where researchers interview participants at length, either in a structured or free-ranging way).

As noted in Chapter 1, the tensions between objective vs subjective; quantitative vs qualitative approaches essentially derive from different philosophies of science (see also Lee, 1999). At one extreme is the natural scientific method (which assumes an empirical world with a single objective reality) and at the other the radical postmodernists who hold a completely subjective view of the world (they assume as many realities as there are people). Lee (1999, pp. 10–11) suggests a middle ground between these extreme views.

> Most [organisational and vocational psychologists likely] accept that organizational members actively engage, at least to some extent, in the social construction of reality and sense making .... By inference, multiple subjective realities can co-exist, and the desirability of qualitative research aimed at understanding these multiple realities is suggested .... Simultaneously, most ... researchers also accept that a vast amount of systematic regularity, though not complete uniformity, occurs within organizational contexts .... It is this systematic regularity in employees' behaviors, interpretations, and agreement on organizational processes that allows the evolution of dominant modes, larger organizational cultures, and a strong, agreed-upon, taken-for-granted, and virtually singular organizational reality.

We agree with Lee's advocacy for a middle position between '(a) the assumptions of an objective reality ... and (b) an ongoing and constant process of interpretation, sense making, and social construction of organizational settings' because it enables a richer understanding of the problem at hand, and increases the potential for innovation.

Pluralism of method in recent work by Heuven and Bakker (2001) shows how useful insights were gained through in-depth interviews (in addition to survey) with cabin attendants about emotional dissonance and burnout. They found that employees of an ostensibly homogenous group showed a great diversity of reactions in relation to emotional labour. They found that emotional display is 'not only defined by organisational demands but is actively regulated by the human service agent on the basis of interaction and reciprocity with recipients'. They noted through interview that healthy and unhealthy groups differed in the extent to which they modified organisational display rules: healthy workers defined the rules more on the basis of reciprocity with passengers, whereas workers who used standardised organisational rules 'perfectly smiling objects' were more burnt out.

Methodological challenges are not confined at the individual, group, and/or organisational level. Cross cultural research in many ways mirrors the problems of research within (unique ~ emic) and between (general ~ etic) occupations/ organisations, and the search of a middle ground – a compromise between an objective reality and the process of interpretation, sense making, and social construction within organisations (local variables). As discussed, theoretical pluralism, multi-methods and participative approaches are frequently advocated in organisational stress research (Theorell, 1998). These generally attempt to uncover emic elements by treating the organisation under study as its own unique culture to at least some degree, while selectively incorporating etic elements from extensions of known theories and results elsewhere into an emerging tailored model of stress in the organisation under study (Narayanan *et al.*, 1999).

> The extension of organisational stress research into another culture redoubles this paradigmatic divide and provides an opportunity to heighten the issue for study. So, lessons learned from the study of organisational stress in a cross-cultural context may provide insights of relevance to the study of organisational stress more generally, and the study of building a culturally-appropriate model of organisational stress (P. Heffernan, personal communication, 2002).

Limitations in the scientific method have lead to more active, participatory designs utilising multiple testing points, continuous feedback from participants, and are problem driven rather than merely theoretically driven.

3.7.3   Comprehensive measurement: Economic indicators of health and well-being

Increasingly stress research is turning to comprehensive measurement of a different kind and contextualising stress in the broader work health productivity framework. Cost benefit analysis (e.g. of stress vs prevention or intervention) and economic analysis of health and well-being are being sought (see Dorman, 2000).

Recent developments call for a *whole of organisational approach* to organisational performance analysis (the bottom line), not just economic interest components. Rather than focusing on just economic factors or health aspects, more versatile performance analysis requirements have led to a multidisciplinary result concept. Liukkonen *et al.* (1999) argue that while historical data regarding financial performance is important (e.g. liquidity, solvency, renumerativeness and profitability), organisational performance analysis should also include a predictive

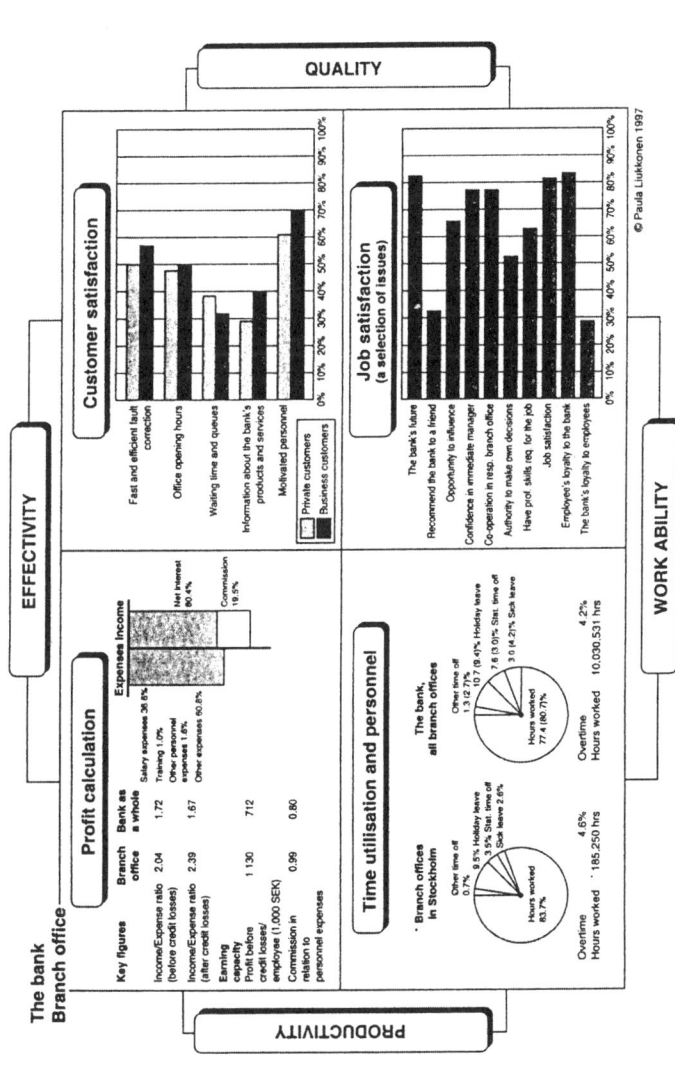

Figure 3.1 Liukkonen, P., Cartwright, S. and Cooper, C., 1999, Costs and benefits of stress prevention in organizations. In M. Kompier and C. Cooper (Eds), *Preventing stress, improving productivity: European case studies in the workplace.* (London: Routledge), p. 346.

element regarding present and future health of employees, to reflect the organisation's collective health and capacity. They propose 4 dimensions of company results: (1) a financial result made up of quantitative measurements from the organisations profit and loss account and balance sheet; (2) time and personnel resources including hours worked, personnel statistics, levels of staffing and competency; (3) customer satisfaction demonstrated by qualitative measures such as how effective a company is at providing value and satisfying the customer; and (4) health and safety, including opinion surveys, and employee health appraisals (see Figure 3.1) (from Kompier and Cooper, 1999).

They propose that the organisational performance analysis method (e.g. Oskar, a Swedish method) requires increased communication between different functional units in the organisation, to find a 'company language' and methods of combined reporting to bring results from different areas together.

> Companies need meaningful information about the quantitative and qualitative values that affect profit development – information about customer satisfaction, employee health, work capacity, staff, competence and time utilisation (p. 40).

In this way companies and researchers are forced to examine cost-benefit analyses of work environments that lead to unhealthy behaviour and low productivity, and health officers are required to examine the cost–benefit analysis of work stress/interventions. Approaches such as these appear to embrace all the value perspectives outlined in Chapter 1 by Levi more evenly: after all our goal is healthy people in productive work.

3.8   CONCLUSION

In conclusion this paper has focused on methodological and measurement issues in work stress research and management. Each of the theoretical models of work stress proposes a particular ideology for conceptualising workplace stress, its antecedents and consequences. Measurement issues that arise from assessment of the theoretical models were discussed. Difficulties emerging from theory testing and identified organisational problems have given rise to more active, participatory research methodologies that use multiple theories and intend to develop new local theory, and multidisciplinary approaches grounded in the site of the struggle. Contemporary approaches to the measurement of work stress impacts in organisations call for multidisciplinary approaches, and the inclusion of all Levi's value positions.

3.9 REFERENCES

Abelson, R.P. and Rosenberg, M.J., 1958, Symbolic psychologic: A model of attitudinal cognition. *Behavioral Science*, **3**, pp. 1–13.

Aldwin, C.A. and Revenson, T.A., 1987, Does coping help? A re-examination of the relationship between coping and health. *Journal of Social Psychology*, **53**, pp. 337–348.

Ashford, S.J., 1988, Individual strategies for coping with stress during organizational transitions. *Journal of Applied Behavioral Science*, **24**, pp. 19–36.

Bandura, A., 1969, *Principles of behavior modification* (New York: Rinehart and Winston).

Baker, D.B., 1985, The study of stress at work. *Annual Review of Public Health*, **6**, pp. 367–81.

Belkic, K., Schnall, P., Landsbergis, P. and Baker, D., 2000, The workplace and cardiovascular health: Conclusions and thoughts for a future agenda. *Occupational Medicine*: State of the art reviews, **15**, pp. 307–321.

Bosma, H. and Marmot, M.G., 1997, Low job control and risk of coronary heart disease in Whitehall 11 (prospective cohort) study. *British Medical Journal*, **314**, pp. 558–565.

Breitmayer, B., Ayres, L. and Knafl, K., 1993, Triangulation in qualitative research: Evaluation of completeness and confirmation purposes. *Image*, **35**, pp. 237–343.

Brief, A.P., Burke, M.J., George, J.M., Robinson, B.S. and Webster, J., 1988, Should negative affectivity remain an unmeasured variable in the study of job stress? *Journal of Applied Psychology*, **73**, pp. 207–214.

Bromet, E.J., Dew, M.A., Parkinson, D.K. and Schulberg, H.C., 1988, Predictive effects of occupational and marital stress on the mental health of a male workforce. *Journal of Organizational Behavior*, **9**, pp. 1–13.

Chen, P.Y. and Spector, P.E., 1991, Negative affectivity as the underlying cause of correlations between stressors and strain. *Journal of Applied Psychology*, **76**, pp. 398–407.

Cherry, N., 1978, Stress, anxiety and work: A longitudinal study. *Journal of Occupational Psychology*, **51**, pp. 259–270.

Costa, P.T. and McCrae, R.R., 1987, Neuroticism, somatic complaints, and disease. *Journal of Personality*, **55**, pp. 299–316.

Costa, P.T., Jr. and McCrae, R.R., 1985, Hypochondriasis, neuroticism and aging. *American Psychologist*, **40**, pp. 19–28.

Cox, T. and Cox, S., 1993, *Psychosocial and organizational hazards at work: Control and monitoring* (Copenhagen: WHO Regional Office).

Cox, T. and Griffiths, A., 1995, The assessment of psychosocial hazards at work. In M. J. Shabracq, J.A.M. Winnubst and C. Cooper (eds), *Handbook of work health psychology* (Chichester: Wiley & Sons).

Cox, T., Griffiths, A. and Rial-Gonzalez, E., 2000, Research on work-related stress. European Agency for Safety and Health at Work. http://agency.osha.eu.int/publications/reports/stress/full.php3.

Daniels, K. and Guppy, A., 1997, Stressors, locus of control, and social support as consequences of affective psychological well-being. *Journal of Occupational Health Psychology*, **2**, pp. 156–174.

de Jonge, J., 1995, *Job Autonomy, Well-Being, and Health: A Study Among Dutch Health Care Workers* (Maastricht, The Netherlands: Datawyse).

de Jonge, J., van Breukelen, G.J.P., Landeweerd, J.A. and Nijhuis, F.J.N., 1999, Comparing group and individual level assessments of job characteristics in testing the Job Demand-Control Model: A multilevel approach. *Human Relations*, **52**, pp. 95–122.

de Jonge, J., Dormann, C., Janssen, P.P.M., Dollard, M.F., Landeweerd, J A. and Nijhuis, F.J.N., 2001, Testing reciprocal relationships between job characteristics and psychological well-being: A cross-lagged structural equation model. *Journal of Occupational and Organizational Psychology*, **74**, pp. 29–46.

Dollard, M.F., Heffernan, P., Winefield, A.H. and Winefield, H.R., 1997, Conducive production: How to produce a PAR worksite proposal. *New Solutions, A Journal of Environmental and Occupational Health Policy*, **7**, pp. 58–70.

Dollard, M.F. and Winefield, A.H., 1998, A test of the Demand-Control/Support model of work stress in correctional officers. *Journal of Occupational Health Psychology*, **3**, pp. 1–23.

Dorman, P., 2000, *The economics of safety, health and well-being at work: An over view*. In Focus Program on Safework (The Evergreen State College International Labour Organisation), pp. 1–42.

Edwards, J.R., 1992, Cybernetic theory of stress, coping, and well-being in organizations. *Academy of Management Review*, **17**, pp. 238–274.

Ekman, P. and Friesen, W.V., 1975, *Unmasking the face* (Englewood Cliffs, NJ: Prentice-Hall).

Elovainio, M., Kivimäki, M., Steen, N. and Kalliomäki-Levanto, T., 2000, Organizational and individual factors affecting mental health and job satisfaction: A multilevel analysis of job control and personality. *Journal of Occupational Health Psychology*, **5**, pp. 269–277.

Festinger, L., 1954, A theory of social comparison processes. *Human Relations*, **7**, pp. 736–741.

French, J.R.P., Jr, Caplan, R.D. and van Harrison, R., 1984, *The mechanisms of job stress and strain* (New York: Wiley).

Frese, M., 1999, Social support as a moderator of the relationship between work stressors and psychological dysfunctioning: A longitudinal study with objective measures. *Journal of Occupational Health Psychology*, **4**, pp. 179–192.

Frese, M. and Zapf, D., 1988, Methodological issues in the study of work stress: Objective vs subjective measurement of work stress and the question of longitudinal studies. In C.L. Cooper and R. Payne (eds), *Causes, coping and consequences of stress at work* (London: Wiley), pp. 375–411.

Fried, Y. and Ferris, G.R., 1987, The validity of the job characteristics model: A review and meta-analysis. *Personnel Psychology*, **40**, pp. 287–322.

Friedman, H.S. and Booth-Kewley, A., 1987, The 'disease prone personality': A meta-analytic view of the construct. *American Psychologist*, **42**, pp. 539–555.

Friedman, M. and Rosenman, R.H., 1974, *Type A behavior and your heart* (New York: Knopf).

Ganster, D.C. and Schaubroeck, J., 1991, Work stress and employee health. *Journal of Management*, **17**, pp. 235–271.

Glick, W.H., Jenkins, G.D. and Gupta, N., 1986, Method versus substance: How strong are underlying relationships between job characteristics and job characteristic outcomes? *Academy of Management Journal*, **29**, pp. 441–464.

Greenhaus, J.H. and Parasuraman, S., 1987, A work and non-work interactive perspective of stress and its consequences. In J.M. Ivancevich and D.C. Ganster (eds), *Job stress: From theory to suggestion* (New York: Haworth), pp. 37–60.

Griffin, R.W., 1983, Objective and social sources of information in task redesign: A field experiment. *Administrative Science Quarterly*, **28**, pp. 184–200.

Griffin, R.W., Bateman, T.S., Wayne, S.J. and Head, T.C., 1987, Objective and social factors as determinants of task perceptions and responses: An integrated perspective and empirical investigation. *Academy of Management Journal*, **30**, pp. 501–523.

Hackman, J.R. and Oldham, G.R., 1975, The job diagnostic survey. *Journal of Applied Psychology*, **60**, pp. 159–170.

Hamilton, D. and Bechtel, G.A., 1996, Research Implications for Alternative Health Therapies. *Nursing Forum*, **31**, pp. 6–11.

Hart, P.M. and Cooper, C.L., 2001, Occupational stress: Toward a more integrated framework. In N. Anderson, D.S. Ones, H. Kepir Sinangil and C. Viswesvaran (eds), *Handbook of Industrial, Work and Organizational Psychology* (Volume 2), (London: Sage Publications), pp. 93–114.

Heuven, E. and Bakker, A.B., 2001, *Emotional dissonance and burnout among cabin personnel*. Manuscript submitted for publication.

James, L.R. and Tetrick, L.E., 1986, Confirmatory analytic tests of three causal models relating job perceptions to job satisfaction. *Journal of Applied Psychology*, **71**, pp. 77–82.

Janman, K., Jones, J.C., Payne, R.L. and Rick, J.T., 1988, Clustering individuals as a way of dealing with multiple predictors in occupational stress research. *Behavioral Medicine*, **14**, pp. 17–29.

Janssen, P.P.M., Nijhuis, F.J.N., Lourijsen, E.C.M.P. and Schaufeli, W.B., 1996, *Healthy work; Less Absenteeism! A manual for work-site health promotion*. (Amsterdam: NIA).

Jex, S.M. and Spector, P.E., 1996, The impact of negative affectivity on stressor–strain relations: a replication and extension. *Work & Stress*, **10**, pp. 36–45.

Jones, E.E., 1979, The rocky road from acts to dispositions. *American Psychologist*, **31**, pp. 107–117.

Karasek, R., 1989, Control in the workplace and its health-related aspects. In S.L. Sauter, J.J. Hurrell and Cooper, C.L. (eds), *Job control and worker health* (New York: Wiley), pp. 129–159.

Karasek, R.A., 1979, Job demands, job decision latitude, and mental strain: Implications for job redesign. *Administrative Science Quarterly*, **24**, pp. 285–308.

Karasek, R.A. and Theorell, T., 1990, *Healthy work: Stress, productivity and the reconstruction of working* life (New York: Basic Books).

Karasek, R.A., Baker, D., Marxer, F., Ahlbom, A. and Theorell, T., 1981, Job decision latitude, job demands, and cardiovascular disease: A prospective study of Swedish men. *American Journal of Public Health*, **71**, pp. 694–705.

Karasek, R.A., Brisson, C., Kawakami, N., Houtman, I., Bongers, P. and Amick, B., 1998, The job content questionnaire (JCQ). An instrument for internationally comparative assessments of psychosical job characteristics. *Journal of Occupational Health Psychology*, **3**, pp. 322–355.

Karasek, R.A., Theorell, T., Schwartz, J., Schnall, P., Pieper, C. and Michela, J., 1988, Job characteristics in relation to the prevalence of myocardial infarction in the U.S. HES and HANES. *American Journal of Public Health*, **78**, pp. 910–918.

Kasl, S., 1986, Stress and disease in the workplace: A methodological commentary on the accumulated evidence. In M.F. Cataldo and T.J. Coates (eds), *Health and industry: A behavioral medicine perspective* (New York: Wiley), pp. 52–85.

Kasl, S.V., 1978, Epidemiological contributions to the study of work stress. In C.L. Cooper and R.L. Payne (eds), *Stress at work* (New York: Wiley), pp. 3–48.

Kiggundu, M.N., 1980, An empirical test of the theory of job design using multiple job ratings. *Human Relations*, **33**, pp. 339–351.

Kirmeyer, S.L. and Dougherty, T.W., 1988, Work load, tension and coping: Moderating effects of supervisor support. *Personnel Psychology*, **41**, pp. 125–139.

Kompier, M. and Cooper, C., 1999, *Preventing stress, improving productivity* (London; Routledge).

Landsbergis, P.A. and Vivona-Vaughan, E., 1995, Evaluation of an occupational intervention in a public agency. *Journal of Organizational Behavior*, **16**, pp. 29–48.

Latack, J.C., 1986, Coping with job stress: Measures and future directions for scale development. *Journal of Applied Psychology*, **71**, pp. 377–385.

Lazarus, R.S. and Folkman, S., 1984, *Stress, appraisal, and coping* (New York: Springer).

Le Blanc, P.M., de Jonge, J. and Schaufeli, W.B., 2000, Job stress and health. In *An Introduction to Work and Organizational Psychology: A European Perspective*, edited by N. Chmiel (Oxford: Blackwell Publishers), pp. 148–177.

Lee, T.W., 1999, *Using qualitative methods in organizational research* (Thousand Oaks, CA: Sage).

Lee, T.W., Mitchell, T.R. and Sablynski, C.J., 1999, Qualitative research in organizational and vocational psychology, 1979–1999. *Journal of Vocational Behavior*, **55**, pp. 161–187.

Linn, M.W., Sandifer, R. and Stein, S., 1985, Effects of unemployment on mental and physical health. *American Journal of Public Health*, **75**, pp. 502–506.

Liukkonen, P., Cartwright, S. and Cooper, C., 1999, Costs and benefits of stress prevention in organizations. In M. Kompier and C. Cooper (eds), *Preventing stress, improving productivity: European case studies in the workplace.* (London: Routledge), pp. 33–51.

Miles, R.H., 1975, Causal inference: Role perceptions and personal outcomes. *Journal of Applied Psychology*, **60**, pp. 334–339.

Minichiello, V., Aroni, R., Timewell, E. and Alexander, L., 1995, *In-depth interviewing* (Sydney: Longman).

Moos, R.H., 1986, *Work Environment Scale manual* (California: Consulting Psychologists Press).

Morrison, D.L., Dunne, M.P., Fitzgerald, R. and Cloghan, D., 1992, Job design and levels of physical and mental strain among Australian prison officers. *Work and Stress*, **6**, pp. 13–31.

Narayanan, L., Menon, S. and Spector, P., 1999, A cross sectional comparison of job stressors and reactions among employees holding comparable jobs in two countries. *International Journal of Stress Management*, **6**, pp. 197–212.

Newton, T.J., 1989, Occupational stress and coping with stress: A critique. *Human Relations*, **42**, pp. 441–461.

NHMRC, 1999, National Health and Medical Research Council. A guide to development, implementation and evaluation of clinical practice guidelines. (Commonwealth of Australia: Canberra).

O'Reilly, C.A. and Caldwell, D.F., 1979, Informational influence as a determinant of perceived task characteristics and job satisfaction. *Journal of Applied Psychology*, **64**, pp. 157–165.

Orth-Gomer, K., Johnson, J.V., Under, A.H. and Edwards, M.E., 1986, Social interaction and mortality in Sweden. Findings in the normal population and in cardiovascular patients. In S.O. Isacsson and L. Janzon (eds), *Social support- health and disease* (Stockholm: Almquist and Wiksell).

Parker, S. and Wall, T., 1998, *Job and Work Design.* (Thousand Oaks, California: Sage Publications).

Parkes, K.R., 1982, Occupational stress among student nurses: A natural experiment. *Journal of Applied Psychology*, **67**, pp. 784–796.

Parkes, K.R., 1990, Coping, negative affectivity, and the work environment: Additive and interactive predictors of mental health. *Journal of Applied Psychology*, **75**, pp. 399–409.

Parkes, K.R., 1991, Locus of control as moderator: An explanation for additive versus interactive findings in the demand-discretion model of work stress? *British Journal of Psychology*, **82**, pp. 291–312.

Payne, R., 1988, A longitudinal study of the psychological well being of unemployed men and the mediating effect of neuroticism. *Human Relations*, **41**, pp. 119–138.

Perrewe, P. and Zellars, K.L., 1999, An examination of attributions and emotions in the transaction approach to the organizational stress process. *Journal of Organizational Behaviour*, **20**, pp. 739–752.

Porac, J.F., 1990, The job satisfaction questionnaire as a cognitive event: First- and second-order processes in affective commentary. In G.R. Ferris and K.M. Rowland (eds), *Theoretical and methodological issues in human resource management* (Greenwich: JAI Press Inc), pp. 85–136.

Quick, J.D., Quick, J.C. and Nelson, D.L., 1998, The theory of preventive stress management in organizations. In C. Cooper (ed.), *Theories of organizational stress* (Oxford: Oxford University Press), pp. 246–268.

Roberts, K.H. and Glick, W.H., 1981, The job characteristics approach to task design: A critical review. *Journal of Applied Psychology*, **66**, pp. 193–217.

Salancik, G.R. and Pfeffer, J., 1978, A social information processing approach to job attitudes and task design. *Administrative Science Quarterly*, **23**, pp. 224–456.

Schaubroeck, J. and Ganster, D.C., 1993, Chronic demands and responsivity to challenge. *Journal of Applied Psychology*, **78**, pp. 73–85.

Schaubroeck, J., Ganster, D.C. and Fox, M.L., 1992, Dispositional affect and work-related stress. *Journal of Applied Psychology*, **77**, pp. 322–335.

Semmer, N., Zapf, D. and Greif, S., 1996, 'Shared job strain': a new approach for assessing the validity of job stress measurements. *Journal of Occupational and Organizational Psychology*, **69**, pp. 293–310.

Siegrist, J., 1996, Adverse health effects of high-effort/low-reward conditions, *Journal of Occupational Health Psychology*, **1**, pp. 27–41.

Snijders, T.A.B. and Bosker, R.J., 1999, *Multilevel analysis: An introduction to basic and advanced multilevel modeling* (Thousands Oaks, California: Sage Publications).

Söderfeldt, B., Söderfeldt, M., Jones, K., O'Campo, P., Muntaner, C., Ohlson, C.G. and Warg, L.E., 1997, Does organization matter? a multilevel analysis of the demand-control model applied to human services. *Social Science and Medicine*, **44**(4), pp. 527–534.

Spector, P.E., 1987, Method variance as an artifact in self-reported affect and perceptions as work: Myth or significant problem? *Journal of Applied Psychology*, **72**, pp. 438–443.

Spector, P.E., Dwyer, J. and Jex, S.M., 1988, Relation of job stressors to affective, health, and performance outcomes: A comparison of multiple data sources. *Journal of Applied Psychology*, **73**, pp. 11–19.

Spector, P.E., Zapf, D., Chen, P.Y., Frese, M., 2000, Why negative affectivity should not be controlled in job stress research: Don't throw the baby out with the bath water. *Journal of Organizational Behaviour*, **21**, pp. 79–95.

Spector, P.E., 1992, A consideration of the validity and meaning of self-report measures of job conditions. In C.L. Cooper and I.T. Robertson (eds), *International Review of Industrial and Organizational Psychology* (New York: Wiley & Sons), pp. 123–151.

Staw, B.M., 1975, Attribution of the causes of performance: A new alternative interpretation of cross-sectional research on organizations. *Organizational Behavior and Human Performance*, **13**, pp. 414–432.

Terry, D.J., Nielson, M. and Perchard, L., 1993, Effects of work stress on psychological well-being and job satisfaction: The stress-buffering role of social support. *Australian Journal of Psychology*, **45**, pp. 168–175.

Theorell, T., 1998, Job characteristics in a theoretical and practical health context. In C. Cooper (ed.), *Theories of organisational stress* (Oxford: Oxford Press), pp. 205–219.

Toch, H. and Klofas, J., 1982, Alienation and desire for job enrichment among correctional officers. *Federal Probation Quarterly*, **46**, pp. 35–44.

Van der Klink, J. Blonk, R., Schene, A.H. and van Dijk, F.J.H., 2001, The benefits of interventions for work-related stress. *American Journal of Public Health*, **91**, pp. 270–276.

Van Yperen, N.W. and Snijders, T.A.B., 2000, A multilevel analysis of the demands-control model: is stress at work determined by factors at the group or at the individual level. *Journal of Occupational Health Psychology*, **5**, pp. 182–190.

Wadsworth, Y., 1998, Participatory Action Research http://www.scu.edu.au/schools/sawd/ari/ari-wadsworth.html

Wall, T.D. and Clegg, C.W., 1981, A longitudinal study of group work redesign. *Journal of Occupational Behavior*, **2**, pp. 31–49.

Wall, T.D. and Martin, R., 1987, Job and work design. In C.L. Cooper and I.T. Robertson (eds), *International Review of Industrial and Organizational Psychology* (New York: John Wiley & Sons), pp. 61–91.

Warr, P., 1987, *Work, Unemployment, and Mental Health* (Oxford: Clarendon Press).

Watson, D. and Clark, L.A., 1984, Negative affectivity: The disposition to experience aversive emotional states. *Psychological Bulletin*, **96**, pp. 465–490.

Watson, D. and Pennebaker, J.W., 1989, Health complaints, stress, and distress: Exploring the central role of negative affectivity. *Psychological Review*, **96**, pp. 234–254.

Watson, D., Pennebaker, J.W. and Folger, R., 1987, Beyond negative affectivity: Measuring stress and satisfaction in the workplace. In J. M. Ivancevich and D. C. Ganster (eds), *Job stress: From theory to suggestion* (New York: Haworth Press), pp. 141–157.

Weick, K.E., 1979, *The social psychology of organizing* (2nd ed.), Reading (Mass: Addison-Wesley).

Zajonc, R.B., 1984, On the primacy of affect. *American Psychologist*, **39**, pp. 117–123.

Zapf, D., Dormann, C. and Frese, M., 1996, Longitudinal studies in organizational stress research: A review of the literature with reference to methodological issues. *Journal of Occupational Health Psychology*, **1**, pp. 145–169.

CHAPTER FOUR

Conventional Wisdom is Often Misleading: Police Stress Within an Organisational Health Framework

Peter M. Hart and Peter Cotton

4.1 INTRODUCTION

Occupational stress among police officers is often viewed as an unfortunate, but inevitable part of police work. Although this view dominates much of the discussion about police stress in scientific, management, and other professional forums, there is no compelling evidence to support the view that police officers are any more or less stressed than other occupational groups (e.g. Hart, Wearing and Headey, 1995; Kirkcaldy *et al.*, 1995).

To address this apparent discrepancy, we drew on the organisational health framework (Hart and Cooper, 2001) to investigate three questions that are central to the debate on the nature and extent of police stress. First, we examined whether the levels of occupational well-being among police officers differed from the levels that are found in other occupational groups. Second, we examined whether police officers' levels of occupational well-being were determined by the work experiences that are peculiar to the nature of police work, or the experiences that are common to most occupational groups. Finally, we examined whether it was the personality characteristics of police officers, the nature of the police organisation, police officers' use of different coping strategies, or their positive and negative work experiences that contributed most to police officers' levels of occupational well-being. By providing answers to these questions, we are able to establish interventions and strategies that are most likely to improve occupational well-being in police organisations.

4.2 OCCUPATIONAL STRESS AMONG POLICE OFFICERS

It goes without saying that police perform an extremely important role in society, particularly through their law enforcement and community service functions. Moreover, the nature of the police role has given rise to a strong stereotypic view about the dangerous and stressful nature of police work. As with most stereotypes, however, it is important to take a careful and systematic look at whether the stereotype actually accords with reality. This was recently noted by Lilienfield (2002), who pointed out that a scientific approach still remains the optimal approach for separating cherished erroneous beliefs from valid knowledge.

Challenging stereotypes and the 'conventional wisdom' through appropriate scientific methods is particularly important in the area of police stress. For many years, researchers and practitioners have tended to assume that police work is inherently stressful (e.g. Anshel *et al.*, 1997; Cacioppe and Mock, 1985; Dantzer, 1987; Sigler and Wilson, 1988). A growing number of researchers have begun to question this assumption, however, arguing that there is little empirical evidence to support the notion that police work is especially stressful (e.g. Anson and Bloom, 1988; Brown and Campbell, 1990; Hart *et al.*, 1994, 1995; Lawrence, 1984; Malloy and Mays, 1984; Terry, 1981).

A number of studies have found that police officers typically regard the organisational aspects of their work (e.g. leader and management practices, appraisal and recognition processes, career opportunities, clarity of roles, coworker relations, goal alignment) to be more stressful than the operational nature of police work. It has been found, for example, that operational experiences, such as being exposed to danger, dealing with victims, and the use of force in the execution of their duties, are not overly stressful for most police officers (Brown and Campbell, 1990; Hart *et al.*, 1994, 1995; Kop and Euwema, 2001).

Instead of the nature of police work itself being stressful, these studies have found that police stress has more to do with the organisational context in which police officers' work. In other words, it seems to be organisational and managerial practices that contribute to police stress, rather than the nature of the job itself. This means that police work may in fact be no different to other jobs, at least in terms of the factors that contribute to levels of occupational stress (e.g. Griffin *et al.*, 2000; Sauter and Murphy, 1995).

The notion that police work is no different to other jobs has also received support from the quality of life and subjective well-being literatures. In a recent longitudinal study, for example, Hart (1999) found that job satisfaction among police officers accounted for between 3% and 13% of the variance in their overall levels of life satisfaction. This was much smaller than the contribution made by the non-work domains of police officers' lives. More importantly, these findings were similar to those that have been found in studies of other occupational (e.g. Adams *et al.*, 1996; Biggam *et al.*, 1997) and community groups (e.g. Heady *et al.*, 1985; Near *et al.*, 1983). Accordingly, these findings suggest that the role of work is no more important for police officers than it is for other occupational groups, and like most employees, police officers' overall levels of psychological well-being are determined more by what happens outside of work, rather than what happens whilst they are at work.

Given that two different lines of enquiry strongly suggest that police officers are very similar to other occupational groups, in terms of the issues that affect their levels of occupational well-being and overall life satisfaction, it is important for researchers and practitioners to question why there is such a strong belief that police work is inherently stressful. One explanation is that the conventional view about the nature and extent of police stress is professionally self-serving. For example, Terry (1981) has argued that police officers are able to set themselves apart from other occupational groups and legitimise the professionalism and value of their occupation by perpetuating the notion that police work is very stressful. This is consistent with arguments that police unions have sometimes put forward during industrial relations and enterprise bargaining negotiations.

It is also easy to see from an outsider's perspective why there is a degree of face validity in the notion that police work is much more stressful than other occupations. Members of the general community are frequently exposed to stereotypic images that show the dangerous aspects of police work. For example, police are often portrayed in risky roles in movies and fictional literature, and these portrayals are reinforced by the stories that are most newsworthy in the print media. When confronted with these images, a person who would not like to find themselves in this type of work situation could easily conclude that it would be quite stressful if they were required to perform these types of work duties. However, this attribution of 'stressfulness' does not necessarily reflect the nature of police work. It merely reflects the fact that many people would find it stressful if they were actually to become police officers themselves.

It should be remembered that police officers choose to become police officers knowing the types of duties and situations they are likely to be involved with and can choose to change jobs if they ultimately find that police work is not to their liking. From an anecdotal standpoint, we have spoken with many police officers that actually enjoy their jobs, suggesting that the perceived 'inherent stressfulness' of police work may have more to do with the views of outsiders, rather than police officers themselves. It seems reasonable to assume that if people are in a job of their choosing they may find it to be quite enjoyable, despite the fact that others may find the job to be quite stressful. The Arnold Schwarzenegger movie entitled 'Kindergarten Cop' highlights this view, albeit from the alternative perspective. In the movie, Arnold plays the role of a tough cop who is not fazed by the dangerous and unsavoury aspects of his work, but finds that it is quite stressful to be placed in the position of teaching a class of young children. The point we are making is that police officers and teachers may not find the nature of their jobs to be particularly stressful, but this may change quite dramatically if they were asked to take on each others' job roles.

There have also been a number of conceptual and methodological problems in the police stress literature that has helped perpetuate the view that police work is inherently stressful. One of the main problems in the police stress literature has been the widespread use of the stressors and strain approach (e.g. Greller *et al.*, 1992). This approach presumes that adverse work experiences (stressors) cause psychological and behavioural strain. Accordingly, research adopting this approach obtains ratings on potential stressors and then correlates these ratings with various indices of strain (e.g. measures of psychological distress and somatic symptoms).

As noted by Hart and Cooper (2001) the stressors and strain approach is an overly simplistic framework that has given rise to four questionable assumptions that pervade much of the occupational stress literature. These assumptions are that: (a) stress is associated with unpleasant emotions (e.g. Klein, 1996; Newton, 1989); (b) employees' feelings of stress occur at the expense of more pleasurable emotions (Quick *et al.*, 1992); (c) stress can be measured by a single variable (e.g. Hurrell, 1998); and (d) stress is caused primarily by adverse work experiences (e.g. Sauter and Murphy, 1995).

A growing body of evidence, however, has called these assumptions into question. For example, the emphasis placed on stressors (e.g. adverse experiences) and strain (e.g. psychological distress) fails to account for the fact that employees' responses to their environment include both positive (e.g. positive affect, psychological morale) and negative (e.g. negative affect, psychological distress) dimensions (e.g. Diener and Emmons, 1985; George, 1996). It has also been found that police officers, like many other employees, are exposed to a range of positive and negative work experiences (Hart *et al.*, 1994). Moreover, these positive and negative experiences make independent contributions to police officers' levels of job satisfaction (Hart, 1999). These findings demonstrate that feelings of stress do not necessarily occur at the expense of more pleasurable emotions, and that stress is not necessarily caused by adverse work experiences. It is possible, for example, that stress is caused by the absence of positive work experiences, rather than the presence of negative work experiences.

Another concern is that many studies of occupational stress rely on the use of context-free, rather than domain-specific, measures of psychological well-being. Context-free measures, such as the General Health Questionnaire (e.g. Goldberg, 1978) assess police officers' overall levels of psychological distress, rather than the levels of distress that are related to the work domain of their lives. Given that the non-work domain of employees' lives tends to be much more important than the work domain in contributing to overall psychological well-being (e.g. Adams *et al.*, 1996; Hart, 1999; Headey and Wearing, 1992), conclusions about occupational stress, based on context-free measures of psychological distress, are likely to be of limited value. This is of particular concern when context-free measures are used to compare the levels of psychological distress among different occupational groups. In these circumstances, it is not possible to determine whether any observed differences are due to the work or nonwork domains of employees' lives.

More recently, researchers have endeavoured to incorporate moderator variables, such as decision latitude and coping processes, into the stressors and-strain framework (e.g. Day and Livingstone, 2001; Sauter and Murphy, 1995). This research, however, continues to be framed by a focus on adverse work experiences and how these relate to negative psychological outcomes (e.g. psychological distress). Moreover, this relatively narrow focus fails to consider the broader organisational context or other important individual and organisational characteristics, such as personality, job skills, leadership, and motivation that are likely to influence occupational well-being (Hart and Cooper, 2001).

A major limitation of the failure to consider the broader organisational context is that the traditional stressor and strain approach tends to motivate interventions that focus primarily on individual employees. Interventions that typically focus on individual employees include coping skills training, employee assistance programmes, and stress inoculation programmes (e.g. Anshel, 1997, 2000; Lowenstein, 1999). Although these types of interventions may be of some value, they focus on changing the employee, rather than changing the conditions or circumstances that may actually be the cause of occupational stress in the first place. In other words, they address the symptoms, rather than the causes of occupational stress.

Moreover, Hurrell (1995) has noted that the stressors and strain approach tends to reinforce the view that occupational stress is an employee problem, rather than an organisational problem that needs to be addressed more systemically. This concern is compounded by the failure of occupational stress researchers to link indices of occupational stress to relevant organisational performance outcomes, such as the cost of absenteeism and workers' compensation claims for stress-related injury, as well as ethical behaviour and complaints about the quality of service delivery. More importantly, the failure to link occupational stress to organisational performance has tended to marginalise the issue of occupational stress in the broader management and organisational behaviour literature (Hart and Cooper, 2001; Wright and Cropanzano, 2000). It may also explain why managers in many police organisations still view occupational stress as an occupational health and safety issue, rather than an issue that is central to the leadership and management practices of the organisation.

Given this overall state of affairs, there is a need for methodologically sound studies that take a broader theoretical and practical perspective. As such, there is a need for occupational stress studies that provide information about the relative contribution of a broader range of organisational and individual factors. For example, research has shown that personality characteristics (Cooper and Payne, 1991; Costa and McRae, 1980), coping processes (Carpenter, 1992), organisational climate (Griffin *et al.*, 2000), and positive and negative work experiences (Hart *et al.*, 1995) are all likely to contribute to indices of occupational stress. It is not known, however, which of these factors is more important in determining occupational stress among police officers.

4.3   MOVING BEYOND THE STRESSORS AND STRAIN APPROACH

We believe that these concerns about the occupational stress literature can be addressed by drawing on the organisational health framework that was recently proposed by Hart and Cooper (2001). The core elements of this framework are shown in Figure 4.1. According to the organisational health framework, it is important for researchers and practitioners to be concerned with the occupational well-being of employees *and* organisational performance. In other words, it is not sufficient to be concerned with occupational well-being in itself, but instead, occupational well-being must be linked to outcomes that affect organisational performance. In a recent empirical study, for example, it was found that satisfaction among employees led to greater discretionary effort which, in turn, contributed to the satisfaction that was being experienced by customers of the organisation (Hart *et al.*, 2002). Although these findings demonstrate that occupational well-being may be related to 'core business' outcomes, it is important to extend this line of research to include a range of performance indicators (e.g. Wright and Cropanzano, 2000).

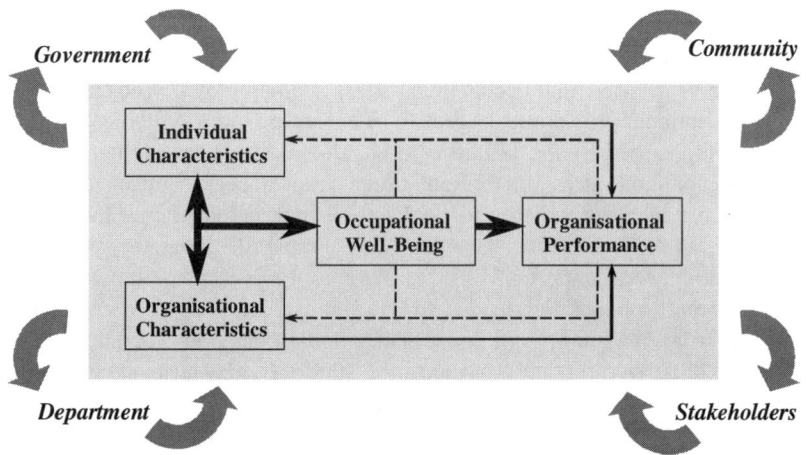

Figure 4.1 A Heuristic Model of Organisational Health.

The focus on occupational well-being is also an important departure from the language of occupational stress. The term 'stress' is typically associated with psychological distress (e.g. Cooper, 1998; Newton, 1989), and forms only one part of the much broader construct of occupational well-being. Drawing on a considerable body of empirical evidence in the quality of life literature that describes the structure of subjective well-being (e.g. Diener, 2000), Hart and Cooper (2001) argue that occupational well-being includes both emotional and cognitive components. The emotional component reflects the two independent dimensions of positive and negative affect (referred to in this study as morale and distress, respectively), whereas the cognitive component reflects employees' judgments about their levels of job satisfaction (Hart, 1999). This means that when assessing occupational well-being it is important to include domain-specific measures that assess all three dimensions.

The organisational health framework also emphasises the role that individual and organisational characteristics play in determining both occupational well-being and organisational performance. Although Hart and Cooper (2001) discuss a range of different individual and organisational characteristics, process theories of occupational stress, such as the cognitive-relational (e.g. Lazarus and Folkman, 1984) and dynamic equilibrium (e.g. Hart, 1999; Hart *et al.*, 1994;

Headey and Wearing, 1989) theories, provide some guidance on the characteristics that are likely to be of most importance. In terms of individual characteristics, the enduring personality characteristics of neuroticism and extraversion (Costa and McCrae, 1989), as well as the use of emotion-focused and problem-focused coping strategies (Latack and Havlovic, 1992), have been related to indices of psychological well-being in both occupational and community studies (e.g. Hart, 1999; Headey and Wearing, 1990; Moyle, 1995). In terms of organisational characteristics, research has shown that organisational climate (Griffin *et al.*, 2000; Michela *et al.*, 1995) and employees' positive (i.e. uplifts or emotionally motivating) and negative (i.e. hassles, pressures, stressors, or emotionally distressing) experiences of work (Hart *et al.*, 1995) strongly influence indices of occupational well-being.

The organisational health framework shown in Figure 4.1 also recognises that the relationship between individual and organisational characteristics on the one hand, and occupational well-being and organisational performance on the other hand, operates in a broader context. The nature of this broader context varies according to the level of analysis that is applied to the core elements of the framework. For example, if the core elements of the model were applied to individual employees and their work teams, then the policies and practices of the wider organisation will form part of the context in which they must operate. If the core elements of the model were applied to the organisation as a whole, however, then other factors, such as government policies, regulatory authorities, and the wider community's expectations, will make up the broader context in which the organisation operates.

According to Hart and Cooper (2001), the organisational health framework provides a theoretical approach that can be used to guide occupational stress research in a way that is of more relevance to organisations and the broader management and work psychology literatures. They acknowledge, however, that the organisational health framework can give rise to a number of competing theoretical propositions and research models. One of the research models proposed by Hart and Cooper is shown in Figure 4.2. This model is based on an integration of the cognitive-relational (e.g. DeLongis *et al.*, 1988; Lazarus and Folkman, 1984) and dynamic equilibrium (e.g. Hart, 1999) theories of stress with the quality of life and subjective well-being literature (e.g. Heady and Wearing, 1989, 1992). Although Hart and Cooper provide a detailed discussion of the theoretical background to this research model, there are six key points that have particular relevance to the current study.

First, the research model shown in Figure 4.2 proposes that the structure of occupational well-being consists of three components that reflect the positive (e.g. energy, enthusiasm, and pride) and negative (e.g. anxiety, depression, and frustration) emotional responses that police officers may have to their work, as well as the judgments that police officers make about their overall levels of job satisfaction (George, 1996; Hart, 1999). Second, it is presumed that these indices of occupational well-being will influence police officers' propensity to seek medical advice, take sickness absence, submit a workers' compensation claim, or

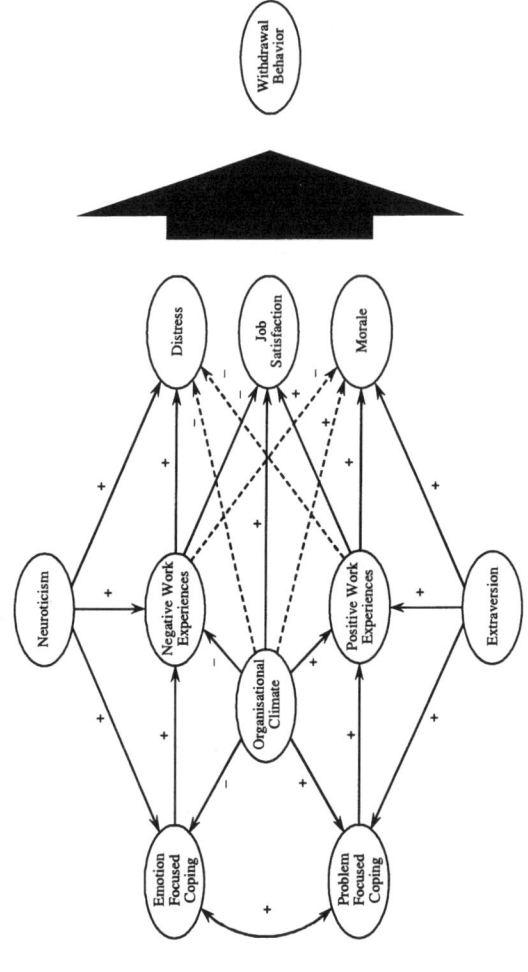

Personal and Organisational Characteristics   **Employee Well-Being**   **Organisational Performance**

Figure 4.2 Example of an Organisational Health Research Model. ['+' indicates a positive relationship and '−' indicates a negative relationship. Dotted lines indicate possible relationships that are expected to be comparatively weak.]

seek another job, as a result of stress-related problems. Given that prior research has not investigated the link between these withdrawal behaviour intentions and the three components of occupational well-being, it is not known which of the three components of occupational well-being are likely to be most influential in determining withdrawal behaviour intentions.

Third, the research model shown in Figure 4.2 suggests that organisational climate will underpin police officers' positive and negative work experiences and that these experiences, in turn, will contribute to their levels of occupational well-being. Consistent with previous research in the quality of life (e.g. Headey and Wearing, 1989) and occupational stress (e.g. Hart, 1994, 1999) literatures, the model suggests that police officers' negative work experiences will contribute more strongly to distress than to morale, whereas their positive work experiences will contribute more strongly to morale than to distress. This gives rise to the following propositions:

P1: Organisational climate contributes to both positive and negative work experiences.
P2: Negative work experiences contribute more strongly than positive work experiences to distress; and,
P3: Positive work experiences contribute more strongly than negative work experiences to morale.

Fourth, the research model suggests that police officers' use of emotion-focused coping strategies will contribute positively to their negative work experiences, whereas the use of problem-focused coping strategies will contribute positively to their positive work experiences. This pattern of relationships is consistent with the results of previous studies (Hart *et al.*, 1995; Heady and Wearing, 1990), and reflects Lazarus' (1990) contention that measures of daily experiences reflect the reappraisals that a person makes after they have attempted to cope (i.e. secondary appraisal) with a situation that was potentially beneficial or harmful (i.e. primary appraisal) to their well-being. Moreover, this pattern of relationships runs contrary to the views expressed in much of the occupational stress literature, where it is often assumed that the use of coping strategies mediates the relationship between employees' work experiences and their levels of psychological well-being. Accordingly, this gives rise to another proposition that was investigated in this study:

P4: Work experiences mediate the relationship between coping and occupational well-being, rather than coping mediating the relationship between work experiences and occupational well-being.

Fifth, the research model shown in Figure 4.2 suggests that the enduring personality characteristics of neuroticism (i.e. emotional reactivity) influences emotion-focused coping, negative work experiences and distress, whereas extraversion (i.e. sociability) influences problem-focused coping, negative work experiences and morale. This pattern of relationships is consistent with prior research (e.g. Hart *et al.*, 1995; Headey and Wearing, 1989; 1990), and reflects the positive and negative affectivity pathways that have often been thought to underpin variables of this nature (e.g. George, 1996). Moreover, a considerable body of empirical evidence has shown that neuroticism exerts a particularly strong influence on other negative emotionally-laden variables (Burke *et al.*, 1993; Moyle, 1995; Williams *et al.*, 1996). Although relatively little is known about the potency of extraversion, recent studies have found that it is less influential. Accordingly, we investigated the following proposition:

P5:   Neuroticism is the strongest determinant of distress, but not of morale and job satisfaction.

Finally, the research model shown in Figure 4.2 suggests that organisational climate underpins police officers' use of emotion- and problem-focused coping strategies, their positive and negative work experiences, and levels of occupational well-being. Although the influence of organisational climate has not been compared directly with the influence of personality characteristics, coping strategies, and positive and negative work experiences, the pattern of relationships shown in the model suggests that organisational climate is likely to have a strong influence on occupational well-being. This is due to the fact that organisational climate reflects the core organisational behaviours that underpin police officers' coping strategies and work experiences. Accordingly, we investigated the following proposition:

P6:   Organisational climate is more influential than coping and positive and negative work experiences in determining police officers' levels of occupational well-being.

4.4 THE PRESENT STUDY

In the present study, we used data obtained from a large sample of employees who worked for an Australian police organisation. Australian police organisations include two types of employees; sworn police officers who perform a range of traditional policing duties, and unsworn public sector workers who perform a range of support functions (e.g, administrative services, communications, and human resources). Both types of employees were invited to participate in the present study. This meant that we were able to compare the views of police officers and other employees who worked in the same organisation. To provide additional comparative data, this study also drew on a normative database that was available for a large heterogeneous sample of public sector employees who did not work in police organisations. The use of these three data sources was a particular strength of the present study, because it enabled us to compare police officers and other employees within the police organisation, and then to compare employees in the police organisation with employees who worked in other organisations.

Another key strength of the present study was that it enabled us to extend the prior work of Hart and his colleagues (e.g. Hart *et al.*, 1994, 1995) in two important ways. First, this is the first study of police officers to include organisational climate and a range of organisational and operational work experiences. Second, we included domain-specific measures of occupational well-being in the current study. Although Hart's prior work with police officers has included measures of job satisfaction (e.g. Hart, 1999), all other outcome measures have been of a non-specific or context-free nature. Moreover, this is one of the first studies to examine the relative influence that police officers' enduring personality characteristics, use of coping strategies, perceptions of organisational climate, as well as their positive and negative work experiences, make to their levels of occupational well-being and withdrawal behaviour intentions.

4.5 METHOD

4.5.1 Participants

Questionnaire booklets were mailed to a random sample of 1550 sworn and unsworn employees who worked for an Australian state police organisation. The sworn employees performed traditional policing duties, whereas the unsworn employees worked in a range of administrative and managerial support areas. This is an important distinction, because the unsworn employees were general public sector employees, rather than police officers.

A total of 793 employees returned completed questionnaire booklets (response rate: 51.16%). Of these, 589 (74.27%) were sworn police officers and 167 (21.06%) were unsworn support personnel. The occupational status of 37 employees was not known. However, the overall ratio of sworn and unsworn employees (77.91% and 22.09%, respectively) was similar to that in the organisation as a whole.

In terms of the sworn police officers, 184 (31.24%) were Constables, 161 (27.33%) were Senior Constables, 174 (29.54%) were Sergeants, 38 (6.45%) were Senior Sergeants, and 32 (5.43%) were Commissioned Officers. Their ages ranged from 20 years to 56 years ($M = 35.00$ years, $SD = 8.56$ years), and their lengths of service ranged from less than 1 year to 38 years ($M = 12.68$ years, $SD = 8.52$ years). Of the 589 sworn police officers, 75 (12.73%) were female.

The unsworn public sector employees represented all occupational levels, from the most junior to the most senior positions. Their ages ranged from 19 years to 63 years ($M = 36.98$ years, $SD = 9.17$ years), and their lengths of service ranged from less than 1 year to 29 years ($M = 5.99$ years, $SD = 5.81$ years). Of the 167 unsworn public sector employees, 115 (68.86%) were female.

4.5.2   Comparative Norm Data

To address some of the research questions posed in this chapter, it was necessary to draw on normative data that was available for Australian employees who did not work for a police organisation. For this purpose, we were able to draw on data that were collected during a different research project. The comparative data was obtained from a sample of 1117 employees who worked for five different public sector organisations (Education, Family Services, Fire and Ambulance Services, Health, and Primary Industries) in the same Australian State as the police organisation that participated in this study. In terms of the employees who provided the comparative data, their ages ranged from 18 years to 65 years ($M =$ 38.72 years, $SD = 10.50$ years), and their lengths of service ranged from less than 1 year to 46 years ($M = 6.22$ years, $SD = 6.97$ years). Of the 1,092 employees, 666 (60.11%) were female (note that the gender of 9 employees was not known).

4.6   MEASURES

4.6.1   Occupational Well-Being

Three separate measures of occupational well-being were used. These measures assessed both the affective and cognitive components of occupational well-being (Hart and Cooper, 2001; Headey and Wearing, 1992). The 14-item Occupational Positive and Negative Affects Scales (Hart *et al.*, 1996) were used to assess the positive (e.g. feeling energised, enthusiastic, cheerful, happy; referred to in this study as Morale) and negative (e.g. feeling anxious, depressed, tense, unhappy; referred to in this study as Distress) emotional responses that employees had to their work (coefficient alpha = .92 and .89, respectively). Employees were asked to rate how often they had experienced 14 positive and negative emotions whilst at work over the past month, on a 7-point scale ranging from 'not at all' to 'all the time' (coefficient alpha = .92 and .87 for Morale and Distress, respectively).

    The 6-item Quality of Work Life Scale (Hart *et al.*, 1996) was used to asses employees' judgments about the quality of their work life. This measure was based on the Life Satisfaction Scale that has been used extensively in the quality of life literature (Pavot and Diener, 1993), by changing the focal point from 'life' more generally to 'life at work.' Example items include, 'The conditions of my life at work are excellent,' and 'I am satisfied with my life at work,' and 'In most ways my work life is close to my ideal.' Employees were asked to rate their level of agreement with each statement on a 7-point scale ranging from 'strongly disagree' to 'strongly agree.' Hart *et al.* (1996) have shown that this measure correlates strongly ($N = 2,655$, $r = .70$, $p < .001$) with job satisfaction (coefficient alpha = .90).

4.6.2   Withdrawal Behaviour

An 8-item scale was used to assess the extent to which employees were likely to withdraw from their jobs, because of stress-related problems. The first four items focused on whether employees had seriously considered withdrawing from their jobs over the past month. The four items were, 'Over the past month, how often have you seriously thought about putting in a worker's compensation claim for a stress-related problem?,' 'Over the past month, how often have you seriously thought about taking sick-leave for a stress-related problem?,' 'Over the past month, how often have you seriously thought about changing jobs because of a stress-related problem?,' and 'Over the past month, how often have you seriously considered seeking medical advice for a stress-related problem?' Employees were asked to rate each item on a 5-point scale ranging from 'very often' to 'rarely or never'. The next four items focused on whether employees seriously intended to withdraw from their jobs in the near future. The four items were, 'Do you seriously believe that in the near future you will put in a worker's compensation claim for a stress-related problem?', 'Do you seriously believe that you will take sick leave in the near future for a stress-related problem?', 'Do you seriously believe that you will change jobs in the near future because of a stress-related problem?', and 'Do you seriously believe that you will seek medical advice in the near future for a stress-related problem?' Employees were asked to rate each of these items on a 5-point scale ranging from 'definitely not' to 'definitely yes'. Principal components analysis showed that the eight items measured a single factor that accounted for 63.70% of the variance (coefficient alpha = .90).

4.6.3 Positive and Negative Work Experiences

Police officers' positive and negative work experiences were assessed with a 109-item measure adapted from the Police Daily Hassles and Uplifts Scales (Hart *et al.*, 1995) and the Positive and Negative Work Experience Scales (Hart *et al.*, 1996). The measure used in this study assessed 8 dimensions of positive work experiences and 12 dimensions of negative work experiences. These dimensions covered both operational (e.g. exposure to danger, dealing with victims, and frustration with the criminal justice system) and organisational (e.g. management behaviour, coworker relations, and decision-making processes) experiences. This meant that we were able to assess experiences that were peculiar to the police role (i.e. job specific experiences) and experiences that were relevant to all occupational groups. As shown later in this chapter, these dimensions could be aggregated to form overall indices of Positive Work Experiences and Negative Work Experiences (coefficient alpha = .95 and .95, respectively). Although the unsworn public sector employees completed a generic version of this measure (i.e. the organisational experiences subscales), in this chapter we do not present the data that were obtained for the unsworn personnel on this measure.

The response format for the Negative Work Experiences Scale required employees to consider whether each experience (item) had occurred 'as a result of their work during the past month' and, if so, how much the experience 'hassled or bothered' them. Employees were then required to rate their response on a five-point scale that ranged from 'definitely does not apply to me' (0) to 'strongly applies to me' (4). The instructions emphasised that employees should indicate (0) if an experience did not occur, or if the experience occurred but was not a hassle or bother. This response format required employees to combine both frequency and intensity when choosing the appropriate response option. The one month time frame ensured that the typical experience of employees was assessed, given the diversity of tasks that employees can be engaged in on a day-to-day basis. Example items included 'Going to dangerous calls' and 'Inadequate feedback on my performance'. The same response format was used for the Positive Work Experiences Scale, except that on this occasion employees were asked to consider the extent to which each experience made them 'feel good' during the past month (Hart, 1999). Example items included 'Helping complainants', and 'Receiving recognition for good work'.

4.6.4 Organisational Climate

Employees' perceptions about seven different aspects of their work environment (appraisal and recognition, coworker interaction, goal congruency, opportunities for development, participative decision-making, role clarity, supportive leadership) were assessed using 35 items from Hart *et al.*'s (1996) Organisational Climate Scale. This scale is based on the components of the School Organisational Health Questionnaire (Hart *et al.*, 2000) that were designed to assess organisational factors that are common to most occupational groups. Employees were asked to rate the extent to which each item (e.g. 'My work objectives are always well defined') described their particular work unit (e.g. police station or work team) on a 5-point scale ranging from 'strongly disagree' to 'strongly agree'. Confirmatory factor analyses showed that the seven separate dimensions could be aggregated at a second-order level to provide an overall index of organisational climate (coefficient alpha = .96).

4.6.5 Coping Strategies

A 24-item Coping Response Inventory (Hart, 1988) was employed to assess the coping strategies used by police officers. This inventory was based on the work of Billings and Moos (1984), and measures six different dimensions of coping: Affective Regulation, Emotional Discharge, Seeking Emotional Support, Information Seeking, Logical Analysis, and Problem Solving. Moreover, Hart *et al.* (1995) have used confirmatory factor analysis to show that the 24 coping items can be grouped at a second-order level to reflect Problem-Focused (Information Seeking, Logical Analysis and Problem Solving) and Emotion-Focused (Affective Regulation, Emotional Discharge, Seeking Emotional Support) Coping (coefficient alpha = .79 and .76, respectively).

In order to assist recall, police officers were asked to nominate the specific work event which had bothered them the most during the preceding six months, and indicate on a 5-point scale the extent to which they had used various coping strategies (items) to manage or deal with this event. Although this procedure is generally used to assess situation specific coping (Carver *et al.*, 1989), Hart (1988) found that police officers reported using similar strategies to cope with bothersome events in their work and non-work lives. Consequently, this procedure seems to assess the typical way in which police officers attempt to cope with events in their daily lives.

4.6.6 Personality characteristics

Costa and McCrae's (1989) NEO Five-Factor Inventory was used to provide a measure of Neuroticism (coefficient alpha = .82) and Extraversion (coefficient alpha = .76). Those who score high on Neuroticism are more likely to worry, and are typically nervous, emotional, insecure, inadequate and hypochondriacal. High scorers on Extraversion tend to be active, talkative, person-oriented, optimistic, fun-loving and affectionate.

4.7 RESULTS

In order to establish whether policing is, in fact, a stressful occupation, it was necessary to approach our analyses from three different perspectives. First, we examined whether there were any mean differences, in the study variables, between the sworn police officers and the unsworn public service employees, as well as between these two groups and public service employees who worked in non-police organisations. These comparisons enabled us to understand whether there were systematic differences in the average levels of occupational well-being, withdrawal behaviours, quality of organisational climate, use of emotion and problem-focused coping strategies, and enduring personality characteristics among police officers and other occupational groups.

Second, we used confirmatory factor analytic techniques to establish whether it was police officers' operational or organisational experiences that contributed most to their levels of occupational well-being. If police work is an inherently stressful occupation, it would be reasonable to assume that negative operational experiences, such as dealing with trauma or being exposed to danger, would have the most deleterious effect on police officers' levels of well-being. If organisational experiences were found to have the most deleterious affect, however, this would indicate that the same types of experiences that are common to all occupational groups is also the primary cause of the 'stress' in police work. This would be a direct empirical challenge to the conventional view that the nature of police work makes the job inherently stressful.

Third, we used structural equation modelling to establish the relative contribution that police officers' personalities, their use of emotion and problem-focused coping strategies, the quality of their workgroup's organisational climate, and their positive and negative work experiences made to their levels of occupational well-being and withdrawal behaviour intentions. This enabled us to assess whether traditional intervention strategies, such as those focusing on police officers' coping skills (e.g. Anshel, 2000), are likely to bring about sustained improvements in police officers' levels of well-being and, if not, what strategies are likely to be most effective.

4.7.1 Mean Differences Between Police Officers and Other Occupational Groups

In Table 4.1, we list the summary statistics for each of the study variables that could be compared, in terms of mean levels, among sworn police officers, unsworn public sector employees, and general public sector employees who did not work for the police organisation. The positive and negative work experience variables were the only ones that could not be compared across the three employee groups. This was due to the fact that the unsworn and general public sector employees could not rate the operational work experiences, because these employees were not involved in operational policing tasks. A series of t-tests were conducted to examine whether the difference in the mean scores listed in Table 4.1 were statistically significant. The results of these tests are shown in Table 4.2.

Table 4.1 Means and Standard Deviations on Study Variables for Police Officers, Unsworn Support Employees, and General Public Sector Employees

Variable	Police Officers			Unsworn Employees			General Public Sector		
	N	*M*	*SD*	*N*	*M*	*SD*	*N*	*M*	*SD*
Quality of Work Life	578	21.77	7.78	165	22.12	7.71	1,074	23.27	7.88
Distress	580	24.86	8.26	167	21.68	9.56	1,087	22.13	8.80
Morale	576	29.13	8.25	164	29.32	8.88	1,074	31.33	7.96
Withdrawal Behaviour	579	14.14	6.36	165	13.18	6.61	N/A	N/A	N/A
Organisational Climate	559	107.55	25.86	163	110.38	26.68	1,020	113.49	24.73
Emotion-Focused Coping	545	18.77	9.26	139	19.18	9.24	991	18.69	9.32
Problem-Focused Coping	545	28.15	8.33	139	30.53	8.48	1,005	30.40	8.44
Neuroticism	573	30.90	7.40	160	31.12	7.44	1,064	31.81	8.07
Extraversion	573	41.08	6.08	156	39.96	6.12	1,056	40.22	6.29

Note: Unsworn employees are public sector employees working in the police organisation, whereas the general public sector data are the results for public sector employees working in non-police organisations. N/A indicates that data were not available.

As shown in Tables 4.1 and 4.2, there were no significant mean differences among the three occupational groups on Emotion-Focused Coping ($p > .10$). This indicates that the extent to which employees use emotion-focused coping strategies is similar across different occupational groups. In comparison to the unsworn public sector employees, police officers did not differ significantly on Morale, Withdrawal Behaviour, Organisational Climate, and Neuroticism ($p > .05$). There was a significant difference, however, on Distress, Problem-Focused Coping, and Extraversion ($p < .05$). These results suggest that when compared to unsworn public sector employees working in the same organisation, police officers experienced higher levels of distress ($M = 24.86$ and 21.68 for sworn police officers and unsworn public sector employees, respectively), engaged less in problem-focused coping strategies ($M = 28.15$ and 30.53 for sworn police officers and unsworn public sector employees, respectively), and tended to be more extraverted ($M = 41.08$ and 39.96 for sworn police officers and unsworn public sector employees, respectively). There was also a significant difference, between the sworn police officers and the general public sector sample, on all study variables except Emotion-Focused Coping. In contrast, Morale was the only variable on which there was a significant difference between the unsworn and general public sector employees.

Table 4.2 Results of t-tests Used to Compare Means Differences
Among Occupational Groups Shown in Table 4.1

Variable	Police Officers v. Unsworn Employees			Police Officers v. General Public Sector			Unsworn Employees v. General Public Sector		
	t	df	p	t	df	p	t	df	p
Quality of Work Life	0.52	741	>.10	3.71	1650	<.001	1.75	1237	>.05
Distress	4.22	745	<.001	6.15	1665	<.001	0.61	1252	>.10
Morale	0.26	738	>.10	5.28	1648	<.001	2.96	1236	<.01
Withdrawal Behaviour	1.69	742	>.05	N/A	N/A	N/A	N/A	N/A	N/A
Organisational Climate	1.22	720	>.10	4.49	1577	<.001	1.47	1181	>.10
Emotion-Focused Coping	0.47	682	>.10	0.16	1534	>.10	0.59	1128	>.10
Problem-Focused Coping	2.99	682	<.01	5.04	1548	<.001	0.17	1142	>.10
Neuroticism	0.33	731	>.10	2.24	1635	<.05	1.02	1222	>.10
Extraversion	2.04	727	<.05	2.67	1627	<.01	0.49	1210	>.10

Note: Unsworn employees are public sector employees working in the police organisation, whereas the general public sector data are the results for public sector employees working in non-police organisations. N/A indicates that no comparative data were available.

The overall pattern of results suggests that police officers may, on average, experience higher levels of distress and lower levels of morale and quality of work life in comparison to other public sector employees. This raises a question about whether these differences are due to the demands of police work, the nature of the police organisation, or the way in which police officers typically deal with difficult and stressful situations. Although not answering this question directly, it is interesting to note that the sworn police officers and unsworn public sector employees both rated the climate of the police organisation as being lower than that experienced by public sector employees working in other organisations. Given that both groups of employees working in the police organisation also had lower levels of morale and quality of work life, in comparison to those who worked in other organisations, the overall pattern of results is consistent with the notion that the police organisation plays a role in determining occupational well-being.

4.7.2   The Influence of Operational and Organisational Experiences

Although a comparison of mean differences can sometimes provide a useful window through which to explore the nature of police stress, this approach often masks the nuances that differentiate the experience of individual employees. Aggregating information to a mean score limits our ability to examine patterns in individual variation. This can be a major methodological limitation when investigating the nature of police stress (Hart and Cooper, 2001). Accordingly, we now turn our attention to understanding the variation among individual employees, and whether systematic patterns in the experience of individual employees can inform our understanding of the determinants of police stress.

Our starting point was to use the data obtained from the 589 sworn police officers to establish the relative potency of the different operational and organisational experiences. The main aim in these analyses was to examine whether it was the operational or organisational experiences that were more important in determining police officers' levels of occupational well-being. This was achieved by using the Linear Structural Relations (LISREL VIII) Program (Joreskog and Sorbom, 1993) to test Hart *et al.*'s (1995) three-factor model of police work experiences. The structural equation analyses reported in this chapter were all based on variance-covariance matrices and employed the maximum likelihood method of estimation. The maximum likelihood method of estimation has been shown to be robust against moderate departures from the skewness and kurtosis of the normal distribution (Cuttance, 1987). The skewness and kurtosis was less than 1.0 in absolute value for most of the study variables.

A series of nested structural equation models was estimated to explain the relations among the 8 Positive Work Experience and 12 Negative Work Experience dimensions. The first model was an independence model that assumed there was no relationship between the 20 dimensions. This model served as a base-line from which to assess the relative fit of the next two models. In the second model, we estimated a 3-factor solution that was consistent with Hart *et al.*'s (1995) theoretical model. As shown in Figure 4.3, this model asserts that the relations among the positive and negative work experiences can be explained by two underlying latent constructs that represent Positive Work Experiences and Negative Work Experiences. Moreover, the model also suggests that the positive and negative operational experiences combine to form a third latent construct that represents police officers' levels of engagement in operation police work. The third model was essentially the same as the second, but on this occasion we estimated two 'cross-loadings' and six correlations among the residual variances for the positive and negative work experience dimensions. The correlations among the residuals were between (a) PWE-Administration and PWE-Supervision; (b) NWE-Victim and NEW-Danger; (c) PWE-Victim and PWE-Offenders; (d) PWE-Management and NWE-Communication; (e) PWE-Coworkers and NEW-Coworkers; and (f) PWE-Victims and NWE-Victims. These additional parameter coefficients were suggested by the modification indices for Model 2 and were defensible on theoretical grounds. The goodness-of-fit statistics for the three models are shown in Table 4.3, and the standardised parameter estimates for Model 3 are shown in Figure 4.3.

Table 4.3 Goodness-of-Fit Statistics for the Structural Equation Models Examining the Relations Among Different Dimensions of Positive and Negative Work Experiences

Model	χ^2	df	p	RMSEA	CFI	SRMSR	GFI
Model 1: Null	4,033.37	190	< .001				
Model 2: 3-Factor	618.69	160	< .001	.08	.88	.07	.88
Model 3: 3-Factor with Modifications	360.17	152	< .001	.06	.95	.05	.93

Note: N = 452 (Listwise). RMSEA = Root Mean Square Error of Approximation, CFI = Comparative Fit Index, SRMSR = Standardised-Root-Mean-Square-Residual, and GFI = Goodness of fit Index.

The goodness-of-fit statistics for Model 3 suggested that there was a good fit between the variance-covariance matrix and the tested model. Moreover, a comparison of the goodness-of-fit statistics for Models 2 and 3 showed that Model 3 was a significantly better fit to the data (χ^2_{diff} = 258.52, df_{diff} = 8, p < .001). The standardised parameter estimates shown in Figure 4.3 support the notion that organisational experiences are more important than operational experiences in determining police officers' occupational well-being. This conclusion is based on the fact that the standardised beta coefficients linking the work experience dimensions to the Positive Work Experiences and Negative Work Experiences latent constructs were noticeable larger for the organisational, rather than the operational experiences. In fact, the results showed that the Negative Work Experiences latent construct explained between 3% and 21% of the variance in the negative operational experiences, but between 26% and 64% of the variance in the negative organisational experiences. Likewise, the Positive Work Experiences latent construct explained between 3% and 10% of the variance in the positive operational experiences, but between 34% and 64% of the variance in the positive organisational experiences. Moreover, the results showed that there was no significant relationship between Positive Work Experiences and Negative Work Experiences ($r = -.01$, p > .10). This finding is consistent with a growing body of empirical evidence in the quality of life (e.g. Headey and Wearing, 1992) and occupational stress (Hart, 1999) literatures, and demonstrates the importance of taking into account both positive and negative work experiences when investigating the job characteristics that contribute to occupational well-being.

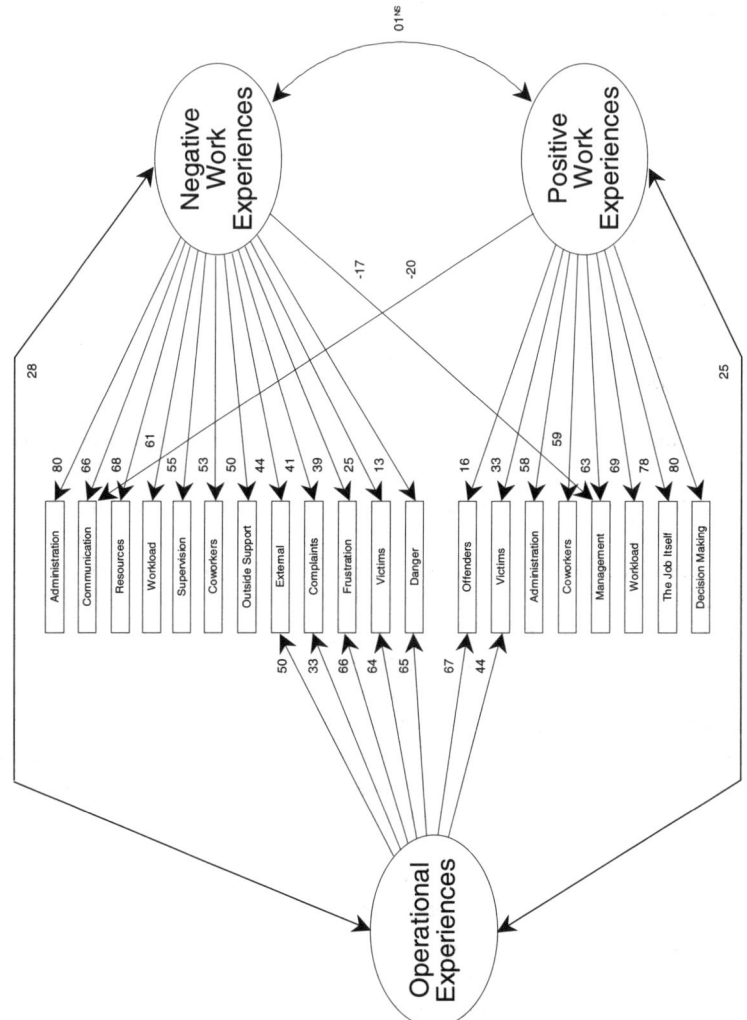

Figure 4.3 Three-Factor Model of Police Work Experiences.
[Standardised parameter estimates are all significant at the .05 level, unless otherwise indicated by [NS].]

In support of previous research by Hart *et al.* (1995), it was also found that the positive and negative operational experiences could be explained, in part, by an underlying Operational Experiences latent construct. The interesting point here is that the sign of the standardised beta coefficients linking Operational Experiences with the seven operational experience dimensions were all positive. This finding calls into question the conventional view that if police officers are experiencing negative operational experiences, such as exposure to danger or traumatised victims, they are unlikely to be enjoying their jobs. These findings demonstrate that police can derive satisfaction from their operational work, even when some of their experiences are causing a degree of discomfort or distress. For example, police officers may become distressed at the trauma they observe at a fatal road accident. However, they may also derive a degree of satisfaction from being able to help injured people who may have survived the accident. In other words, the one job may have a mix of negative and positive experiences. This is also consistent with the positive correlations that were found between Operational Experiences and Negative Work Experiences ($r = .28$, $p < .01$) and Operational Experiences and Positive Work Experiences ($r = .25$, $p < .01$).

The next question to be addressed in our analyses was how do these experiences contribute to police officers' levels of distress and morale, as well as their withdrawal behaviour intentions? To address this question, we estimated two nested structural equation models that examined the relations among Operational Experiences, Positive Work Experiences, Negative Work Experiences, Distress, Morale and Withdrawal Behaviour, after controlling for the effects of Neuroticism and Extraversion. We estimated the Operational Experiences, Negative Work Experiences, and Positive Work Experience latent constructs using the confirmatory factor technique that was adopted in the previous analysis (see Figure 4.3). The unit weighted composite scores for Neuroticism, Extraversion, Distress, Morale and Withdrawal Behaviour were used as single indicators of their respective latent constructs, and measurement error was taken into account by setting the percentage of error variance at (1 – alpha). This approach has been used in previous studies (e.g. Hart, 1999). As Bagozzi and Heatherton (1994) note, this form of measurement model is satisfactory when global questions are being asked about the relations among the constructs of interest. Moreover, it was necessary to adopt this procedure for practical reasons, because each construct was measured by a large number of items.

The first model was an independence model that assumed there was no relationship between the observed variables. In the second model, we regressed Distress, Morale and Withdrawal Behaviour onto Negative Work Experiences and Positive Work Experiences. In turn, all of these latent constructs and Operational Experiences were regressed onto Neuroticism and Extraversion. The goodness-of-fit statistics for the two models are shown in Table 4.4, and the standardised parameter estimates are shown in Figure 4.4. Note that in the interests of diagrammatic clarity, the standardised beta coefficients linking Neuroticism and Extraversion to all other latent constructs in the model have been shown in Table 4.5.

Table 4.4 Goodness-of-fit Statistics for the Structural Equation Models Examining the
Relations Between Positive Work Experiences, Negative Work Experiences, Distress, Morale,
Withdrawal Behaviour, Neuroticism and Extraversion

Model	χ^2	df	p	RMSEA	CFI	SRMSR	GFI
Model 1: Null	5,283.30	300	<.001				
Model 2: Theoretical Model	554.73	240	<.001	.05	.94	.05	.91

Note: $N = 452$ (Listwise). RMSEA = Root Mean Square Error of Approximation, CFI = Comparative Fit
Index, SRMSR = Standardised-Root-Mean-Square-Residual, and GFI = Goodness-of-fit Index.

Table 4.5 Standardised Beta Coefficients Linking Neuroticism and Extraversion to withdrawal Behaviour,
Distress, Morale, Negative Work Experiences, Positive Work Experiences,
and Operational Experiences

Dependent Variable	Neuroticism	Extraversion
Withdrawal Behaviour	.36	.18
Distress	.49	.04[NS]
Morale	-.01[NS]	.26
Negative Work Experiences	.29	.03[NS]
Positive Work Experiences	-.11[NS]	.40
Operational Experiences	.25	.20

Note: $N = 452$ (Listwise). [NS] = Nonsignificant at the .05 level.

The goodness-of-fit statistics for Model 2 suggested that there was a good fit
between the variance-covariance matrix and the tested model. The standardised
parameter estimates showed Negative Work Experiences and Positive Work
Experiences both contributed to Distress, Morale and Withdrawal Behaviour. The
overall pattern of results was consistent with the notion that negative work
experiences contribute more strongly than positive work experiences to distress
($\beta = .31$ and $-.27$, $p < .05$, for Negative Work Experiences and Positive Work
Experiences, respectively), whereas positive work experiences contribute more
strongly than negative work experiences to morale ($\beta = .45$ and $-.32$, $p < .05$, for
Positive Work Experiences and Negative Work Experiences, respectively). These
results were found after controlling for the effects of Neuroticism and
Extraversion, and were consistent with the pattern of results that have been found
in previous research (Hart *et al.*, 1995; Headey and Wearing, 1992). In terms of
police officers' withdrawal behaviour intentions, however, it was found that their
positive and negative work experiences made similar contributions ($\beta = .33$ and
$-.31$, $p < .05$, for Negative Work Experiences and Positive Work Experiences,
respectively). These findings again demonstrate the importance of taking into
account both positive and negative work experiences when investigating the
determinants of occupational stress.

As shown in Table 4.5, it was found that Neuroticism contributed significantly to Withdrawal Behaviour ($\beta = .36$, $p < .05$), Distress ($\beta = .49$, $p <.05$),
Negative Work Experiences ($\beta = .29$, $p < .05$), and Operational Experiences ($\beta = .25$,

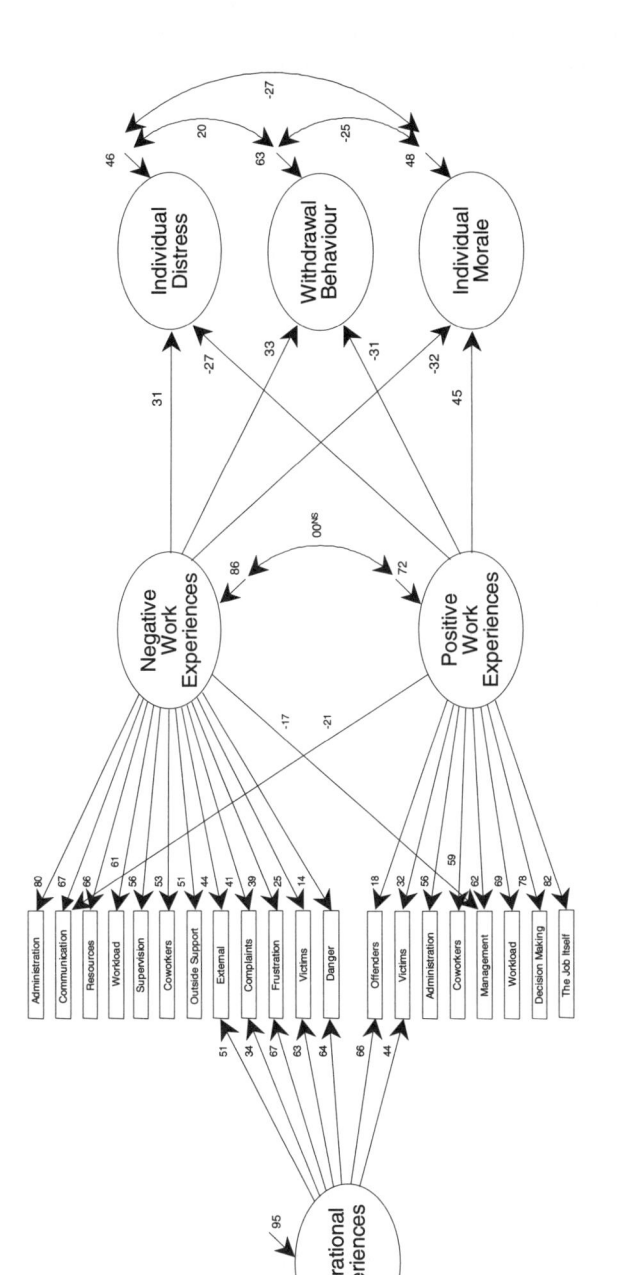

Figure 4.4 Standardised Parameter Estimates Showing the Relationship Between the Three-Factor Model of Police Work Experiences and Indices of Occupational Well-Being. [Standardised parameter estimates are all significant at the .05 level, unless otherwise indicated by [NS].]

$p < .05$), whereas Extraversion contributed significantly to Withdrawal Behaviour ($\beta = .18$, $p < .05$), Morale ($\beta = .26$, $p < .05$), Positive Work Experiences ($\beta = .40$, $p < .05$), and Operational Experiences ($\beta = .20$, $p < .05$). The pattern of relationships between the two personality constructs, occupational well-being, and work experience variables was consistent with the notion that separate positive and negative affectivity paths tend to explain the relationships among these types of variables (Hart, 1999; Hart and Cooper, 2001).

Overall, the results of the two sets of structural equation analyses demonstrated that organisational experiences were more important than operational experiences in determining police officers' levels of occupational well-being. This calls into question the assumptions that are often made about the nature of police stress. In particular, these results clearly demonstrate that the nature of police work is not inherently stressful. Rather, it is the organisational context in which police officers work that explains the differences between those police officers' who experience lower or higher levels of occupational well-being. The role of the organisational context also appears to be more important than operational experiences in determining police officers' intentions to explore different job options, seek medical advice, take sickleave, or submit workers' compensation claims, as a result of stress-related problems.

4.7.3 Comparing the Influence of Personality, Coping Strategies, Organisational Climate and Work Experiences

Our attention now turns to examining the influence of positive and negative work experiences within a broader context. A series of structural equation analyses were conducted to examine the influence that personality, coping strategies, organisational climate, and positive and negative work experiences had on police officers' levels of occupational well-being and withdrawal behaviour intentions. For the purpose of these analyses, the unit weighted composite scores for Neuroticism, Extraversion, Emotion-Focused Coping, Problem-Focused Coping, Organisational Climate, Negative Work Experiences, Positive Work Experiences, Distress, Morale, Quality of Work Life and Withdrawal Behaviour were used as single indicators of their respective latent constructs, and measurement error was taken into account by setting the percentage of error variance at $(1 - \text{alpha})$. It was necessary to use this approach so that the number of parameters estimated during each of the analyses was kept within acceptable limits (Bagozzi and Heatherton, 1994; Hart, 1999).

The first model was an independence model that assumed there was no relationship between the 11 latent constructs. The second model is shown in Figure 4.5. For reasons of diagrammatic clarity, this Figure only illustrates the paths that were hypothesised in the theoretical model shown in Figure 4.2. In this model, we regressed Withdrawal Behaviour onto Distress, Morale and Quality of Work Life. These, in turn, were regressed onto Negative Work Experiences, Positive Work Experiences and Organisational Climate. Negative Work Experiences and Positive

Work Experiences were both regressed onto Emotion-Focused Coping and Problem-Focused Coping. Negative Work Experiences, Positive Work Experiences, Emotion-Focused Coping and Problem-Focused Coping were all regressed onto Organisational Climate. Finally, Withdrawal Behaviour, Distress, Morale, Quality of Work Life, Negative Work Experiences, Positive Work Experiences, Organisational Climate, Emotion-Focused Coping and Problem-Focused Coping were all regressed onto Neuroticism and Extraversion. We also estimated the correlations among the residuals for the three occupational well-being constructs, the residual for the work experiences constructs, and the residuals for the coping constructs. The correlation between Neuroticism and Extraversion was also estimated.

In Model 3, we reversed the order of the work experience and coping constructs. In other words, in this model we regressed Distress, Morale and Quality of Work Life onto Emotion-Focused Coping and Problem-Focused Coping. Emotion-Focused Coping and Problem-Focused Coping were both regressed onto Negative Work Experiences and Positive Work Experiences. This model reflected the traditional way in which the 'causal' order of work experiences, coping strategies and occupational well-being is often viewed. In other words, this model reflected the assumption that work experiences influence the use of coping strategies, and that coping strategies, in turn, influences occupational well-being.

Models 4 and 5 imposed equality constraints on Model 2 to test Propositions 3 and 4. In Model 4, we imposed an equality constraint on the paths that linked Negative Work Experiences and Positive Work Experiences to Distress. This enabled us to test the hypothesis that Negative Work Experiences made a stronger contribution than Positive Work Experiences to Distress. In Model 6, we imposed an equality constraint on the paths that linked Negative Work Experiences and Positive Work Experiences to Morale. This enabled us to test the hypothesis that Positive Work Experiences made a stronger contribution than Negative Working Experiences to Morale. Where necessary, the observed variables were reverse-scored before we estimated the models that contained the equality constraints. This was necessary to ensure that the equality constraints were imposed on paths of the same sign. The standardised beta coefficients for the key hypothesised paths in Model 2 are shown in Figure 4.3. The goodness-of-fit statistics for all models are shown in Table 4.6, and the standardised beta coefficients linking Neuroticism and Extraversion to all other latent constructs in the model are shown in Table 4.7.

The goodness-of-fit statistics for Model 2 suggested that there was an excellent fit between the variance–covariance matrix and the tested model. Moreover, a comparison of the goodness-of-fit statistics for Models 2 and 3 showed that the hypothesised model was a better fit to the data than the model depicting the more traditional way of viewing the relationships between work experiences, coping strategies, and occupational well-being. This conclusion is based on the fact that the chi-square statistics showed that there was no significant discrepancy between Model 2 and the observed data ($p > .05$), whereas there was a significant discrepancy between Model 3 and the observed data ($p < .05$). It was not possible to use the chi-square difference test to compare the relative fit of

Models 2 and 3, because the two models were not nested. These results were consistent with Proposition 4, and support earlier findings (Hart *et al.*, 1995) that suggest employees reconstruct history when completing work experience measures (Lazarus, 1990) and, as such, these should mediate the relationship between measures of coping and occupational well-being (Hart and Cooper, 2001). Within the context of the cognitive-relational theory of stress (Lazarus and Folkman, 1984), this implies that measures of work experiences actually assess employees' reappraisals of potentially stressful situations, rather than the primary appraisals that trigger the stress process. This is consistent with the notion that employees have already coped with the events, to some degree, before they are asked to report their experience of these events.

Table 4.6 Goodness-of-Fit Statistics for the Structural Equation Models Examining the Relations Between Withdrawal Behaviour, Occupational Well-Being, Work Experiences, Organisational Climate, Coping Strategies, and Personality

Model	χ^2	df	p	RMSEA	CFI	SRMSR	GFI
Model 1: Null	1,921.38	55	<.001				
Model 2: Theoretical Model	16.22	11	0.13	.03	1.00	.02	.99
Model 3: Traditional Model	35.09	11	<.001	.07	.99	.02	.99
Model 4: Distress	16.21	12	0.18	.03	1.00	.02	.99
Model 5: Morale	20.90	12	0.05	.03	1.00	.02	.99

Note: $N = 420$ (Listwise). RMSEA = Root Mean Square Error of Approximation, CFI = Comparative Fit Index, SRMSR = Standardised-Root-Mean-Square-Residual, and GFI = Goodness-of-fit Index. Model 3 was the traditional model were coping strategies mediated the relationship between work experiences and occupational well-being, Model 4 tested whether Negative Work Experiences contributed more strongly than Positive Work Experiences to Distress, and Model 5 tested whether Positive Work Experiences contributed more strongly than Negative Work Experiences to Morale.

Table 4.7 Standardised Beta Coefficients Linking Neuroticism and Extraversion to Withdrawal Behaviour, Occupational Well-Being, Work Experiences, Organisational Climate and Coping Strategies

Dependent Variable	Neuroticism	Extraversion
Withdrawal Behaviour	.32	.29
Quality of Work Life	−.18	.03[NS]
Distress	.49	.02[NS]
Morale	−.04[NS]	.31
Negative Work Experiences	.17	.14
Positive Work Experiences	.03[NS]	.25
Organisational Climate	−.21	.22
Emotion-Focused Coping	.47	.30
Problem Focused Coping	.16	.27

Note: $N = 420$ (Listwise). [NS] = Nonsignificant at the .05 level.

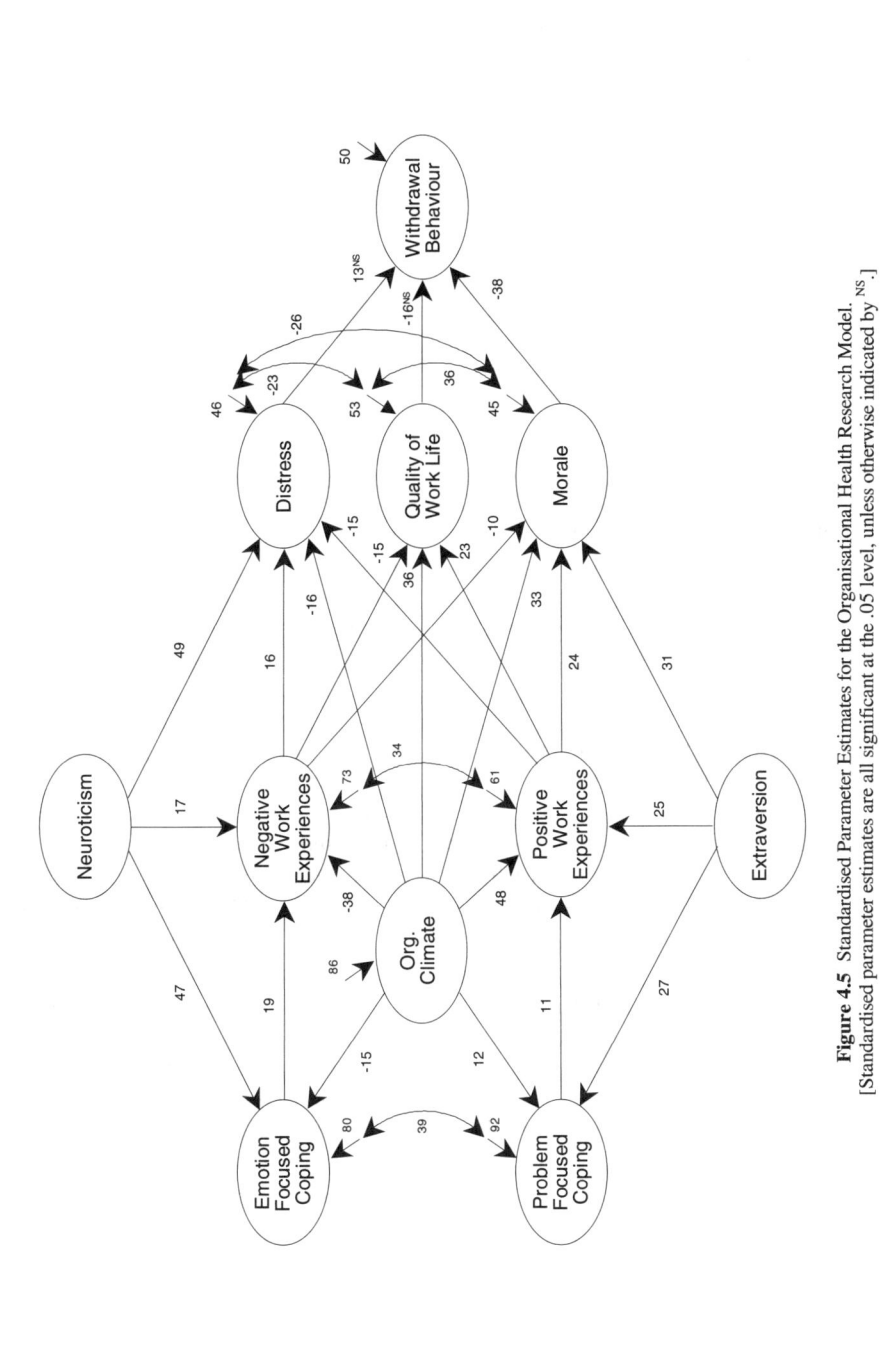

Figure 4.5 Standardised Parameter Estimates for the Organisational Health Research Model. [Standardised parameter estimates are all significant at the .05 level, unless otherwise indicated by NS.]

Given that Models 4 and 5 were nested within Model 2, it was possible to use the chi-square difference test to examine their relative fit. According to the chi-square difference tests, there was no significant difference between the fit of Models 2 and 4 ($\chi^2_{diff} = 0.01$, $df_{diff} = 1$, $p > .05$). These results failed to support Proposition 2, in as much as the beta coefficient linking Negative Work Experiences to Distress ($\beta = .16$) was not significantly different from the beta coefficient linking Positive Work Experiences to Distress ($\beta = -.15$). A chi-square difference test showed, however, that there was a significant difference between the beta coefficient linking Positive Work Experiences to Morale ($\beta = .24$) and the beta coefficient linking Negative Work Experiences to Morale ($\beta = -.10$) ($\chi^2_{diff} = 4.68$, $df_{diff} = 1$, $p < .05$). This results supported Proposition 3.

The standardised beta coefficients for Model 2 also supported Proposition 1. It was found that Organisational Climate contributed to Positive Work Experiences and Negative Work Experiences. Moreover, the results showed that Organisational Climate explained 23% of the variance in Positive Work Experiences and 14% of the variance in Negative Work Experiences. This was noticeably more than the variance explained by stable personality characteristics. As shown in Table 4.7, it was found that Extraversion explained 6% of the variance in Positive Work Experiences and 2% of the variance in Negative Work Experiences, whereas Neuroticism explained 3% of the variance in Negative Work Experiences and was not related significantly to Positive Work Experiences. This supports Hart's (1999) contention that measures of daily work experiences reflect conditions in the work environment, more than they reflect employees' stable personality characteristics and, as such, are not overly contaminated by the negative and positive affectivity bias that has often been raised as a major limitation of self-report measures of work experiences (Costa and McCrae, 1990; Brief *et al.*, 1993; Williams *et al.*, 1996).

These results also showed the central role that was played by organisational climate. Consistent with the theoretical model proposed by Hart and Cooper (2001), it was found that Organisational Climate was also related significantly to Emotion-Focused Coping ($\beta = -.15$), Problem-Focused Coping ($\beta = .12$), Distress ($\beta = -.16$), Morale ($\beta = .33$), and Quality of Work Life ($\beta = .36$). The overall pattern of results suggests that police officers' engage in both emotion and problem-focused coping strategies to help them manage or deal with day-to-day organisational experiences. These coping strategies, in turn, influence police officers' positive and negative work experiences, and it is these experiences, along with organisational climate that contributes to police officers' levels of occupational well-being. In terms of the cognitive-relational theory of stress, this pattern of relationships is consistent with the notion that organisational climate reflects primary appraisals, emotion and problem-focused coping reflect secondary appraisals, and that positive and negative work experiences reflect police officers' reappraisals.

Interestingly, it was found that there was a significant positive relationship between Emotion-Focused Coping and Negative Work Experiences. This implies that when police officers engage in emotion-focused coping strategies, it is likely to make matters worse, rather than better. The results showed, however, that the use of problem-focused coping strategies was more adaptive, in that the use of these strategies resulted in more positive work experiences. An example, based on an overload of paperwork, may help to illustrate these findings. A police officer may arrive at work to find that they have a large amount of paperwork that should have been attended to a few days earlier. This may cause them to experience a degree of discomfort or distress. If the officer chose to engage in emotion-focused coping strategies to deal with this situation, the results suggest that this would increase the level of discomfort and distress associated with their incomplete paperwork and level of work demands. This could be explained by the fact that while the officer is focusing on managing his or her emotions (e.g. by using relaxation techniques), the paper work is not being addressed and is becoming more and more overdue. If the officer were to deal with the problem, however, by completing the paper work and, maybe, examining their time-management or prioritisation of work tasks, the results suggest that this would lead the officer to feel positive about their work. This example also explains why police officers can experience both negative and positive emotions associated with their workloads (see Figure 4.3). Of course, this example assumes that police officers choose one strategy over another. The results shown in Figure 4.5, however, show that if police officers use one type of coping strategy, they are also likely to use the other type of strategy. This conclusion is based on the fact that the correlation between Emotion-Focused Coping and Problem-Focused Coping was .39 ($p < .01$). Moreover, consistent with previous research (Hart *et al.*, 1995; Heady and Wearing, 1990), it was found that there was no significant relationship found between Emotion-Focused Coping and Positive Work Experiences ($\beta = .07$, $p > .05$), nor between Problem-Focused Coping and Negative Work Experiences ($\beta = .04$, $p > .05$).

It was also noteworthy that Withdrawal Behaviour was determined by Morale, rather than Distress. Whereas the results showed that Morale explained 14% of the variance in Withdrawal Behaviour, there was no significant relation linking Withdrawal Behaviour to Distress ($\beta = .13$, $p > .05$) and Quality of Work Life ($\beta = -.16$, $p > .05$). This finding runs contrary to the conventional view that psychological distress contributes to the absenteeism and workers' compensation claims that result from occupational stress. Instead, this finding suggests that stress-related absenteeism and workers' compensation claims are driven more by the absence of morale (e.g. employees lack of energy, enthusiasm and pride in their work), rather than the presence of distress.

In order to test Propositions 5 and 6, we examined the total effects that were based on the pattern of relationships shown in Model 2. The standardised total effects are shown in Table 4.8 and showed that Neuroticism was the strongest determinant of Distress (standardised total effect = .60) and Withdrawal Behaviour (standardised total effect = .51). Organisational Climate, however, was found to be the strongest determinant of Morale (standardised total effect = .49) and Quality of Work Life (standardised total effect = .54). Moreover, Organisational Climate was the second strongest determinant of Distress (standardised total effect = −.36) and the third strongest determinant of Withdrawal Behaviour (standardised total effect = −.32). Interestingly, examination of the total effects showed that Emotion-Focused

Coping and Problem-Focused Coping failed to make a significant contribution to Distress, Morale, Quality of Work Life and Withdrawal Behaviour ($p > .01$). Overall, the total effects supported Propositions 4 and 6.

Table 4.8 Standardised Total Effects Showing the Relative Contribution
of the Predictor Variables to Police Officers' Levels of
Occupational Well-Being and Withdrawal Behaviour Intentions

Predictor Variable	Withdrawal Behaviour	Quality of Work Life	Distress	Morale
Quality of Work Life	$-.16^{NS}$			
Distress	$.13^{NS}$			
Morale	$-.38$			
Negative Work Experiences	$.10$	$-.19$	$.16$	$-.13$
Positive Work Experiences	$-.15$	$.25$	$-.16$	$.24$
Organisational Climate	$-.32$	$.54$	$-.36$	$.49$
Emotion-Focused Coping	$.01^{NS}$	$-.02^{NS}$	$.02^{NS}$	$-.01^{NS}$
Problem-Focused Coping	$-.01^{NS}$	$.02^{NS}$	$-.01^{NS}$	$.02^{NS}$
Neuroticism	$.51$	$-.32$	$.60$	$-.16$
Extraversion	$.08^{NS}$	$.18$	$-.07^{NS}$	$.46$

Note: $N = 420$ (Listwise). ^{NS} = Nonsignificant at the .05 level.

4.8 DISCUSSION

The results of this study clearly demonstrate that organisational factors, particularly those related to organisational climate, were the most influential determinants of occupational well-being among police officers. Accordingly, these findings indicate that police officers' levels of occupational well-being will only be improved if there is a focus on improving the leadership and management practices, as well as the organisational structures and processes (e.g. appraisal and recognition processes, decision-making styles, clarity of roles, goal alignment, etc.), that underpin the climate of police stations and other work teams within police organisations. In fact, the results shown in Table 4.8 demonstrate that a 10% improvement in the organisational climate of this particular police organisation would lead to a 3.6% decrease in distress, a 4.9% improvement in morale, a 5.4% improvement in quality of work life, and a 3.2% decrease in police officers' intentions to seek medical advice, leave their current job, take sickness absence, or submit a workers' compensation claim for stress-related problems.

Moreover, the central role of organisational climate is supported by the results showing that organisational experiences are more important than operational experiences (i.e. the specific nature of police work, including dealing with danger and victims) in determining police officers' levels of well-being. These findings are consistent with the results that have been found in other occupational groups (e.g. Griffin *et al.*, 2000; Hart, 1994; Hart *et al.*, in press), and demonstrate that police officers are really no different to employees in other occupational groups, when it comes to the issue of occupational well-being.

The notion that police organisations are similar to many other organisations, in terms of the issues that contribute to the well-being of employees, suggests that the strategies used to improve organisational climate and well-being in other organisations will also have relevance in police organisations. This point is worth noting, because police officers have often informed us that their situation is quite unique and different to what goes on in other organisations. This view is clearly not supported by the results of the present study.

One of the strategies for improving occupational well-being that has been used in other organisations is the use of regular employee opinion surveys that are used to build accountability among mangers for the development of their people management practices. For example, many Australian organisations in the public and private sector are now conducting regular employee opinion surveys to obtain information that is linked to performance management systems (e.g. balanced score cards) and used to inform ongoing improvements in people management practices. Longitudinal evidence is starting to emerge, showing that when regular employee opinion surveys are linked to an organisation's accountability (i.e. performance management) and development (e.g. improvements in leadership behaviour) frameworks, it is possible to achieve sustained improvements in leadership and managerial practices, organisational climate, and occupational well-being among employees (e.g. Hart, 2000).

The results of the present study also call into question the value of many traditional stress management training programmes that focus on teaching employees how best to cope with stressful situations in the workplace. These types of programmes often focus on a range of emotion- and problem-focused coping strategies, such as time management and stress relaxation training. The findings of this study, however, suggest that on the whole, coping skills exert a negligible influence over police well-being outcomes. Considering the relatively much stronger contribution of organisational factors, individual stress management approaches are likely to be of much less benefit overall than organisationally oriented interventions. It should be noted, however, that we have not explored the potential moderating effects of emotion- and problem-focused coping strategies, and have examined a normative, rather than clinical sample of police officers. Nevertheless, these findings caution against the use of organisation-wide stress management training programmes, unless there is evidence from within the organisation to demonstrate that they will add value.

Our results also showed that neuroticism was the strongest determinant of distress among police officers. This result is consistent with the findings in the negative affectivity literature (e.g. Burke *et al.*, 1993; Moyle, 1995; Williams *et al.*, 1996), and raises the question about what can be done, from an organisational perspective, to address the influence of personality factors. Although the temptation is to consider personality screening during selection and placement processes, this may not necessarily be the most appropriate response. It has been shown for example, that the enduring personality characteristic of neuroticism has little, if any, effect on performance outcomes (Barrick and Mount, 1991). There is also little empirical evidence to support the notion that distress contributes to sickness absence and workers' compensation claims for stress-related injuries (e.g. George, 1989; Hart *et al.*, in press). This is consistent with the current findings that showed that withdrawal behaviour intentions were related to the absence of morale, rather than the presence of distress. Accordingly, there are no strong grounds for adopting personality screening as a method of reducing distress among employees.

The central importance of organisational climate, as well as the findings that withdrawal behaviour (intention to submit a compensation claim for stress) was more strongly influenced by the lack of morale rather than the presence of distress, raises issues about how best to support police officers' who are exposed to traumatic and distressing operational incidents. This is also an important issue to address in light of the current controversy in the clinical literature over the status of workplace critical incident debriefing practices. A growing number of studies have found that employee involvement in debriefing following exposure to distressing operational incidents is not associated with any positive impact on clinical outcomes (e.g. Carlier *et al.*, 2000; Hobbs *et al.*, 1996; Bisson *et al.*, 1997). Such findings must be reconciled with the fact that participants in debriefing processes typically report high levels of satisfaction (e.g. Armstrong *et al.*, 1998). The current findings suggest that debriefing practices may actually be most effective when they are oriented towards the provision of social and organisational support, rather than focusing on clinical issues. In other words, we can reconceptualise debriefing as an organisational rather than a clinical intervention; as a gesture of support to employees which likely impacts on morale rather than on distress levels. As such, appropriately re-oriented forms of debriefing practice may still have a useful role to serve in the workplace, albeit not as a clinical intervention. Such a reconceptualisation, we suggest, opens up a promising avenue for further research investigating the efficacy of various components of debriefing practice and their role in the workplace management of trauma.

Similarly, these results may help to clarify the conflicts that the second author has frequently observed in clinical contexts between treatment providers and clinicians assessing stress-related workers compensation claims. For example, a common presenting 'trauma' profile involves a police officer with a history of attendance at various operational incidents. The assessment process reveals that the major trigger for the claim is actually disgruntlement over a management decision, career progression issues, or other related contemporary organisational concerns. However, these concerns have been inadvertently obscured by treating stress professionals who presume that operational experiences exert the strongest influence on occupational well-being. As such, it is common for an officer's employment history to be reframed through an iatrogenically fostered re-attribution process. In other words, the clinician, through encouraging a focus on reviewing past operational incidents, unwittingly influences the officer to recontextualise and link his or her current distress to past incidents that did not have any particular negative impact at the time. We suggest that the assessment and clinical management of work-related stress conditions may be advanced through better integration of the work psychology, stress and occupational clinical literatures, and by treating providers paying closer attention to research findings in all of these areas.

Our results further suggest that stressors and strain studies can be misleading in terms of guiding priorities for intervention resources. The focus on negative experiences, coping and other individual level factors does not allow the relative potency of these factors to be adequately assessed. The present study has demonstrated that there is a need to take a broader perspective so that a system of variables can be examined to determine which factors are the most important determinants of individual well-being outcomes. Such an approach provides a much more comprehensive basis for prioritizing interventions that will actually make a difference.

Although our results were consistent with prior research among police officers (e.g. Hart *et al.*, 1995), as well as other occupational and community groups (e.g. Griffin *et al.*, 2000; Heady and Wearing, 1992), there are a number of limitations that should be considered when interpreting our results. First, it should be noted that the present study utilised a cross-sectional design. As such, it is not possible to draw conclusions about causation. At best, we are able to state that our data are consistent with a causal theory. Nevertheless, it is important to replicate these findings with longitudinal data that enables an investigation of stability and change among the variables that were included in the research model (e.g. Hart, 1999). Second, the present study used a self-report survey methodology, and this always raises concerns about common method bias. However, common method bias is not likely to have played a major role in the current study, because the pattern of relationships among the study variables showed that there was strong discrimination in police officers' responses. For example, the zero correlation shown in Figure 4.3 between Positive Work Experiences and Negative Work Experiences cannot be consistent with common method bias. Nevertheless, future studies should consider the use of third-party ratings and the use of objective data about withdrawal behaviours (e.g. actual sickness absence data). Third, the sample size for the unsworn public service employees working in the police organisation was relatively small. This prevented us from testing the research model with this occupational group. Given our claim that the results suggest police officers' are no different to other employees, in terms of the issues that affect occupational well-being, it is important to replicate the results for the research model with data obtained from other occupational groups (e.g. Hart *et al.*, 2002).

Notwithstanding these limitations, the present study has provided strong empirical evidence that calls into question a number of common assumptions about police stress. We found that the weight of evidence runs counter to the stereotypic views of police stress in that police officers appear to be just the same as any other occupational group in terms of the issues that affect occupational well-being. This demonstrates the importance of conducting research and challenging conventional views, because a reliance on conventional views and folklore can be quite misleading and lead to the implementation of policies and practices that are unlikely to bring about sustained improvements in occupational well-being among police officers.

4.9   ACKNOWLEDGEMENTS

This research was supported by a National Health and Medical Research Council Public Health Fellowship (Grant No. 954208) awarded to the first author. We also thank Ruth Tyman who played a lead role in collecting the data within the police organisation that was used during this study, and Mark A. Griffin who assisted with data collection for the normative sample used during this study.

Correspondence concerning this article should be addressed to Peter M. Hart, Department of Psychology, University of Melbourne, Parkville, Victoria, 3052, Australia. Electronic mail may be sent via Internet to hart@insightsrc.com.au.

4.10   REFERENCES

Adams, G.A., King, L.A. and King, D.W., 1996,. Relationships of job and family involvement, family social support, and work-family conflict with job and life satisfaction. *Journal of Applied Psychology*, **81**, pp. 411–420.

Agho, A.O., Price, J.L. and Mueller, C.W., 1992, Discriminant validity of measures of job satisfaction, positive affectivity and negative affectivity. *Journal of Occupational and Organizational Psychology*, **65**, pp. 185–196.

Anshel, M.H., Robertson, M. and Caputi, P., 1997, Police stress and coping, *Journal of Occupational and Organizational Psychology*, **70**, pp. 337–356.

Anshel, M.H., 2000, A conceptual model and implications for coping with stressful events in police work. *Criminal Justice and Behaviour*, **27**, pp. 375–400.

Anson, R.H. and Bloom, M.E., 1988, Police stress in an occupational context. *Journal of Police Science and Administration*, **16**, pp. 229–235.

Armstrong, K., Zatzick, D., Metzler, T., Weiss, D.S., Marmar, C.R., Garma, S., Ronfelt, H. and Roepke, L., 1998, Debriefing of American Red Cross personnel. *Social Work in Health Care*, **27**, pp. 33–50.

Bagozzi, R.P. and Heatherton, T.F., 1994, A general approach to representing multifaceted personality constructs: Application to state self-esteem. *Structural Equation Modeling*, **1**, pp. 35–67.

Barrick, M.R. and Mount, M.K., 1991, The Big Five Personality Dimensions and Job Performance: A Meta-Analysis. *Personnel Psychology*, **44**, pp. 1–25.

Bisson, J., Jenkins, P., Alexander, J. and Bannister, C., 1997, Randomised controlled study of psychological debriefing for victims with acute burns trauma. *British Journal of Psychiatry*, **171**, pp. 78–81.

Biggam, F.H., Power, K.G. and Macdonald, R.R., 1997, Coping with the occupational stressors of police work: a study of Scottish police officers. *Stress Medicine*, **13**, pp. 109–115.

Billings, A.C. and Moos, R.H., 1984, Coping, stress and social resources among adults with unipolar depression. *Journal of Personality and Social Psychology*, **46**, pp. 877–891.

Brief, A.P., Burke, M.J., George, J.M., Robinson, B.S. and Webster, J., 1988, Should negative affectivity remain an unmeasured variable in the study of job stress? *Journal of Applied Psychology*, **73**, pp. 193–198.

Brown, J.M. and Campbell, E.A., 1990, Sources of occupational stress in the police. *Work and Stress*, **4**, pp. 305–318.

Burke, M.J., Brief, A.P. and George, J.M., 1993, The role of negative affectivity in understanding relations between self-reports of stressors and strains: a comment on the applied psychology literature. *Journal of Applied Psychology*, **78**, pp. 402–412.

Cacioppe, R.L. and Mock, P., 1985, The relationship of self-actualisation, stress and quality of work experience in senior Australian police officers. *Police Studies*, **8**, pp. 173–186.

Carpenter, B.N. (ed.), 1992, *Personal Coping: Theory, Research, and Application* (Westport, CT: Praeger).

Carlier, I.V.E., Voerman, A.E. and Gersons, B.P.R., 2000, The influence of occupational debriefing on post-traumatic stress symptomatology in traumatise police officers. *British Journal of Medical Psychology*, **73**, pp. 87–98.

Carver, C.S., Scheier, M.F. and Weintraub, J.K., 1989, Assessing coping strategies: A theoretically based approach. *Journal of Personality and Social Psychology*, **56**, pp. 267–283.

Cohen, S. and Hoberman, H.M., 1983, Positive events and social supports as buffers of life change stress. *Journal of Applied Social Psychology*, **13**, pp. 99–125.

Cooper, C.L. (ed.), 1998, *Theories of Organisational Stress* (New York: Oxford).

Cooper, C.L. and Payne, R. (eds), 1991, *Personality and stress: Individual Differences in the Stress Process* (Chichester: John Wiley).

Costa, P.T. Jr and McCrae, R.R., 1980, Influence of extraversion and neuroticism on subjective well-being. *Journal of Personality and Social Psychology*, **38**, pp. 668–678.

Costa, P.T. Jr and McCrae, R.R., 1984, Personality as a lifelong determinant of well-being. In Cmalatesta and C. Izard (eds), *Affective processes in adult development and aging* (pp. 141–157).

Costa, P.T. Jr and McCrae, R.R., 1985, *The NEO Personality Inventory, Manual*. Odessa (FL: Psychological Assessment Resources).

Costa, P.T. Jr and McCrae, R.R., 1989, *The NEO PI/FFI Manual Supplement, l*. Odessa (FL: Psychological Assessment Resources).

Costa, P.T. Jr and McCrae, R.R., 1990, Personality: Another 'hidden factor' in stress research. *Psychological Inquiry, An International Journal of Peer Commentary and Review*, **1**, pp. 22–24.

Cox, T., 1988, Organisational health. *Work and Stress*, **2**, pp. 1–2.

Cox, T., 1992, Occupational health: past, present and future. *Work and Stress*, **6**, pp. 99–102.

Cuttance, P., 1987, Issues and problems in the application of structural equation models. In P. Cuttance and R. Ecob (eds), *Structural Modeling by Example: Applications in Educational, Sociological and Behavioral Research* (New York: Cambridge), pp. 241–279.

Day, A.L. and Livingstone, H.A., 2001, Chronic and acute stressors among military personnel: do coping styles buffer their negative impact? *Journal of Occupational Health Psychology*, **6**, pp. 348–360.

Dantzer, M.L., 1987, Police related stress: a critique for future research. *Journal of Police Criminal Psychology*, **3**, pp. 43–48.

DeLongis, A., Folkman, S. and Lazarus, R.S., 1988, The impact of daily stress on health and mood: Psychological and social resources as mediators. *Journal of Personality and Social Psychology*, **54**, pp. 486–495.

Diener, E., 2000, Subjective wellbeing: The science of happiness and a proposal for a national index. *American Psychologist*, **55**, pp. 34–43.

Diener, E. and Emmons, R.A., 1985, The independence of positive and negative affect. *Journal of Personality and Social Psychology*, **47,** pp. 1105–1117.

George, J.M., 1989, Mood and absence. *Journal of Applied Psychology*, **74**, pp. 317–324.

George, J.M., 1996, Trait and state affect. In K.R. Murphy (ed.), *Individual Differences and Behavior in Organizations* (San Francisco: Jossey-Bass), pp. 145–171.

Griffin, M.A., Hart, P.M. and Wilson-Evered, E., 2000, Using employee opinion surveys to improve organisational health. In L.R. Murphy and C.L. Cooper (eds), *Health and productive work: an international perspective* (London: Taylor and Francis), pp. 15–36.

Goldberg, D., 1978, *Manual of the General Health Questionnaire.* Windsor: National Foundation for Educational Research

Greller, M.M., Parsons, C.K. and Mitchell, D.R.D., 1992, Additive effects and beyond: Occupational stressors and social buffers in a police organisation. In J.C. Quick, L.R. Murphy and J.J. Hurrell, Jr (eds), *Stress and Well-Being at Work, Assessments and Interventions for Occupational Mental Health,* (Washington DC: American Psychological Association), pp. 33–47.

Hart, P.M., 1988, *Stress measurement and perceived quality of life amongst police officers.* Unpublished Master's Preliminary thesis, University of Melbourne.

Hart, P.M., 1994, Teacher quality of work life: Integrating work experiences, psychological distress and morale. *Journal of Occupational and Organizational Psychology*, **67**, pp. 109–132.

Hart, P.M., 1999, Predicting Employee Life Satisfaction: A Coherent Model of Personality, Work and Nonwork Experiences, and Domain Satisfactions. *Journal of Applied Psychology*, **84**, pp. 564–584.

Hart, P.M., 2000, *Understanding Organisational Health: Benchmarking and Improvement in Victoria Schools.* Melbourne: Department of Education, Employment and Training.

Hart, P.M. and Cooper C.L., 2001, Occupational Stress: Toward a more integrated framework. In N. Anderson, D.S. Ones, H.K. Sinangil and C. Viswesvaran (eds),

Handbook of Industrial, Work and Organizational Psychology, Volume 2. (Sage: London), pp. 93–114.

Hart, P.M., Cotton, P., Wearing, A.J. and Cooper C.L., 2002, *Stress in the public service: A test of the organizational health framework*. Manuscript in preparation, University of Melbourne.

Hart, P.M., Griffin, M.A., Wearing, A.J. and Cooper, C.L., 1996, *Manual for the QPASS Survey* (Brisbane: Public Sector Management Commission).

Hart, P.M., Palmer, R.H., Christie, S. and Lander, D., 2002, April, *Linking Climate, Job Satisfaction and Contextual Performance to Customer Experience*. Paper presented at the 17th Annual Conference of the Society for Industrial and Organizational Psychology, Toronto, Canada.

Hart, P.M., Wearing, A.J. and Headey, B., 1993, Assessing police work experiences: development of the Police Daily Hassles and Uplifts Scales. *Journal of Criminal Justice*, **21**, pp. 553–572.

Hart, P.M., Wearing, A.J. and Headey, B., 1994, Perceived quality of life, personality and work experiences: construct validation of the Police Daily Hassles and Uplifts Scales. *Journal of Criminal Justice and Behavior*, **21**, pp. 283–311.

Hart, P.M., Wearing, A.J. and Headey, B., 1995, Police stress and well-being: Integrating personality, coping and daily work experiences. *Journal of Occupational and Organizational Psychology*, **68**, pp. 133–156.

Hart, P.M., Wearing, A.J., Conn, M., Carter, N.L. and Dingle, R.K., 2000, Development of the School Organizational Health Questionnaire: a measure for assessing teacher morale and school organisational climate. *British Journal of Educational Psychology*, **70**, pp. 211–228.

Headey, B., Glowacki, T., Holmstrom, E. and Wearing, A.J., 1985, Modelling change in perceived quality of life. *Social Indicators Research*, **17**, pp. 267–298.

Headey, B. and Wearing, A.J., 1989, Personality, life events, and subjective wellbeing: towards a dynamic equilibrium model. *Journal of Personality and Social Psychology*, **57**, pp. 731–739.

Heady, B. and Wearing, A.J., 1990, Subjective well-being and coping with adversity. *Social Indicators Research*, **22**, pp. 327–349.

Headey, B. and Wearing, A.J., 1992, *Understanding Happiness: A Theory of Subjective Well-Being* (Melbourne: Longman Cheshire).

Hobbs, M., Mayou, R., Harrison, B. and Worlock, P., 1996, A randomised controlled trial of psychological debriefing for victims of road traffic accidents. *British Medical Journal*, **313**, pp. 1438–1439.

Hurrell, J.J., 1995, Police work, occupational stress and individual coping. *Journal of Organisational Behaviour*, **16**, pp. 27–28.

Hurrell, J.J., Jr, Nelson, D.L. and Simmons, B.L., 1998, Measuring job stressors and strains: where we have been, where we are, and where we need to go. *Journal of Occupational Health Psychology*, **3**, pp. 368–389.

James, L.R. and McIntyre, M.D., 1996, Perceptions of organisational climate. In K. Murphy (ed.), *Individual Differences and Behavior in Organisations* (San Francisco: Jossey-Bass) pp. 416–450.

Joreskog, K.G. and Sorbom, D., 1993, *LISREL 8, User's Reference Guide* (Chicago, IL: Scientific Software).

Judge, T.A., Thoresen, C.J., Pucik, V. and Welbourne, T.M., 1999, Managerial coping with organizational change. *Journal of Applied Psychology*, **84**, pp. 107–122.

Kasl, S.V. and Rapp, S.R., 1991, Stress, health and well-being: The role of individual differences. In C.L. Cooper and R. Payne (eds), *Personality and Stress: Individual Differences in the Stress Process* (Chichester: Wiley), pp. 269–284.

Kirkcaldy, B., Cooper, C.L. and Ruffalo, P., 1995, Work stress and health in a sample US police. *Psychological Reports*, **76**, pp. 700–702.

Klein, G., 1996, The effect of acute stressors on decision-making. In J.E. Driskell and E. Salas (eds), Stress and human performance (Mahwah, NJ: Lawrence Erlbaum), pp. 49–88.

Kop, N. and Euwema, M.C., 2001, Occupational stress and the use of force by Dutch police officers. *Criminal Justice and Behaviour*, **28**, pp. 631–652.

Latack, J.C. and Havlovic, S.J., 1992, Coping with job stress: A conceptual evaluation framework fro coping measures. *Journal of Organisational Behaviour*, **13**, pp. 479–508.

Lawrence, R.A., 1984, Police stress and personality factors: A conceptual model. *Journal of Criminal Justice*, **12**, pp. 247–263.

Lazarus, R.S., 1990, Theory-based stress measurement. *Psychological Inquiry: An International Journal of Peer Commentary and Review*, **1**, pp. 3–13.

Lazarus, R.S. and Folkman, S., 1984, *Stress, Appraisal, and Coping* (New York: Springer).

Lilienfield S.O., 2002, When worlds collide: Social science, politics, and the Rind *et al.* (1998) child sexual abuse meta-analysis. *American Psychologist*, **57**, pp. 176–188.

Lowenstein, L.F., 1999, Treating stress in the police force. *The Police Journal,* January, pp. 65–74.

Malloy, T.E. and Mays, G.L., 1984, The police stress hypothesis. *Criminal Justice Bulletin*, **11**, pp. 197–224.

Michela, J.L., Lukaszewski, M.P. and Allegrante, J.P., 1995, In S.L. Sauter and L.R. Murphy (eds), *Organisational Risk Factors for Job Stress* (Washington, DC: American Psychological Association), pp. 61–80.

Moyle, P., 1995, The role of negative affectivity in the stress process: Tests of alternative models. *Journal of Organisational Behaviour*, **16**, pp. 647–670.

Near, J.P., Smith, C.A., Rice, R.W. and Hunt, R.G., 1983, Job satisfaction and nonwork satisfaction as components of life satisfaction. *Journal of Applied Social Psychology*, **13**, pp. 126–144.

Newton, T.J., 1989, Occupational stress and coping with stress: a critique. *Human Relations*, **42**, pp. 441–461.

Pavot, W. and Diener, E., 1993, Review of the satisfaction with life scale. *Psychological Assessment*, **2**, pp. 164–172.

Quick, J.C., Murphy, L.R. and Hurrell, J.J. Jr (eds), 1992, *Stress and Well-Being at Work: Assessments and Interventions for Occupational Mental Health* (Washington, DC: American Psychological Society).

Sauter, S.L. and Murphy, L.R., 1995, *Organisational Risk Factors for Job Stress* (Washington, DC: American Psychological Society).

Schneider, B. (ed.), 1990, *Organisational climate and culture* (San Francisco: Jossey-Bass).

Sigler, R.T. and Wilson, C.N., 1988, Stress in the workplace: Comparing police stress with teacher stress. *Journal of Police Science and Administration*, **16**, pp. 151–162.

Terry, W.C. III, 1981, Police stress: the empirical evidence. *Journal of Police Science and Administration*, **9**, pp. 61–75.

Terry, W.C. III, 1983, Police stress as an individual and administrative problem: Some conceptual and theoretical difficulties. *Journal of Police Science and Administration*, **11**, pp. 156–165.

Watson, D. and Tellegen, A., 1985, Towards a consensual structure of mood. *Psychological Bulletin*, **98**, pp. 219–235.

Williams, L.J., Gavin, M.B. and Williams, M.L., 1996, Measurement and nonmeasurement processes with negative affectivity and employee attitudes. *Journal of Applied Psychology*, **81**, pp. 88–101.

Wright, T.A. and Cropanzano, R., 2000, Psychological well-being and job satisfaction as predictors of job performance. *Journal of Occupational Health Psychology*, **5**, pp. 84–94.

CHAPTER FIVE

Burnout Among Oncology Care Providers: Radiation Assistants, Physicians and Nurses

Pascale M. Le Blanc and Wilmar B. Schaufeli

5.1 INTRODUCTION

Oncology can be an exciting and challenging specialty for those who work in it. Care providers generally give many important reasons for choosing to be involved in the direct care of cancer patients, e.g. a compassion for patients who face a life threat, a desire to be closely involved in the treatment of chronically ill patients, and the challenges presented by the complex tasks of diagnosis and clinical care. In a qualitative study by Haberman *et al.* (1994), an oncology nurse described the essence of her practice as follows: 'We (i.e. *the patient and herself; the authors*) walk down the same road, but we wear different shoes.' And in a study by Peteet *et al.* (1989) among staff members of a comprehensive cancer centre in the USA, the ideal described by the largest number of clinicians was to be 'a friend within a professional relationship'. However, in their daily routines, oncology care providers are also faced with a host of psychosocial problems that may drain their excitement and damage their commitment to ideals that initially drew them to the specialty (Flint Sparks, 1989). Eventually, this may lead to burnout, a form of chronic job stress that is characterised by emotional exhaustion, depersonalisation and reduced personal accomplishment (Maslach, 1982; Schaufeli and Enzmann, 1998). In the best light, burnout can be seen not as a condemnation of the professional activity per se but rather a reflection of the total quantity of the emotional stresses of practice, which dominate the majority of professional time in the practice of oncology (Whippen and Canellos, 1991). However, left untreated, burnout might reduce a person's ability to use the excellent capabilities that may have taken years of training to develop (Flint Sparks, 1989), and eventually might result in the care provider leaving the profession.

In this chapter, we would like to give a systematic overview of work-related factors which are associated with work stress in general, and burnout in particular, among oncology care providers. Next, the results of a national, questionnaire-based survey among Dutch oncology care providers with respect to (de)motivating aspects of their working situation and their levels of burnout and psychiatric morbidity will be presented and discussed. The last part of this chapter will address interventions to combat work stress and burnout in these professions, including a brief description of team based burnout prevention training for oncology care providers, developed in continuation of our national survey.

5.2   WORK STRESSORS IN ONCOLOGY

In this paragraph, an overview of the main sources of work stress for oncology care providers is presented. In total, six categories of work stressors are distinguished and will be discussed in more detail below.

5.2.1   Patient-related/Emotional Stressors

The diagnosis of cancer is a major stressful life event that affects an increasing number of people in the industrialised countries. The initial period after diagnosis may be regarded as a life crisis, as cancer patients are confronted with a life-threatening disease, its treatment, and severe physical side effects. Patients may also experience feelings of uncertainty, a diminished self-image, and changes in social relationships (Moos and Schaefer, 1984). Not surprisingly, about a quarter of cancer patients experience depressive symptoms during this period (McDaniel *et al.*, 1995). These, and many other patient reactions to cancer (e.g. regressive behaviour, numbness and inappropriate denial, panic and grief, a need to propitiate and bargain, disappointment and anger) are difficult to handle with professional demeanour. The required steadfastness of personal behaviour demands much of oncology care providers' strength and maturity. On the one hand, 'difficult' patients may make staff become angry and unempathic, or depersonalised (Maslach, 1982). Yet, the illness still endows the patient with pathos, and caregivers may feel guilty about their hostile feelings. On the other hand, some patients may become 'special' to a caregiver – usually because of personal experiences – and these patients may arouse over-involvement (Lederberg, 1998). Cancer patients' need for psychological security may be extreme. Patients and their families often develop highly dependent relationships with oncology care providers and place great trust and faith in them. Though the intimacy and closeness of these interpersonal relationships can evoke feelings of accomplishment and unique importance (Meyer, 1992), their intensity may also put a heavy emotional burden on care providers. Patients' verbalisations of their fears may be quite helpful for themselves, but will be stressful for care providers who feel that they should somehow reassure the patient (Vachon *et al.*, 1978). Moreover, continuously having to satisfy the emotional needs of cancer patients and their families may become emotionally exhausting in the long run. In fact, in caring for cancer patients, the question is how to remain remote enough to think, function and prevent burnout, yet close enough to relate (Himmelsbach, 1978).

Despite the fact that some malignancies are now curable even in advanced stages, a large proportion of cancer patients still have a poor prognosis, as their illness remains unpredictable in its course and its response to treatment. Foreseeable declines, patients' struggles to come to grips, and family desperation may create a climate of dull pain, suppressed sadness and regret, if not outright guilt, on oncology wards. Intense and repeated exposures to mutilation and suffering may not only lead to pain and sadness among oncology care providers, but also to fear and revulsion, and this makes it impossible for them to maintain a simplistic belief in 'terrestrial justice' (Lederberg, 1998). In these situations, they also have to acknowledge the limits of modern medicine. They may feel incompetent, 'helpless helpers', when confronted with an irresistibly advancing disease and ultimately death (Hürny, 1988). Frequent exposure to death may lead to a loss of their sense of invulnerability and a painful awareness of personal mortality. Even seasoned staff will react with tension and the need to escape, when deaths are too frequent, unexpected, or difficult (Lederberg, 1998). Moreover, the lack of time to grieve after each patient also increases the stress of repeated and successive deaths (Adams *et al.*, 1991; Cooper and Mitchell, 1990; Gray-Toft and Anderson, 1987; Vachon, 1987).

5.2.2 Environmental/Physical Stressors

In addition to these stressors that are more or less inherent in the nature of patient responses to cancer, Lederberg (1998) discusses some other stressors that impinge on oncology care providers and are inherent to the nature of treatment. Oncology staff are required to apply toxic treatments, i.e. radiation and chemotherapy, which can also provoke feelings of anxiety among them. They often have to 'make their patients sick' for an expected benefit later on. The management of these and other ever-present side effects is a constant reminder to patient and staff of the cytotoxic nature of cancer treatment. Even transient side-effects, such as nausea and hair loss, can be distressing for staff because they are so difficult for some patients to tolerate. Less obvious long-term effects such as infertility are very significant for caregivers in their childbearing years. Treatment procedures that carry a significant risk of mortality, like bone marrow transplantation, have an ever-present emotional tension. Surgery remains a crucial and often successful form of treatment, but when it is disfiguring or extremely radical, it may also be experienced as fundamentally damaging (Lederberg, 1998).

5.2.3 Mental/Cognitive Stressors

Third, oncology care providers are confronted with a variety of complicated treatment regimens and a rapidly growing amount of knowledge in their subspecialty that is not easy to assimilate (Hürny, 1988). 'High tech' oncology settings demand a high level of intellectual acuity for rapid decision making based on a thorough understanding of complex techniques. The technical demands of medical oncology and radiation therapy may be so great that the care provider misses out on the more enjoyable aspects of patient care (Herrera, 1986). In addition, decision making has become more complicated, as treatment modalities have multiplied and patients have become more active participants.

5.2.4   Social/Interpersonal Stressors

As treatment moves through successive phases, the shift from curative to palliative intent may be very difficult to negotiate. It engenders many negative reactions in patients and families, which are often diverted onto staff (Lederberg, 1998). Physicians, in particular, may be repeatedly confronted with a conflict between the curative goals on which most of their training is based and the palliative goals of much cancer care (Delvaux *et al.*, 1988). A descriptive survey conducted in the USA has indeed suggested that American oncologists experience high levels of burnout that are particularly related to this conflict (Whippen and Canellos, 1991). For nurses, some authors have shown that their level of stress is higher when they are asked to meet simultaneously acute and palliative clinical care objectives (Benoliel, 1969; Bene and Foxal, 1991). In addition, both professions have to deal with complex ethical issues of clinical research, where benefits for the individual patient may not be directly evident (Hürny, 1988). The need to maintain a balance between human concern and scientific objectivity can be a constant struggle (Flint Sparks, 1996).

5.2.5   Discipline-specific Stressors

So, in general, working in oncology can be considered stressful and may give rise to ambivalent feelings among care providers. Unfortunately, in the formal training of oncology care providers, no solid basis of psychosocial knowledge and skills is given to facilitate coping with the more or less specialty-specific issues that were just discussed. There is still a lack of education in interviewing skills, needs and coping mechanisms of patients with a chronic, potentially fatal disease, care of the terminally ill, and the impact of terminal illness on the caregiver–patient relationship (Hürny, 1988). Lack of staff know-how on communication issues often leads to patient and family dissatisfaction, which in turn may give rise to tension, anxiety and avoidance among staff members as well as low self-esteem and decreased job satisfaction (Ford *et al.*, 1994; Ramirez *et al.*, 1995; Ullrich and Fitzgerald, 1990). Also, these type of stressors may eventually lead to dysfunctional staff responses like anxiety, depression, burnout, and psychiatric problems (Lederberg, 1998).

   Especially in this field of medicine, the (interdisciplinary) team can be an important (re)source of physical and emotional support. Moreover, it allows for the flexibility needed to work in such an emotionally charged field (Hürny, 1988).

However, it can also be a source of stress, as each discipline, or each individual may focus on its own facet of care with little sympathy for the other's viewpoint. This can be exacerbated in research settings where ethical and philosophical conflicts between research and clinical care commitments can readily occur, even in objectively well-managed cases. Whether acknowledged or not, these conflicts may intensify other sources of division. Some of these differences are discipline-based. As doctors have the most detailed information about the natural history of the disease, and its likely course and prognosis in a given patient, it is most natural for them to assume an intellectual approach. However, some nurses do not identify with this approach and view themselves more as 'nurturers'. There is definitely no incompatibility between the two stances, but the primary identification with one or the other dictates a set of emotional responses and sources of professional satisfaction. Failure to acknowledge the existence of these two stances, or valuing one at the expense of the other, may create tensions on units or within a team (Lederberg, 1998). As a result of these differences in approach, the different disciplines in oncology can be more or less characterised by their own specific 'stress-profiles', depending on their role in the process of care giving. We will return to this issue in more detail in the next section.

5.2.6   General Stressors

In addition to the above, literature shows that care providers in oncology not only experience many stressors unique to their specialty, e.g. the frequent confrontation with death and dying, but are also exposed to work stressors that are more common to other health care workers, e.g. high workload and lack of autonomy (Hürny, 1988; Schaufeli, 1999).

5.3   JOB STRAIN AND STRESSORS IN ONCOLOGY

5.3.1   Strain

Though this overview of potential sources of stress in oncology may lead to the impression that oncology care providers experience more severe job strain than professionals in other medical specialties, the few empirical studies in which stress and burnout levels of care providers in oncology and non-oncology settings are compared yield contradictory results. Some empirical studies found a lower level of job strain or burnout among nurses working in oncology as compared to those working in other medical specialties. For instance, in a study by Yasko (1983), oncology clinical nurse specialists' mean scores on the Staff Burnout Scale for Health Professionals turned out to be significantly lower than those of nurses of eight other medical specialties. Jenkins and Ostchega (1986), who replicated Yasko's study among oncology nurses, found similar results.

On the other hand, there are also a few empirical studies that found higher levels on aspects of burnout among oncology care providers than among care providers in other disciplines. In a study among a representative sample of all active physicians in Finland (Olkinuora *et al.*, 1990), oncology was listed among the 'high burnout specialties'. Catalan *et al.* (1996), who compared levels of burnout, assessed by the Maslach Burnout Inventory (MBI), and levels of psychiatric morbidity, assessed by the General Health Questionnaire (GHQ-28) (Goldberg, 1978) found no differences between staff working with AIDS and those working with cancer patients with respect to emotional exhaustion and depersonalisation scores. However, AIDS staff reported both lower intensity and frequency of personal accomplishment than oncology staff. A possible explanation for the finding of reduced personal accomplishment in AIDS staff may be the fact that AIDS care workers deal with patients whose prognosis is generally regarded as more gloomy and for whom therapeutic options are more limited than for cancer patients. With respect to overall GHQ-morbidity, similar proportions scored above the cut-off for caseness in each group: AIDS 40 per cent and oncology 44 per cent. Also, Lopez-Castillo *et al.* (1999) found no significant differences in psychiatric morbidity, assessed by the GHQ-28, between oncology, internal medicine and AIDS care providers, but overall levels of psychiatric morbidity were high: 38 per cent of the total sample reported levels of psychological distress at or above the caseness level. With respect to burnout, again assessed by the MBI, levels of emotional exhaustion were highest in oncology, and were significantly higher in both oncology and general internal medicine than in the other two groups.

However, most recent empirical studies show no significant differences between oncology and non-oncology professionals with respect to the total level of job-strain, burnout or psychiatric morbidity. In a study by Van Servellen and Leake (1993), the MBI was administered to a sample of nurses from 18 units in seven different hospitals, including special care units managing people with AIDS, cancer, medical intensive care, and general medical diseases. This study showed no significant differences in burnout scores across nurse samples representing variations in patient diagnosis and delivery method. Papadatou *et al.* (1994) compared the scores of nurses working in public oncology hospitals on the MBI to those of nurses working in general hospitals and also found no significant differences between these two groups. Also, Tyler and Ellison (1994) found similar total amounts of stress, assessed by different subscales of the Nursing Stress Scale and the Nursing Stress Index, and psychiatric morbidity, assessed by the original GHQ, for nurses in four different high dependency areas: theatres, the liver unit, haematology/oncology, and elective surgery.

So, the results of studies in which the level of strain that accompanies the care of cancer patients is compared to the level of strain accompanying other medical specialties are inconclusive. However, literature is quite consistent in claiming that oncology differs from other medical specialties with respect to the *nature* of work-related stressors.

5.3.2 Stressors

Herschbach (1992) compared the sources of work-related stress experienced by physicians and nurses working with cancer patients to those experienced by care providers working with patients in cardiac, intensive care or surgical units. The oncology group turned out to suffer more from feelings of emotional involvement and self-doubt, whereas they suffered less from stress connected with institutional factors than did the comparison groups. In the study by Tyler and Ellison (1994), also some specific differences between departments in sources of stress were found. The workload was most stressful for nurses in theatres. In the haematology/oncology unit the major problem was the close nurse–patient relationship resulting from long-term one-to-one nursing. Oncology nurses in the study of Papadatou *et al.,* (1994) reported significantly less stress due to lack of personnel and increased workload than nurses in general hospitals, but more stress because of the lack of support by senior co-workers. Oncology nurses also seemed to have more difficulties in finding a balance between their professional and family life than general nurses. Catalan *et al.* (1996) reported that similar proportions of care providers in AIDS care and oncology found it difficult to work with people who were dying, with people with a life-threatening illness, to be involved with people with deteriorating health, to work with young patients, and to deal with patients with psychological problems. However, oncology workers differed from AIDS staff in the frequency with which they endorsed three specific situations concerning difficulties: they experienced less lack of time and less lack of resources for their work, but more difficulties in working with the patient's family. Finally, in the study of Lopez-Castillo *et al.* (1999) the number of stressful interpersonal circumstances (e.g. deteriorating health of patients, working with patients' partners, working with young patients) was significantly higher in oncology as compared to internal medicine and AIDS care. So, in general stressors for oncology care providers appear to be more strongly related to the social and interpersonal aspects of the job, including relationships with patients and co-workers in the team, and less strongly to institutional working conditions as compared to care providers in other specialties.

Besides the above comparative studies, some studies have focused on samples made up of oncology care providers exclusively. From the results of these studies, it again becomes quite clear that, besides the 'general' stressors pertaining to interpersonal relationships at work, there are some particular problems for physicians and nurses, which relate to differences in their professional roles.

For physicians, a major discomfort is the inability to help patients, or to provide optimal care. Their stress may partly be related to a lack of confidence when faced with their limited ability to alter the course of the illness (Ullrich and Fitzgerald, 1990). As there is an increasing tendency for medical oncologists to serve the role of a primary or general physician to cancer patients, more time is devoted to symptomatic or palliative care rather than active therapy. According to Whippen and Canellos (1991), who conducted a qualitative survey among a random sample of 1000 American oncologists, training programmes in medical oncology have not been well oriented to the physician's role in the palliative care of advanced cancer, and this may contribute to the incidence of burnout in actual practice. If the oncologist's perception of his or her role is primarily directed to the remission of disease without an appreciation of his/her positive contributions to the palliative care of cancer patients, then the stress of this type of practice would appear to be unavoidable. Results of a national questionnaire based survey among 476 cancer clinicians by Ramirez *et al.* (1995) also showed that – in addition to feeling overloaded and not having adequate resources – burnout was especially related to high levels of patient-related stress (e.g. dealing with patients' suffering, dealing with treatment toxicity and errors) and low levels of patient-related satisfaction (e.g. satisfaction from dealing well with patients and relatives, and from your professional status). Moreover, those clinicians who felt insufficiently trained in communication and management skills had significantly higher levels of distress, in terms of burnout (MBI) and psychiatric morbidity (GHQ-12) than those who felt sufficiently trained.

Nurses, on the other hand, are the principal link between the patient and his/her family, and doctors, who are often absent when nurses need most of their expertise and their support in various clinical situations (Hinds *et al.*, 1990). Therefore, it is not surprising that nurses are most concerned about ethical issues, e.g. patients are adequately informed (Peteet *et al.*, 1989). Moreover, their identification with the suffering of the patient and their over-commitment may lead to undue tiredness, which closely resembles the emotional exhaustion dimension of the burnout syndrome (Ullrich and Fitzgerald, 1990).

To conclude, though it is not yet clear whether the *level* of job strain in oncology is actually higher than in other medical specialties, there is quite strong (empirical) evidence for a difference between oncology and other fields of medicine with respect to its main causes. Whereas oncology care providers are mainly confronted with social/interpersonal stressors, care providers in other medical specialties mostly have to deal with more 'general' stressors related to their working conditions (e.g. high workload). Within oncology, different 'stress-profiles' for physicians and nurses can be distinguished, that are mainly discipline-based and can be reduced to the basic distinction between curing (physicians) and caring (nurses).

5.4   THE DUTCH NATIONAL SURVEY ON WORK AND WELL BEING OF ONCOLOGY CARE PROVIDERS

Though a review of the most recent literature gives a good impression of the main sources of work stress in oncology, most empirical studies on this subject have used small 'convenience' samples of oncology care providers (sometimes combined with care providers of other medical specialties). The only two exceptions are the studies by Whippen and Canellos (1991) and by Ramirez *et al.* (1995), but their samples consisted of physicians only.

In the Netherlands, the incidence of work stress or burnout among oncology care providers had never been quantitatively assessed. As there were some signs 'from the field' indicating that Dutch oncology care providers might also be at risk for burnout, in 1996 a research project called 'Burnout in cancer care' was started at the Department of Social and Organizational Psychology of Utrecht University financed by a grant from the Dutch Cancer Society. The aims of the first part of this project were to:

(1) assess the prevalence of burnout among Dutch oncology care providers, and
(2) identify (de)motivating aspects of their working situation ('stressors' and 'satisfiers') and the relationship of these aspects with burnout.

The distinction between 'stressors' and 'satisfiers' can be traced to the general distinction between two dimensions that can be distinguished in any kind of job, i.e. job demands and job resources. Jones and Fletcher (1996, p. 34) define demands as '... the degree to which the environment contains stimuli that peremptorily require attention and response. Demands are the "things that have to be done" '. Clearly, in every job something has to be done. More specifically, we refer to job demands as those physical, psychological, social, or organisational aspects of the job that require sustained physical and/or psychological (i.e. cognitive or emotional) effort and are therefore associated with certain physiological and/or psychological costs. Job demands are not necessarily negative, according to the Job Demands Control model (Karasek and Theorell, 1990). The model assumes that so-called active jobs, characterised by a combination of high demands and high control, foster growth, learning and development at the job. However, job demands may turn into job stressors when meeting those demands requires extremely high effort and is associated with very high costs that elicit negative responses such as anger, anxiety, or burnout. Job demands may be either quantitative (i.e. workload, high work-pace, meeting tight deadlines) or qualitative (i.e. emotional demands such as confrontation with suffering patients or complaining customers; physical demands such as noise, heat, and crowding; mental demands, such as attention, vigilance, and concentration).

Job resources, on the other hand, refer to those physical, psychological, social, or organisational aspects of the job that either/or: (1) reduce job demands and the associated physiological and psychological costs; (2) are functional in achieving work goals; and (3) stimulate personal growth, learning and development. Hence, not only resources are necessary to deal with job demands and in order 'to get things done', but they also are important in their own right because they foster employee's growth, learning and development. Job resources may be located at the level of the task (e.g. feedback, variety, significance, identity, autonomy – cf. Hackman and Oldham's, 1976, classical task characteristics), the organisation of work (e.g. role clarity, participation in decision making, safe environment), interpersonal and social relations (e.g. supervisor and co-worker support, opportunity for interpersonal contact), and the organisation at large (e.g. pay, career opportunities, job security, supervisory coaching, supportive organisational climate, valued social position).

5.4.1 Procedure and Participants

In the Netherlands, almost all oncology care providers are affiliated with one or more professional associations. In January 1997, a random sample of members of five Dutch associations of oncology care providers was asked to fill out an extensive questionnaire on their work and well being. Questionnaires were sent to the home addresses of 1585 care providers: 750 nurses, 432 physicians, and 403 radiation assistants (i.e. care providers who carry out radiation treatments). An explanatory letter from the researchers and a statement of support from the respective professional association were also enclosed. In order to guarantee anonymity, the completed questionnaire could be returned in a pre-stamped envelope. Three weeks after the initial mailing, a reminder was sent out.

5.4.2   Instruments

The scales that were used to assess *(de)motivating aspects of the working situation* have been constructed by the first author. Based on a review of the literature on work stress and burnout in oncology and in-depth pilot interviews with 20 randomly selected oncology care providers about (de)motivating aspects of their working situation, 68 items about sources of stress ('stressors') and 30 items about sources of satisfaction ('satisfiers') were formulated. Each item assesses how prevalent a particular 'stressor' or 'satisfier' is in a respondent's working situation, and is scored on a five-point scale ranging from (1) 'not at all' to (5) 'extremely often'. By means of principal components analysis, items were clustered into 18 different scales. With the exception of 'unfair salary' (as compared to one's effort), all scales had acceptable internal consistencies (Cronbach's α; see Table 5.1).

Burnout was assessed by the three subscales of the Dutch version (UBOS; Schaufeli and Van Dierendonck, 2000) of the Maslach Burnout Inventory (MBI; Maslach and Jackson, 1986). In this chapter, we will restrict ourselves to the emotional exhaustion and depersonalisation dimensions of burnout, as they are generally considered as the 'core of burnout' whereas personal accomplishment reflects a personality characteristic like self-efficacy rather than a genuine burnout component (e.g. Cordes and Dougherty, 1993; Demerouti *et al.*, 2001; Shirom, 1989). Scores on these subscales range from (0) 'never' to (6) 'every day'. In the UBOS, one item of the original MBI has been eliminated because of insufficient factorial validity (Byrne, 1993; Schaufeli and Van Dierendonck, 1993). Recently, Schaufeli and Van Dierendonck (1993; 2000) demonstrated that the reliability and construct validity of the Dutch version are comparable to the original American version.

5.4.3   Results

Response Rate and Demographic Characteristics of the Sample

A total of 816 oncology care providers returned the questionnaire (response rate 52 per cent): 410 nurses (response rate 55 per cent), 179 physicians (response rate 42 per cent) and 227 radiation assistants (response rate 56 per cent). Age ranged from 21 to 63 years: the mean age was 38.5 years ($SD = 8.9$). Eighty-Six per cent of the

Table 5.1 Descriptive statistics for the key study variables ($N = 816$ oncology care providers)

Scale	N items	Cron-bach's alpha	M	SD	Correlations EE	DP
Stressors						
Negative aspects of relation with patients (e.g. distrustful patients, uncooperative patients)	12	.92	2.54	.84	.15**	.22**
Negative aspects of relation with colleagues (e.g. conflicts, lack of appreciation)	11	.90	2.57	.86	.29**	.20**
Time pressure	10	.84	2.71	.82	.39**	.25**
Organisational problems (e.g. bureaucracy, unclear policy)	10	.84	2.43	.90	.37**	.15**
Confrontation with death and dying	6	.81	2.27	1.08	.13**	.10**
Negative aspects of oncology (e.g. side-effects of treatment, limited curative possibilities)	4	.84	2.88	.83	.11**	.11**
Emotional demands (e.g. identification with patients)	4	.80	2.53	.83	.16**	.12**
Shift work	2	.85	1.31	1.38	.05	.13**
Scientific demands (e.g. publications, clinical trials)	3	.74	1.28	1.13	.13**	.18**
Physical demands	1		2.22	1.26	.20**	.05
Unfair salary	2	.50	2.45	1.90	.15**	.07*
Satisfiers						
Lack of autonomy	6	.86	3.35	1.13	−.30**	−.20**
Positive aspects of relation with patient (e.g. openness, appreciation)	10	.89	3.71	.73	−.09**	.00
Positive aspects of relation with colleagues (e.g. cooperativeness)	5	.79	3.88	.63	−.02	−.03
Positive aspects of oncology (e.g. fast developments, multidisciplinary)	4	.77	3.56	.76	−.04	−.10**
Contribution to scientific knowledge	3	.70	2.44	1.13	.00	.03
Skill (task) variety	2	.68	3.65	.96	−.05	−.02
Developmental/career prospects	5	.71	3.47	.79	−.07*	−.03

*$p < .05$,   **$p < .01$.

respondents were women. The overwhelmingly female response can be explained by the fact that 73 per cent of the questionnaires was distributed among nurses and radiation assistants, i.e. professional groups in which the majority of employees are women. On average, the participants had 10.1 years (SD = 6.4) of working experience in oncology.

Further analyses showed that there are some significant differences in demographics between the three professional groups. First, there is a significant difference in mean age of the three groups (F = 89.30; p < .001), with physicians having the highest mean age (44.9 years) and radiation assistants the lowest (34.2 years). Second, the number of years of working experience in oncology of physicians (13.5 years; SD = 7.4) was significantly higher (F = 34.91; p < .001) than that of nurses and radiation assistants (9.1 and 9.4 years, respectively). Third, most of the physicians who participated in this study were male (77 per cent), whereas the majority of the nurses (88 per cent) and radiation assistants (68 per cent) were female.

Stressors and Satisfiers

In Table 5.1, mean item scores and standard deviations of the different 'stressors' and 'satisfiers' for the total sample of 816 oncology care providers are presented. This table shows that, relatively speaking, oncology care providers in our sample feel most stressed by negative aspects of oncology (e.g. negative side-effects of treatment, limited curative possibilities), and by having to work under time pressure, whereas they are most satisfied with positive aspects of the relationship with colleagues (e.g. cooperativeness) and positive aspects of the relationship with patients (e.g. openness of relationship with patients and their families, patients' and families' appreciation of your efforts).

By means of one way ANOVAs we tested whether there are significant differences between the three subgroups of oncology care providers with respect to their perception of (de)motivating aspects of the working situation. The differences in (sub)group means were tested by a Scheffe' test with a significance level of .05.

With respect to the '*stressors*', it was found that radiation assistants score significantly lower than the other two groups on having to work under time pressure (F = 40.12, p < .001), confrontation with death and dying (F = 162.26, p < .001), and negative aspects of oncology (F = 7.61, p < .001). Nurses score significantly higher than the other two groups on organisational problems such as bureaucracy and unclear policy (F = 7.86, p < .001), and on unfair salary (F = 11.17, p < .001). Physicians score significantly lower than the other two groups on physical demands (F = 28.06, p < .001), and also significantly lower than nurses on emotional demands such as having to deal with patients' suffering and identification with patients (F = 4.05, p < .01). Finally, there are significant differences between all three groups in shift work (lowest: radiation assistants; highest: physicians; F = 94.13, p < .001), and in scientific demands such as having to perform clinical trials and having to write scientific publications (lowest: radiation assistants; highest: physicians; F = 163.75, p < .001).

With respect to the '*satisfiers*', it was found that radiation assistant's score significantly lower than the other two groups on positive aspects of the relationship with patients ($F = 11.48, p < .001$), and positive aspects of oncology such as rapid developments in knowledge and the opportunities for multidisciplinary collaboration ($F = 16.07, p < .001$). Physicians score significantly higher than the other two groups on 'being able to contribute to scientific knowledge' ($F = 70.53, p < .001$). Finally, there is a significant difference between the three professional groups in the perception of developmental and career prospects (lowest: radiation assistants; highest: nurses; $F = 34.99, p < .001$) and in autonomy (lowest: radiation assistants; highest: physicians, $F = 14.99, p < .001$).

These results again support the idea, already put forward in our literature review, that the different disciplines within the field of oncology can be characterised by different 'stress-profiles', that are primarily related to (differences in) their role in the caring process.

Burnout

Table 5.2 shows mean scores on the two burnout dimensions for (a) the total group of 816 Dutch oncology care providers, (b) the subgroup of 410 nurses, (c) the subgroup of 179 physicians, (d) the subgroup of 227 radiation assistants, and (e) the Dutch health care normative sample (Schaufeli and Van Dierendonck, 2000), consisting of almost 4,000 health care providers from different occupational groups.

Table 5.2 Mean scores on emotional exhaustion and depersonalisation for the total sample of oncology care providers (OCP), radiation assistants, physicians, nurses and the Dutch normative sample

Group	Total sample OCP	Radiation assistants	Physicians	Nurses	Dutch normative sample
	($N = 816$)	($N = 227$)	($N = 179$)	($N = 410$)	($N = 3892$)
Emotional Exhaustion	20.7	19.9	20.0	21.4	15.5
Depersonalisation	9.7	9.3	10.5	9.4	7.4

From Table 5.2, it becomes clear that Dutch oncology care providers are, on average, more burned out than Dutch health care providers in general, as they score significantly higher on the two core components of the burnout-syndrome than the Dutch normative sample (emotional exhaustion, $t = -18.84, p < .001$; depersonalisation, $t = -16.42, p < .001$).

Comparison of the mean burnout scores of the three different subgroups of oncology care providers by means of one-way ANOVAs, reveals some significant differences. Nurses' emotional exhaustion scores are significantly higher than those of physicians and radiation assistants ($F = 4.60, p < .01$), whereas physicians' depersonalisation scores are significantly higher than those of the other two groups ($F = 5.20, p < .01$).

Relationship between (De)motivating Aspects of the Working Situation and Burnout

Inspection of Table 5.1 makes clear that, relatively speaking, having to work under time pressure and organisational problems show the highest correlations with feelings of emotional exhaustion, whereas having to work under time pressure also shows the highest correlation with depersonalisation.

To find out which of the (de)motivating aspects of the working situation had the strongest relationships with burnout, hierarchical multiple regression analyses were performed for each burnout dimension separately. In the first step of these analyses, we controlled for potential confounding effects of 'gender', 'age' and 'years of working experience in oncology'. In the second step, the 12 'stressors' and 6 'satisfiers' were entered into the equation.

In Table 5.3, the results of the multiple regression analyses for emotional exhaustion are presented for each of the three groups of oncology care providers.[1]

Table 5.3 Hierarchical multiple regression analyses of the stressors and satisfiers on emotional exhaustion

Professional group	Radiation assistants ($N = 227$)		Physicians ($N = 179$)		Nurses ($N = 410$)	
	B	R^2 Change	B	R^2 Change	B	R^2 Change
1. Age		.02	−.39**	.05	.15**	.04
Gender					−.12*	
2. **Stressors**:						
Lack of autonomy	.18*	.29		.35		.27
Negative aspects of relation with colleagues					.17**	
Time pressure	.24**		.29**		.28#	
Emotional demands	.19*		.24**		.14*	
Scientific demands			.20*			
Physical demands	.21**					
Unfair salary	.15*					
Satisfiers:						
Positive aspects of relation with patient	−.21**				−.18**	
F		4.11#		4.72#		7.52#
R^2 Total		.31		.40		.31

$*p <.05, **p <.01, {}^{\#}p < .001$.

----

[1] For reasons of economy, only significant regression coefficients are presented in Tables 5.3 and 5.4.

Inspection of Table 5.3 makes clear that, after controlling for potential confounding effects of gender, age and years of working experience in oncology, having to work under time pressure shows the strongest relationship with feelings of emotional exhaustion for all three groups: the more they have to work under time pressure, the stronger oncology care providers' feelings of emotional exhaustion. Also, for all three groups a positive relationship between emotional demands and this burnout-dimension is found. Whereas for both radiation assistants and nurses an increase in positive aspects in the relationship with patients' is associated with a decrease in emotional exhaustion, only for nurses is an increase in negative aspects in the relationship with colleagues (e.g. conflicts, lack of mutual respect and appreciation) related to an increase in feelings of emotional exhaustion. For physicians, an increase in scientific demands is related to an increase in emotional exhaustion. Finally, for radiation assistants the 'stressors' lack of autonomy, unfair salary, and physical demands are also positively related to feelings of emotional exhaustion.

In Table 5.4, the significant results of the multiple regression analyses for depersonalisation are presented for each of the three groups of oncology care providers.

Table 5.4 Hierarchical multiple regression analyses of the stressors and satisfiers on depersonalisation

Professional group	Radiation assistants (N = 227)		Physicians (N = 179)		Nurses (N = 410)	
	B	R^2 Change	B	R^2 Change	B	R^2 Change
1. Gender		.02	−.17*	.02	−.25**	.05
2. *Stressors*:						
Lack of autonomy		.16		.22	.19**	.15
Time pressure	.22*				.19**	
Emotional demands			.21*			
Satisfiers:						
Positive aspects of oncology					−.14*	
F		1.96*		2.23**		4.05#
R^2 Total		.18		.24		.20

$*p < .05, **p < .01, {}^#p < .001.$

Table 5.4 shows that the relationships between this burnout-dimension and the 'stressors' and 'satisfiers' are less strong than for emotional exhaustion. Whereas the percentages of explained variance range from 31 per cent to 40 per cent for emotional exhaustion, they are between 18 per cent and 24 per cent for depersonalisation. For both nurses and radiation assistants, again, an increase in having to work under time pressure is related to an increase in depersonalisation. Moreover, for nurses a significant positive relationship between lack of autonomy and this burnout-dimension is found. Quite interestingly, for nurses an increase in positive aspects of oncology is related to a decrease in depersonalisation towards patients. Finally, for physicians, an increase in emotional demands is related to an increase in depersonalisation.

So, whereas there are two stressors that are significantly related to burnout in all three professional groups, i.e. time pressure and emotional demands, there are also some more 'discipline-specific' relationships between certain stressors and (aspects of) burnout.

5.4.4 Conclusions

Level of Burnout

From this study on (de)motivating aspects of the working situation of Dutch oncology care providers and the relationship of these 'stressors' and 'satisfiers' with burnout, we can draw the following conclusions.

First, mean levels of burnout (i.e. emotional exhaustion and depersonalisation) in our sample are quite unfavorable in comparison to Dutch norm scores for health care providers.

Moreover, we found significant differences in mean burnout-scores between the three subgroups of oncology care providers. Nurses' emotional exhaustion scores are significantly higher than those of physicians and radiation assistants. This might be explained by the fact that, relatively speaking, of all three professional groups nurses spend the most time with patients, and, as a consequence, are exposed more frequently to high (emotional) demands than physicians and radiation assistants. The results of the oneway ANOVAs indeed show that nurses' exposure to emotional demands is significantly higher than that of physicians. Especially, once a patient is acknowledged to be terminally ill, the 'burden of care' shifts to the nurse (Bram and Katz, 1989). Physicians score significantly higher on depersonalisation than the other two professional groups. As already indicated by Whippen and Canellos (1991) and Lederberg (1998), due to the focus of their professional training, physicians might be more 'cure-oriented', i.e. oriented to the medical-technical aspects of patient-care, compared to nurses and radiation assistants. Their professional attitude might therefore be more 'distant' than that of nurses and radiation assistants, who are perhaps more 'care-oriented', i.e. oriented to the (psychosocial) aspects of patient care. Moreover, physicians might feel personally responsible when their (medical) treatment fails. A possible way to deal with the accompanying feelings of guilt and frustration might be to adopt a more distant attitude towards patients, i.e. treating them in a depersonalised way. Indeed, it is consistently found that levels of depersonalisation are particularly high in physicians as compared to other health care professionals (Schaufeli, 1999). This explanation is also supported by the fact that only for physicians an increase in emotional demands in the relationship with patients (e.g. dealing with patients' suffering) is related to an increase in depersonalisation (see Table 5.4). Other possible explanations include that physicians are more highly educated, have higher status and are male, all possible moderators of burnout level.

Stressors and Satisfiers

According to the oncology care providers in our sample, negative aspects of oncology; such as limited curative possibilities and negative side effects of treatment, and having to work under time pressure are the most prevalent 'stressors'. Both 'stressors' were also reported as being important in the survey among cancer clinicians in UK by Ramirez *et al.* (1995). Positive aspects of the relationship with patients as well as positive aspects of the relationship with colleagues were rated as the most prevalent 'satisfiers'. These findings are again in line with those of Ramirez *et al.* (1995) and underline the importance of good interpersonal relationships at work.

As was already indicated several times in the previous paragraphs, the results of the one way ANOVAs show specific profiles of 'stressors' that correspond quite well to hospital practice and the positions in the 'oncology care system' held by the corresponding professionals. Relatively speaking, nurses have the highest scores on organisational problems, unfair salary, and emotional demands. Physicians have the highest scores on scientific demands and shift work, and the lowest scores on physical demands, emotional demands and lack of autonomy. Finally, radiation assistants have the highest scores on lack of autonomy, and the lowest scores on having to work under time pressure, shift work, confrontation with death and dying, negative aspects of oncology, and scientific demands.

Correlates of Burnout

With respect to the relationship between the 'stressors' and 'satisfiers', and burnout, our results show that having to work under time pressure is the most important risk factor – which is in line with the bulk of research on burnout (Schaufeli and Enzmann, 1998) – immediately followed by emotional demands in the relationship with patients. Moreover, reducing time pressure would also enhance the quality of patient care, as for both radiation assistants and nurses, having to work under time pressure is positively related to depersonalisation. For nurses, feelings of emotional exhaustion are strengthened by negative aspects in the relationship with colleagues, whereas for both nurses and radiation assistants feelings of emotional exhaustion are lessened by positive aspects in the relationship with patients. However, of all three professional groups, radiation assistants have the lowest score on this 'satisfier'! This might be explained by the fact that their contacts with patients are of a relatively short duration (only during radiation sessions, when they walk in and out of the radiation room), and they have to work on schedule. As a result, there is (too) little time for rewarding contacts with patients. Also, during the pilot interviews, radiation assistants were complaining about the lack of (positive) feedback on the results of their work they receive from physicians (who are doing the periodical check-ups of patients).

Practical Implications

As the results of our national survey show that the level of burnout among oncology care providers is relatively high compared to that of other groups of health care workers, there is a strong need to develop, implement and evaluate stress management interventions for these professional groups. This issue will be addressed in more detail in the next section. But first, some practical points of action to reduce feelings of work stress and burnout will be derived from the results of our survey.

First, our findings stress the importance of good collaborative practice between the different disciplines in oncology. Ramirez *et al.* (1995) suggest the value of communication skills training to enhance the satisfaction (and at the same time reduce the stress) of dealing with patients and collaborating as colleagues. If necessary, this could be supplemented with management skills training to reduce feelings of time pressure or overload, which also emerged from our results as a major stressor. Another way to reduce the stress of being overloaded, which can be derived from a leading theoretical model on work stress, is to increase care providers' level of job autonomy (Karasek, 1979). Empirical studies among different occupational groups have demonstrated that a high level of job autonomy helps workers to cope with a high workload (Karasek and Theorell, 1990).

Second, the specific differences in 'stress-profiles' between different professional groups that were found in our study should be taken into account when designing and implementing stress management interventions for oncology care providers. Ideally, care providers should be made aware of the different stances of different professional groups within the interdisciplinary team, as this awareness might be conducive to their collaboration.

For nurses, a negative relationship between the positive aspects of oncology and depersonalisation is found. So, an increase in motivating aspects such as multidisciplinary collaboration, and new developments in the treatment of cancer positively affect nurses' quality of patient care. By letting nurses participate in the design and carrying out of multidisciplinary scientific research projects, e.g. clinical trials, these aspects could be increased.

On the other hand, for physicians a significant, positive relationship between scientific demands and both burnout-dimensions is found. Due to the high caseload of many physicians, tasks such as reading and writing scientific articles and designing clinical trials often have to be carried out in their free time (e.g. at night at home, or during the weekends). As a result, theyx are unable to recover from their work, fatigue accumulates, and eventually may result in feelings of burnout. If possible, it is therefore advisable to reduce physicians' tasks in direct patient-care, e.g. by increasing the number of staff-members, so that they can devote more time to scientific tasks during regular working hours. Having more time to perform the latter type of tasks might also be a powerful antidote to the stress associated with high emotional demands in the relationship with patients.

Finally, in order to reduce feelings of exhaustion among radiation assistants, physical demands that are probably related to lifting and repositioning patients during radiation sessions should be reduced, e.g. by introducing special lifting devices. Moreover, increasing their level of job autonomy can also reduce their exhaustion levels. This seems especially important, as radiation assistants have significantly lower autonomy-scores than the other two groups of oncology care providers. Finally, if possible, their salary should be raised in order to make them feel that financial rewards are in proportion to their work-related efforts. The idea is consistent with the central hypothesis of the Effort–Reward Imbalance model (see also Siegrist, 1996).

5.5 INTERVENTIONS

Research over the past two decades has shown that burnout is not only related to negative outcomes for the individual worker, but also for the organisation, including absenteeism, turnover rates, and lowered productivity (for recent reviews, see Cordes and Dougherty, 1993; Schaufeli and Enzmann, 1998). So, both from the individual and from the organisational point of view, efforts to combat this form of chronic job stress can be considered important. The results of our national survey yielded some practical insights into ways to reduce the (high) levels of work stress and burnout among oncology care providers by changing aspects of job design and/or the organisation of work. However, as our review of the literature as well as the results of the national survey showed that the quality of the (working) relationships with colleagues is of great importance for the well being of care providers in this field, the second part of our research project 'Burnout in cancer care' was focused on this aspect of the working situation. More specifically, in this part of our project, we tried to gain a detailed insight into collaborative practice between oncology care providers and into ways of improving it. The idea behind this was that a well-functioning and supportive work group could be a powerful antidote to the high levels of stress oncology care providers have to deal with in their daily working life.

This idea was supported by the results of a literature review on stress management in oncology, which showed that the most frequently employed means to assist oncology care providers in coping with work-related stressors is a so-called staff support group. A staff support group consists of regular meetings in which care providers have the opportunity to share personal, work-related experiences and feelings with colleagues in a supportive, nonjudgmental environment (Ryerson and Marks, 1982).

The availability of social support at work is very crucial in the adaptation of the care provider to the care of cancer patients. Empathic concern and active care by one's co-workers can greatly reduce the effects of the accompanying stress and help prevent burnout (Flint Sparks, 1989). Moreover, as early as 1978, Maslach suggested that a person should frequently analyse his or her personal feelings related to work. She found that burnout-rates are lower in health care workers who actively express, analyse, and share their personal feelings with their colleagues. In addition, support groups can defuse tension and aid in problem solving, as new perspectives and solutions to perceived and real problems can ensue from such peer interaction. Also, sharing responsibility for the quality of the working environment and for the mutual support of staff is important in maintaining staff morale (Cull, 1991). Finally, Kash and Holland (1989) found that increasing sensitivity, support and communication for staff members also increased the patient's positive perception of care.

In cooperation with two experienced team counsellors, a burnout prevention training for oncology care providers that combines the just described advantages of a support group with those of a participatory action research approach (Murphy and Hurrell, 1987) was developed and implemented on nine oncology wards (see for more details: De Geus *et al.*, 2000). A *team-based* stress management approach was chosen, which means that complete, 'functional' teams, with members from different professional groups who work together on a daily basis, participated in the training programme. In total, 29 oncology wards participated in this part of our research project.

Before the training programme started (T1), an extensive questionnaire survey was conducted among all staff members of the participating wards. The questionnaire assessed oncology care providers' perceptions of the most important (potential) work 'stressors' that were identified previously by means of our national survey, as well as their perceptions of social and working relationships within their team. Next, 9 training wards were randomly selected from the total number of wards participating in this study; the other 20 wards functioned as a comparison group. Participants were care providers (physicians, nurses, and radiotherapy assistants) working in direct care with cancer patients. At the training wards, staff were offered the opportunity to participate in the burnout intervention programme. Directly after the end of the training programme (T2) as well as six months later (T3), staff of the training and the comparison wards again filled out the questionnaire. The total number of respondents at the different measurements were 774 (T1), 466 (T2) and 391 (T3), respectively.

To familiarise themselves with the training wards, the team counsellors held extensive intakes with the management (e.g. head nurses, physicians, co-ordinators, and team leaders) of each of the wards where the training programme was to be implemented. During these conversations, the protocol of the training was clarified, and potential intervention effects ('gains') were discussed. The counsellors also inquired after the ward management's reasons for participation in the training programme, their main objectives, and their 'criteria' for successfulness of the intervention. Moreover, they also gathered information on the structure and policies of the larger organisation. Finally, the ward management's perception of the working situation, including the main sources of job stress was discussed. By means of these intakes, the team counsellors tried to increase the ward management's motivation for the implementation of organisational change processes.

Next, a 'kick-off' meeting for the entire staff of each of these wards was organised. During this meeting, the team counsellors presented the protocol of the training programme, whereas the researchers once more explained the intervention study design. Staff were encouraged to ask questions about the intervention protocol and/or the study design. By means of these meetings, we tried to increase staff's commitment to participate and to promote positive anticipatory attitudes towards the training programme.

The information that was gathered during the intakes and the kick-off meetings was written down in a so-called 'take off'-document, which was the first in a series of reports about the progress and results of the programme. These reports formed a sort of 'log-book', to keep all participants informed during the periods in between the programme sessions.

The training programme itself consisted of six monthly sessions of three hours each, which were supervised by the two team-counsellors. The starting points for action were the results of the T1 measurement on care providers' perceptions of 'stressors' and relationships with their colleagues. During the first training session, these results were fed back to the staff members. This was done to help them to structure their subjective feelings by providing them with relevant topics for discussion and for their plans to reduce work stress. However, participants were only informed about the ward's scores on the above-mentioned (perceptions of) aspects of the working situation, because these formed the starting point for later actions; info about the teams' burnout levels was presented to the team counsellors and the participants after the last questionnaire measurement (T3). This was done because we wanted to avoid the potential effects of 'labelling' (low versus high-risk profile).

During the following sessions, small problem-solving teams were formed that collectively designed, implemented, evaluated, and re-formulated plans of action to cope with the most important stressors in their working situation. In addition, the team counsellors trained them in some more general communication and cooperation skills (how to give and receive feedback, social support etc.) that might be lacking. So, the participants were their own 'agents of change', whereas the team counsellors acted as their 'coaches'. They gave feedback on the plans of action that were formulated by the different teams, and assisted and advised them in the process of implementing these plans.

With respect to the effect of the training programme on care providers' burnout levels, it was found that staff of the comparison wards showed an increase in levels of both emotional exhaustion and depersonalisation during the 1-year study period, whereas staff of the training wards showed a stabilisation in their levels of both burnout components. So, sharing responsibility for the quality of the working environment and mutual support of co-workers turned out to be an effective means to – at least – prevent feelings of work stress from increasing. Because of the relatively short duration of the training programme, it may have been unrealistic to expect an actual decrease in burnout-levels. This may perhaps require a longer, more intensive period of active intervention. However, we feel that the first results with respects to the effects of this programme are quite promising.

5.6   EPILOGUE

Though this chapter has focused on sources of work stress in oncology, we would also like to emphasise all the positive experiences which keep caregivers working in this specialty and will relish for many years. As Lederberg (1998) states, many of them thrive on emotional excitement and intensity, many love the intellectual challenge especially in the context of interacting with people in a helpful way. They gain satisfaction from being part of a valued social enterprise, from playing a positive role in profound human experiences, and from having survived a form of existential initiation. Moreover, caregivers derive sustenance from being surrounded by people who share these commitments and experiences with them, even if it is unexpressed much of the time.

5.7   REFERENCES

Adams, P.J., Hershatter, M.J. and Moritz, D.A., 1991, Accumulated loss phenomenon among hospice caregivers. *The American Journal of Hospice and Palliative Care*, **8**, pp. 29–37.

Bene, B. and Foxal, M., 1991, Death anxiety and job stress in hospice and medsurgical nurses. *Hospice Journal*, **7**, pp. 25–41.

Benoliel, Q.J., 1969, The threath of death: some consequences for patients and nurses. *Nursing Forum*, **8**, pp. 295–296.

Bram, P.J. and Katz, L.F., 1989, A study of burnout in nurses working in hospice and hospital oncology settings. *Oncology Nursing Forum*, **16**, pp. 555–560.

Byrne, B.M., 1993, The Maslach Burnout Inventory: testing for factorial validity and invariance across elementary, intermediate, and secondary teachers. *Journal of Occupational and Organizational Psychology*, **66**, pp. 197–212.

Catalan, J., Burgess, A., Pergami, A., Hulme, N., Gazzard, B. and Phillips, R., 1996, The psychological impact on staff of caring for people with serious diseases: the case of HIV infection and oncology. *Journal of Psychosomatic Research*, **40**, pp. 425–435.

Cooper, L.C. and Mitchell, S., 1990, Nursing the critically ill and dying. *Human Relations*, **43**, pp. 297–311.

Cordes, C.L. and Dougherty, T.W., 1993, A review and integration of research on job burnout. *Academy of Management Review*, **18**, pp. 621–656.

Cull, A., 1991, Staff support in medical oncology: a problem solving approach. *Psychology and Health*, 1991, **5**, pp. 129–136.

De Geus, A.C., Son, A.M. van, Le Blanc, P.M. and Schaufeli, W.B., 2000, *Take Care: een teamgerichte interventie ter bevordering van welzijn op het werk*

[Take care: a team-based burnout intervention training], (Houten/Diegem: Bohn Stafleu Van Loghum).

Delvaux, N., Razavi, D. and Farvacques, C., 1988, Cancer care: a stress for health professionals. *Social Science and Medicine*, **27**, pp. 159–166.

Demerouti, E., Bakker, A.B., Nachreiner, F. and Schaufeli, W.B., 2001, The job demands resources model of burnout. *Journal of Applied Psychology*, **86**, pp. 499–512.

Flint Sparks, T., 1989, Coping with the psychosocial stresses of oncology care. *Journal of Psychosocial Oncology*, **6**, pp. 165–179.

Ford, S., Fallowfield, L.J. and Lewis, S., 1994, Can oncologists detect distress in their out-patients and how satisfied are they with their performance during bad news consultations? *British Journal of Cancer*, **70**, pp. 767–770.

Gray-Toft, P.A. and Anderson, J.G., 1987, *Sources of stress in nursing terminal patients in a hospice. Omega*, **17**, pp. 21–41.

Goldberg, D.P., 1978, *Manual of the General Health Questionnaire*. Windsor: Nfer-Nelson.

Haberman, M.R., Germino, B.B., Maliski, S., Stafford-Fox, V. and Rice, K., 1994, What makes oncology nursing special? Walking the road together. *Oncology Nursing Forum*, **21**, pp. 41–47.

Hackman, J.R. and Oldham, G.R., 1976, Motivation through the design of work: test of a theory. *Organizational Behavior and Human Performance*, **16**, pp. 250– 279.

Herrera, H., 1986, Interpersonal dimensions of illness. *Journal of Psychosocial Nursing*, **24**, pp. 33–35.

Herschbach, P., 1992, Work-related stress specific to physicians and nurses working with cancer patients. *Journal of Psychosocial Oncology*, **10**, pp. 79–99.

Himmelsbach, K., 1978, Social work with the cancer patient. *Proceedings of the American Cancer Society Second National Conference on Human Values and Cancer* (New York: American Cancer Society Inc.).

Hinds, P., Fairclough, D., Dobos, C., Grier, R., Herring, P., Mayhall, J., Arheart, L.K., Day, L.A. and McAulay, L., 1990, Development and testing of the stressor scale for pediatric oncology nurses. *Cancer Nursing*, **13**, pp. 354–360.

Hürny, C., 1988, Psychosocial support of cancer patients: a training programme for oncology staff. *Recent Results in Cancer Research*, **108**, pp. 295–300.

Jenkins, J.F. and Ostchega, Y., 1986, Evaluation of burnout in oncology nurses. *Cancer Nursing*, **9**, pp. 108–116.

Jones, F. and Fletcher, B.C., 1996, Job control and health. In *Handbook of Work and Health Psychology*, M.J. Schabracq, J.A.M. Winnubst, J.A,M, and C.L. Cooper (eds) (Chichester: Wiley), pp. 33–50.

Karasek, R.A., 1979, Job demands, job decision latitude, and mental strain: implications for job redesign. *Administrative Science Quarterly*, **24**, pp. 285–308.

Karasek, R.A. and Theorell, T., 1990, *Healthy work* (New York: Basic Books).

Kash, J. and Holland, J.C., 1989, Special problems of physicians and house staff in oncology. In *Handbook of Psycho-oncology*, J.C. Holland and Rowland, J.R. (eds) (New York: Oxford University Press), pp. 647–657.

Lederberg, M.S., 1998, Oncology staff stress and related interventions. In *Psycho-Oncology*, edited by J.C. Holland (New York/Oxford: Oxford University Press), pp. 1035–1048.

Lopez-Castillo, J., Gurpegui, M., Ayuso-Mateos, J., Luna, J.D. and Catalan, J., 1999, Emotional distress and occupational burnout in health care professionals serving HIV-infected patients: a comparison with oncology and internal medicine services. *Psychotherapy and Psychosomatics*, **68**, pp. 348–356.

Maslach, C., 1978, The burnout syndrome and patient care. In *Stress and survival*, edited by C. Garfield (St. Louis: Mosby Press), pp. 111–120.

Maslach, C., 1982, *Burnout: the cost of caring* (Englewood Cliffs, NJ: Prentice Hall).

Maslach, C. and Jackson, S.E., 1986, *MBI: Maslach Burnout Inventory* (Palo Alto CA: Consulting Psychologists Press).

McDaniel, J.S., Musselman, D.L., Porter, M.R., Reed, D.A. and Nemeroff, C.B., 1995, Depression in patients with cancer. Diagnosis, biology, and treatment. *Archives of General Psychiatry*, **52**, pp. 89–99.

Meyer, C., 1992, The richness of oncology nursing. *American Journal of Nursing*, **92**, pp. 71–78.

Moos, R.H. and Schaefer, J.A., 1984, The crisis of physical illness: an overview and conceptual approach. In *Coping with physical illness 2: New perspectives*, edited by R.H. Moos (New York: Plenum Medical Book Company), pp. 3–25.

Murphy, L. and Hurrell, J.J., 1987, Stress management in the process of organizational stress reduction. *Journal of Managerial Psychology*, **2**, pp. 18–23.

Olkinuora, M., Asp, S., Juntunen, J., Kauttu, K., Strid, L. and Äärima, 1990, Stress symptoms, burnout and suicidal thoughts among Finnish physicians. *Social Psychiatry and Psychiatric Epidemiology*, **25**, pp. 81–86.

Papadatou, D., Anagnostopoulos, F. and Monos, F., 1994, Factors contributing to the development of burnout in oncology nursing. *British Journal of Medical Psychology*, **67**, pp. 187–199.

Peteet, J.R., Murray-Ros, D., Medeiros, C., Walsh-Burke, K., Rieker, P. and Finkelstein, D., 1989, Job stress and satisfaction among the staff members at a cancer center. *Cancer*, **64**, pp. 975–982.

Ramirez, A.J., Graham, J., Richard, M.A., Cull, A., Gregory, W.M., Leaning, M.S., Snashall, D.C. and Timothy, A.R., 1995, Burnout and psychiatric disorder among cancer clinicians. *British Journal of Cancer*, **71**, pp. 1263–1269.

Ryerson, D. and Marks, N. (1982). Career burnout in the human services: strategies for intervention, In *The burnout syndrome: current research, theory and intervention*, J.W. Jones (ed.) (Park Ridge: London House Press), pp. 151–164.

Schaufeli, W.B., 1999, Burnout. In *Stress in Health Professionals: Psychological and Organizational Causes and Interventions*, J. Firth-Cozens and R. Payne, (eds) (Chichester: Wiley & Sons), pp. 17–32.

Schaufeli, W.B. and Van Dierendonck, D., 1993, The construct validity of two burnout measures. *Journal of Organizational Behavior*, **14**, pp. 631–647.

Schaufeli, W.B. and Enzmann, D., 1998, *The burnout companion to study and practice: a critical analysis* (Washington D.C.: Taylor & Francis).

Schaufeli, W.B. and Van Dierendonck, D., 2000, *UBOS: De Utrechtse Burnout Schaal, Handleiding [UBOS: The Utrecht Burnout Scale, Test Manual]*, (Lisse: Swets & Zeitlinger).

Shirom, A., 1989, Burnout in organizations. In *International Review of Industrial and Organizational Psychology 1989*, edited by C.L. Cooper and I.T. Robertson (eds) (Chichester: John Wiley & Sons), pp. 25–48.

Siegrist, J., 1996, Adverse health effects of high-effort/low-reward conditions. *Journal of Occupational Health Psychology*, **1**, pp. 27–41.

Tyler, P.A. and Ellison, R.N., 1994, Sources of stress and psychological well being in high dependency nursing. *Journal of Advanced Nursing*, **19**, pp. 469– 476.

Ullrich, A. and Fitzgerald, P., 1990, Stress experienced by physicians and nurses in the cancer ward. *Social Science and Medicine*, **31**, pp. 1013–1022.

Vachon, M., 1987, Team stress in palliative hospice care. *The Hospice Journal*, **3**, pp. 75–103.

Vachon, M.L.S., Lyall, W.A.L. and Freeman, S.J.J., 1978, Measurement and management of stress in health professionals working with advanced cancer patients. *Death Education*, **1**, pp. 365–375.

Van Servellen, G. and Leake, B., 1993, Burnout in hospital nurses: a comparison of acquired immunodeficiency syndrome, oncology, general medical, and intensive care unit nurse samples. *Journal of Professional Nursing*, **9**, pp. 169–177.

Whippen, D.A. and Canellos, G.P., 1991, Burnout syndrome in the practice of oncology: results of a random survey of 1,000 oncologists. *Journal of Clinical Oncology*, **9**, pp. 1916–1920.

Yasko, J.M., 1983, Variables which predict burnout experienced by oncology clinical nurse specialists. *Cancer Nursing*, **6**, pp. 109–116.

CHAPTER SIX

Senior Nurses: Interventions to Reduce Work Stress

Amanda Griffiths, Raymond Randall, Angeli Santos and Tom Cox

6.1 INTRODUCTION

This chapter outlines an approach to assessing and improving the working conditions of a particular group of employees from the service professions: senior hospital nurses. The process employed for this investigation – a risk management approach – can be adapted to suit all working groups, and can also be simplified so as to be undertaken by workers themselves. The aim of this process is to produce a reasonable account of the major likely stressors for a particular group – an account that is 'good enough' to enable employers and employees to decide on possible improvements. The version of the technique described in this chapter is not a process designed primarily for research purposes, nor does it employ sophisticated statistical techniques, but rather is designed to facilitate practical steps towards organisational change. In the study reported here, senior nurses played an active role in the assessment of their work, but particularly in the design and implementation of several interventions devised to reduce and prevent work stress.

6.1.1 The Legislative Background

The European Commission, in 1989, published the *Framework Directive on the Introduction of Measures to Encourage Improvements in the Safety and Health of Workers at Work*. These requirements had to be made law ('transposed') in each of the Member States of the European Union, within their respective national legislative frameworks within a specified time frame. The Directive required employers to avoid risks to the health and safety of their employees, evaluate the risks which cannot be avoided, to combat the risks at source *(Article 6:2)*, to keep themselves informed of the 'latest advances in technology and scientific findings concerning workplace design' *(Article 10:1)* and to 'consult workers and/or their representatives and allow them to take part in discussions on all questions relating to safety and health at work' *(Article 11:1)*. Employers were also charged to develop a 'coherent overall prevention policy which covers technology, organization of work, working conditions, [and] social relationships' *(Article 6:2)*.

In addition, employers were required to 'be in possession of an assessment of the risks to safety and health at work' and to 'decide on the protective measures to be taken' *(Article 9:1)*.

In Britain, many of these provisions were already catered for under the *Health and Safety at Work etc Act 1974* (Health and Safety Executive, 1990) but some of the requirements, such as the duty to undertake assessments for all risks to health, were introduced in the *Management of Health and Safety at Work Regulations 1992* (Health and Safety Commission, 1992) and their revision (Health and Safety Commission, 1999). Managers were advised to consider stress when undertaking their general risk assessments (Health and Safety Executive, 1995). A risk assessment involves 'a systematic examination of all aspects of the work undertaken to consider what could cause injury or harm, whether the hazards could be eliminated, and if not what preventive or protective measures are, or should be, in place to control the risks' (European Commission, 1996). In other words, employers have a responsibility to take reasonable and practicable steps to protect their employees from those aspects of work or the working environment that are foreseeably detrimental to safety and health.

By the early 1990s, it had become clear in Britain and in other European countries, that (i) work was widely thought to be giving rise to significant levels of stress, (ii) stress was therefore a foreseeable risk, and (iii) employers were legally required to undertake risk assessments for known causes of ill-health.

6.1.2   Nursing in UK

Among the service professions, nursing is one of several disciplines contributing to a considerable body of research into the causes and effects of work-related stress. In many countries nursing is frequently cited as a 'stressful' profession. For example, a recent national survey of work-related ill-health in UK reported nursing to be among the professions where work stress was most commonly reported as a cause of illness (Jones *et al.*, 1998). Not surprisingly therefore, at the time of writing this chapter, there is a serious national shortage of qualified nurses in UK. This is due to many nurses leaving, as well as a shortage of trained nurses entering the profession. Work stress is perceived as a major problem, resulting in high levels of staff sickness, absenteeism, a decline in quality of care, and growing job dissatisfaction (Clegg, 2001; Aiken *et al.*, 2001). Evidence also suggests that nurses experiencing stress are more likely to have conflicts with colleagues, experience feelings of inadequacy, self-doubt, low self-esteem, irritability, depression, somatic disturbance, sleep disorders and burnout (Callaghan *et al.*, 2000; MacNeil and Weisz, 1987). In UK, recent estimates suggest the direct cost of absent nurses to the National Health Service to be about 1 billion pounds sterling (approximately 2.5 billion Euro, or 2.1 billion US dollars) per annum (Bourbonnais *et al.*, 1998; Kunkler and Whittick, 1991).

6.1.3   Sources of Nursing Stress

For nurses, stress has been traditionally been associated with the design, organisation and management of work and staff relationships (psychosocial factors) as well as factors particular to the job such as dealing with dying patients and their relatives (Guppy and Gutteridge, 1991; Hemingway and Smith, 1999). A review of the recent literature suggested the primary sources of stress for nurses to be work overload, inadequate staffing levels, unsociable work schedules, poor managerial and supervisory support, poor communication, lack of autonomy, role conflict and role ambiguity (Edwards *et al.*, 2000; Hemingway and Smith, 1999; Hillhouse and Adler, 1997; Stordeur *et al.*, 2000). Some of this work has focused on organisational culture (Newton and Keenan, 1987). Such research suggests that high levels of autonomy, peer cohesion and supervisor support accompanied by low levels of work pressure are consistently associated with lower levels of perceived stress (Hemingway and Smith, 1999). Other studies reveal a rise in role ambiguity as nurses increase their scope of professional practice – a recent trend designed to broaden nursing skills and responsibilities, but which may also increase workload (Castledine 1998; Chapman, 1998; Dunn 1997).

6.1.4 Stress in Senior Nurses

In response to today's rapidly changing work environment, competition between hospitals, and the current shortage of hospital nurses in many industrialised societies, senior nurses are commonly faced with the challenge of managing a workforce that is hard-pressed and experiencing high levels of job dissatisfaction (Aiken *et al.*, 2001; Edwards *et al.*, 2000; Stordeur *et al.*, 2001). In a recent multi-national study of 43,000 nurses from 711 hospitals, problems with the design and management of their work were also identified as undermining the provision of quality care to patients (Aiken *et al.*, 2001). This decline in service provision has been attributed largely to organisational reengineering and restructuring initiatives that have been designed to emulate models of efficiency and productivity, rather than to address nurses' particular concerns.

Senior nurses, many of whom are ward managers, frequently administer a delicate balance between demands and resources. The amount of work that can be satisfactorily accomplished is determined by many factors, but notably by staffing levels, the competence of individual nursing staff and by the availability of suitable equipment (Guppy and Gutteridge, 1991; Muncer *et al.*, 2001). In today's hospital setting, low staffing levels are perceived not only as a direct cause of stress in terms of increasing job demands, but also as leading to the adoption of multiple roles. For ward managers in particular, the interface between clinical and managerial workloads becomes challenging.

There is evidence to suggest that with increasing seniority, certain aspects of nursing that were once stressful, are no longer experienced as such. For example, some studies have reported that junior grade staff find dealing with patients and relatives and managing home-work conflicts to be the most significant stressors, whereas senior nurses are more concerned about the overall management of their workload (Butterworth *et al.*, 1999). Thus, with seniority, nurses may adapt to specific task-related aspects of the job, whilst at the same time being exposed to, and finding stressful, more organisational, context-related aspects of the job

(Guppy and Gutteridge, 1991). Senior nurses have also been found to report more concern with staff relationships and resources and higher levels of job satisfaction than their junior colleagues (Butterworth *et al.*, 1999; Guppy and Gutteridge, 1991) – findings which may well be explained by the different nature of their responsibilities and perceived control over work. It is clear that qualitative differences between the work of junior and senior nurses, and their different responses to it, merit a tailored approach to audit and intervention.

6.1.5   Interventions to Reduce Stress in Senior Nurses

The scientific literature is replete with generic accounts of the causes and effects of stress in nurses. Whilst there is a considerable amount of information to draw from the traditional academic approach to the study of work stress in this group, there are significant limitations. Many studies provide an indication of a researcher's view of an association between various broadly defined work characteristics and particular employee health indices, usually driven by theoretical rather than practical interest. The measures used are often derived from macro-theories and models drawn from large populations in order to identify general laws, and are often used to provide comparisons across different forms of work. Few studies have defined their samples adequately enough for the purposes of risk assessment, and few provide sufficient detail in their assessment of problems to enable any reduction of problems (risk management) to be effected. Thus, there are few publications of an evaluated, apparently successful, context-based and tailor-made approach to interventions. There are even fewer that embody a democratic, participatory approach where those employees, as the experts in their own jobs, play a central role in the re-design of their work. It is now increasingly recognised that such an approach may offer a promising way forward, although at the same time raising some challenging methodological issues for researchers (Cox, Griffiths and Randall, 2002; Griffiths, 1999). We suggest that the way forward is to use an approach that provides sufficiently detailed information, with a meaningful analysis of contextual issues, for the subsequent design of interventions. In this way, employers may begin to fulfil their legal duty of care to their employees, as outlined above, with respect to work stress.

The aim of this chapter is to describe and discuss one approach that attempts address these issues. It outlines the process and outcome of a risk management approach to reducing stress among a group of approximately 80 senior grade nurses working across 15 wards from Children's Services in a large hospital in UK. This group of senior nursing staff, known as F, G and H grade nurses, have both clinical and managerial responsibilities. G grade nurses have overall responsibility for a ward, and are more involved in strategic decisions about the running of the ward than F grades. But F grade nurses may often be the most senior nurse on a ward at any given moment. The most senior nurses, H grade, have an advisory role and are responsible for a group of wards. The majority of this group were F grade nurses, and the minority were H grade. The study involved the design, implementation and evaluation of a collection of interventions designed to improve (i) the design and management of the senior (F, G and H Grade) nursing staff's working conditions, and (ii) the well-being of that group. The interventions were designed and implemented by staff themselves in part as a response to an initial risk assessment for work stress carried out by the researchers almost a year earlier, and as part of general service development. Before the study is outlined, we will first outline the process adopted.

6.2 METHOD: A RISK MANAGEMENT APPROACH

A risk-management approach to the prevention and reduction of work-related stress has been developed by staff from the Institute of Work, Health and Organisations (I-WHO) at the University of Nottingham over the past ten years. Originally proposed by Cox (1993), it is an evidenced-based, problem solving approach consisting of four stages: (i) identifying the problems (risk assessment), (ii) prioritisation of problems and the design of feasible interventions (translation), (iii) intervention implementation (risk reduction) and (iv) evaluation (Cox, Griffiths and Rial González, 2000; Cox *et al.*, 2000; Cox, Griffiths and Randall, 2002; Cox, Randall and Griffiths, 2002). At the time of writing, this approach has been used by the authors and their colleagues in various forms, with approximately 50 different working groups from both manufacturing and service sectors.

Risk management is a familiar term to many managers, but one that is traditionally employed to refer to relatively well-known and quantifiable physical hazards at work. The adaptation of the traditional risk-management paradigm to deal with work stress is not 'rocket science', particularly in terms of the accuracy and specificity of its measures. In finding a practical way forward for managers, the objective is not to seek an exhaustive and precise account of all stressors and their associated health risks for all individuals. Instead, the objective is to produce a reasonable account, with a sound scientific basis, of the more salient likely risk factors for stress in a given work group – an account that is 'good enough' to enable employers and employees to decide on possible improvements, and will enable managers to comply with their legal duty of care (Griffiths *et al.*, 1995). In this work on risk management, particular emphasis lies on employees as 'experts' in relation to their own jobs. In this respect, data collection is a process of knowledge elicitation and modelling. For the system to work well, employees need to be made aware of process issues, to develop reasonable expectations and to participate actively through its entirety. Before the process begins however, there has to be considerable consultation between stakeholders (including employee representatives) and experts (researchers or consultants). Typically, at the beginning of the risk management process the researchers take the lead, whereas by its completion, successful projects are usually led and managed by the employees themselves.

6.2.1 Risk Assessment

The aim of the risk assessment stage is to identify, for a defined employee group, any significant potential sources of stress relating to its work and working conditions (Cox, Randall and Griffiths, 2002). This can be achieved by several methods – interviews, focus groups, and organisational documentation – depending on the size of the group. Tailor-made questionnaires, based on interviews or focus group data, quantify and summarise the problems specific to that group, and provide sufficient detail to allow for the design and planning of specific interventions. Risk assessment is not an organisation-wide approach: such an approach would miss important details. The identification of risks relies on the expert judgement of groups of relevant working people about the design and management of their work. This information is treated at the group level and consensus is measured in terms of their expert judgements on working conditions. A group is most usefully defined by its function and the inter-relatedness of its members rather than simply by location. Such an approach has worked well with employees who are dispersed such as train drivers or home helps, as well as with those, like nurses, who work in closer proximity.

Information about the possible outcomes of work-related stress is gathered both from the risk assessment survey and from other sources such as absence data or occupational health referrals. The measure of employee well-being used by I-WHO is the General Well-being Questionnaire (Cox *et al.*, 1983; Cox and Griffiths, 1995), developed in Nottingham for this purpose and used on a wide variety of occupational groups. The exercise of relating stressful hazards to their possible effects on health can be formally investigated using simple statistical techniques such as odds ratios (Cox, Randall and Griffiths, 2002; Wang *et al.*, 1995). In some cases, however, particularly with small samples, it is possible to rely on more informal analyses of association.

The first stage involves the researchers' familiarisation with the organisation, its work environment, procedures, key stakeholders and other interested parties. This stage is a two-way process however, whereby the organisation in turn learns about the researchers, the processes involved, and their own involvement in those processes. Senior management support is crucial. Following this, the researchers conduct a series of work analysis interviews with a representative sample of the population, to elicit qualitative information about employees' perceptions of problematic aspects of their working conditions and how such problems might affect them. In addition, an audit of existing management control and employee support systems is conducted. This aims to identify organisational policies and procedures that are relevant to the management of work-related stress, and to establish what the organisation in question is already doing (and not doing) to prevent or manage such stress. The audit would typically include an examination of health and safety policies, management training, organisational culture, occupational health provision, employee assistance programmes, and so on. These initial steps are designed to build a model of the working conditions of the assessment group that is good enough to support the design, and subsequently the implementation, of a bespoke survey.

The risk assessment survey is administered anonymously to the entire population. In this particular study of senior nurses, 58 completed questionnaires were returned (72% response rate). The survey allows simple quantification of significant problematic working conditions and an exploration of other potentially significant indicators. In this study five indicators of individual and organisational health were used: (i) general well-being ('worn-out' scores), (ii) overall job satisfaction, (iii) intention to leave, (iv) absence, and (v) musculo-skeletal pain. The 12 items from the General Well-being Questionnaire (Cox *et al.*, 1983) that measure the frequency and variety of symptoms of tiredness, confusion and fluctuating emotions (feeling 'worn-out') were used. Example items are 'How often have you...become easily annoyed or irritated?....found it hard to make up your mind?...found your feelings easily hurt?' Absence was measured in self-reported days absent in the previous year. The information thus gathered allowed the key features or likely risk factors to be identified and to be prioritised for action in the light of their likely consequences for employees and the organisation.

6.2.2 Design and Implementation of Interventions

The way in which the information from the risk assessment is discussed, explored and used to develop interventions is termed 'translation' (Cox, Randall and Griffiths, 2002). In effect, it is the translation of the information obtained into a reasonable and practicable action plan (a package of interventions) aimed to reduce likely risk factors for stress at work. This stage often takes some time to accomplish satisfactorily. The development of the action plan involves deciding upon what is to be targeted, the methods used, those responsible, the proposed time schedule, the resources required, and how the interventions will be evaluated.

Often, several risk factors can be dealt with by means of a single intervention. To use a medical analogy: by examining a patient's various symptoms, a doctor may diagnose an underlying disease which in turn, allows for treatment of the underlying pathology rather than the symptoms. In our experience, for example, improving communication systems within an organisation or particular functional working group often deals effectively with many specific stress-related problems. Successful interventions are not necessarily costly. Examples of these in the hospital setting are the introduction of regular team meetings or open forums, newsletters, reviewing administrative procedures, (e.g. assigning specific clerks to particular consultants' clinics) and adjustments to rota systems. Others, such as increasing staffing levels, or installing new equipment, may incur significant initial costs, but may 'pay their way' indirectly further down the line. Preventing the loss of one key member of staff, for example, may save the organisation considerable on-costs. Interventions, or 'risk-reduction' measures need not be expensive, disruptive or even revolutionary. Risk reduction measures are often simply examples of innovative good management practice.

6.2.3 Evaluation of Interventions

The evaluation of interventions is an important step, but one that is often overlooked. Not only does it inform the organisations about the effectiveness of the interventions implemented, but allows for the situation to be re-assessed, thereby providing a basis for wider organisational learning. Essentially, it initiates a process for continuous improvement. Managing work-related stress should not be a singular, isolated activity viewed only from within a health and safety framework, but rather, part of an on-going cycle of good management practice. In our view, good management *is* stress management.

Three methods of data collection were employed in this evaluation: (i) interviews with management and key stakeholders, (ii) interviews with staff, and (iii) a questionnaire survey. Together, these three methods of data collection yielded considerable information about the interventions and their impact. Group data collected at the intervention evaluation stage were compared with data collected from the survey (carried out just under a year earlier) in the initial risk assessment stage. Information from staff involved in specific interventions was compared with that from staff not involved in those interventions. This facilitated a clearer picture of impact. Both outcome (change or no change) and process (implementation) issues for each intervention were evaluated. Qualitative data from interviews frequently enabled apparently conflicting results from the questionnaire survey to be resolved, as will be explained further in the Results section. Such information also provided valuable material for the design of future interventions within the organisation. Details about data collection methods are provided below.

(i) Interviews with Management and Key Stakeholders

Seven managers and key stakeholders were interviewed in order to catalogue the interventions and changes that had occurred since the initial risk assessment. Information about how the interventions had been implemented was also gathered from this group.

(ii) Interviews with Staff

Semi-structured interviews were carried out, on a one-to-one basis, with a representative cross-section of 21 nurses from all three grades. Interviews were voluntary and their confidentiality emphasised. Nurses were asked about their experience of the interventions and to describe their impact. They were also questioned about their working conditions and general well-being. These interviews provided a rich source of information about the interventions. They explored (i) the intervention's implementation, (ii) the detail of exactly how the intervention made or did not make a difference, (iii) barriers to the success of an intervention, and (iv) facilitators of the intervention. Thus, a consensus view of the process and outcome of each intervention was reached. In this way it could also be established whether any intervention's failure might be due, for example, to implementation problems or to unforeseen events.

(iii) Questionnaire Survey

Items in the intervention evaluation questionnaire were based on those used in the risk assessment survey so as to facilitate comparisons. In all 51 staff completed questionnaires (64% response rate). The evaluation survey included an additional section asking for information about interventions. In particular, staff were asked three specific questions for each intervention: (i) whether or not they were aware of the intervention (and whether or not they had been actively involved in it), (ii) how much of an impact they thought the intervention had on them and on their work, and (iii) whether the intervention had made things better for them. These questions allowed some estimation of the separate impact of each of the package of interventions.

6.3 RESULTS: INTERVENTIONS WITH SENIOR NURSES

This section provides a summary of the results of the initial risk assessment, the design of interventions and details of the changes that followed interventions. Full details and complete data for this study, and several such studies in the hospital environment, are available in Cox, Randall and Griffiths (2002).

6.3.1 Risk Assessment Results – Time 1

In the initial risk assessment, senior nurses demonstrated high levels of job satisfaction, noting in particular strong working relationships among colleagues and a supportive environment for clinical, managerial and emotional issues. Also valued was role clarity, having transparent targets, and being able to exercise sufficient autonomy. Nonetheless, the overall well-being of the group was lower than average (worn-out scores were high), a high proportion of G and H grade staff intended to leave, and various specific issues emerged as being problematic. Achieving a balance between clinical and managerial responsibilities was particularly challenging, particularly with little time available, and support provided, for the completion of complex managerial tasks. A lack of funding opportunities and inadequate time for training and professional development were also identified as issues. Problems in communication systems were thought to create tensions both within and between wards. Many of the most senior nurses also reported taking substantial amounts of work home, a factor that caused them problems with work-life balance. The most senior grade of nurses also reported an inadequate understanding of their roles by other staff.

    The strength of the link between inadequate aspects of work and various other measures (such as high worn-out scores or intention to leave) was determined using odds ratios to indicate how much more likely the group reporting a problematic working condition was to report a high worn-out score or intention to leave, when compared to the group who did not report the problematic working condition. This information, together with information on resources, time constraints and practical considerations, was used by staff to design and prioritise interventions. These data are not presented here, since they are quite voluminous but are available in Cox, Randall and Griffiths, (2002). However, the items presented in the tables below give some indication of the problematic aspects of the work of this group of nurses. G and H grade nurses shared similar views about their work; in much of this section, data relevant to those groups have been combined.

6.3.2   Intervention Design and Implementation

After the risk assessment stage, and a period of feedback and discussion, the design of a package of interventions was driven by intensive consultation with the nurses themselves. Five of these are presented in detail in this chapter. Most interventions were suggested, developed and implemented by staff themselves. The interventions had only been in place for a short time (six months) when they were subject to preliminary evaluation, necessitated by the constraints of external research funding arrangements.

In order to give staff more time to cope with both their clinical and managerial workload, a review of office hours was undertaken, resulting in an increase in the time allocated for office work. New computer facilities were placed on most wards to facilitate the completion of administrative work. Many wards appointed housekeeping staff to ease peripheral demands on staff in their clinical roles. Policies on study leave were updated, ensuring that feedback on training courses was given by staff who had attended those courses, that approval for attendance on training courses was speeded up, and that allocation of training was more equitable. A staff open forum was set up to allow all staff from Children's Services to meet with other staff and their managers on a monthly basis to address issues of concern. Regular problem-solving workshops for groups of same-grade staff continued. A staff newsletter was implemented in order to facilitate the flow of information in the Service, particularly between wards. A new problem-solving and service development strategy ('Shared Governance') was also in the early stages of its implementation.

6.3.3   Exploring Changes between Time 1 and Time 2

The data in this section are presented in three sections. First, differences in reported working conditions between Time 1 (the initial risk assessment) and Time 2 (the intervention evaluation) were explored. Second, differences in individual and organisational health at these two time points were examined. Such comparisons are useful for identifying large, significant changes in the group as a whole. However, these data may inform us that changes have occurred, but not about *how* such changes may have transpired. Further, such group data may also mask potentially important changes; working conditions and well-being may have improved significantly for some staff, but not for others. In particular, this may be true for those staff involved in particular interventions: not all staff experienced all interventions. In the third section therefore, the shortcomings of the broad picture are explored by examining the possible impact of each intervention individually.

This third section also serves to explore the reasons behind some of the overall changes reported between Time 1 and Time 2.

6.3.4 Perceived Changes in Working Conditions between Time 1 and Time 2

Staff responses to questions about potentially problematic aspects of their working conditions were compared between the initial risk assessment and intervention evaluation stage. Overall, some impressive changes were observed, as demonstrated in Table 6.1.

Table 6.1 Comparison of Reported Problems at Time 1 and Time 2

	% Staff Reporting the Problem	
	Time 1	Time 2
All Nurses (F, G and H Grade)		
Lack of praise and recognition from line management	60	31
Inadequate time for study leave and training	66	41
Inadequate information about ward performance	79	56
Poor advice and support on long-term planning issues	59	37
Lack of sharing of new ideas and good practice	80	49
Inadequate support from line management	57	21
F Grade Nurses		
Difficulty extracting important information from extensive written communications	67	40
Poor quality of support and advice from senior staff	61	36
G and H Grade Nurses		
Too little time spent with consultants discussing ward management and future plans and ideas	67	37
Lack of support for staff after distressing events	54	32
Other staff's lack of understanding of this group's workload and priorities	79	50

Table 6.1 demonstrates in particular, that relationships with management improved considerably. Support from management, advice on long-term planning, and the amount of praise and recognition nurses received showed marked improvements in all three grades. The lack of time for study leave was also less problematic. And although still a problem for approximately half the group, far fewer staff reported problems concerning a lack of sharing of ideas and good practice between wards. For F grade staff in particular, many found information to be more readily available, and the quality of support and advice from their more senior staff had improved dramatically. G and H Grade staff reported that support for staff after distressing situations had improved, and that there was more time for discussion with consultants.

In addition, not shown in Table 6.1, there were modest improvements the amount of time available to tackle managerial tasks, in the availability of administrative support, in balancing clinical and managerial workloads, and in the amount of control staff had over the way that balance was achieved. Nurses also reported that new ideas were more readily allowed to develop and were implemented more quickly.

6.3.5   Changes in Individual and Organisational Well-being

Table 6.2 compares this group of nurses' worn-out scores at the time of the initial risk assessment survey (Time 1) with those after the intervention stage (Time 2). Also indicated in the table, are scores typically observed in other groups of employees studied by the authors over the last 20 years. Overall, worn-out scores decreased marginally (nurses reported being slightly less worn-out). The change is slightly more pronounced for G and H Grade staff (down nearly 2 points) than for the F Grade staff (down just under a point).

Table 6.2 Comparison of Worn-out Scores at Time 1 and Time 2

	Typical Scores in Other Groups	F Grade Nurses		G and H Grade Nurses	
		Time 1	Time 2	Time 1	Time 2
Worn-out	16–17	19.5	18.7	20.0	18.4

Although not dramatic, in our experience large-scale improvements in well-being for the whole group would not be expected given the short time period between the intervention and the evaluation (six months). If interventions are successful, such changes typically emerge in the medium to long-term. In addition, change may only be apparent for those staff involved in particular interventions (and not all staff participated in all interventions). Perhaps more interesting, as an indication of possible changes to come, is the consideration of whether nurses perceived any changes in the working conditions that they themselves had previously identified as being problematic for them, as described in the section above.

Table 6.3 reveals a slight decrease in numbers of G and H grade nurses intending to leave, but a slight increase in the number of F grade nurses intending to leave. Overall the numbers intending to leave are still relatively high (approximately a third of the group). No doubt factors other than those explored in this study were operative in this respect. For example, higher grade nurses were much in demand at the time of this study and any of this group could readily have obtained employment elsewhere. Nonetheless, the group remained satisfied with their jobs and absence remained low. A higher proportion of this group reported work-related musculo-skeletal pain at Time 2 than Time 1. This still represents a relatively low figure for nursing staff. Surveys repeatedly show nurses to be amongst the top professions for musculo-skeletal disorders (Hodgson *et al.*, 1993; Jones *et al.*, 1998). However, the possible reasons for the increase in this group were not clear. None of the interventions, however, targeted musculo-skeletal pain and many of the wards did not use drop-side cots; therefore staff often lifted from an awkward position. The fact that more nurses had regular access to new computing facilities may have been relevant, but we have no evidence currently to support this link.

As with worn-out scores, changes in such measures usually emerge in the medium to long-term. The changes are also likely to be localised to the group involved in particular interventions, rather than the group as a whole. The following section explores this suggestion further.

Table 6.3 Comparison of Individual and Organisational Well-being at Time 1 and Time 2

	Typical Scores in Other Groups	F Grade Nurses		G & H Grade Nurses	
		Time 1	Time 2	Time 1	Time 2
Job satisfaction (% staff satisfied)	40–50%	64%	68%	58%	60%
Intention to leave (% staff wanting to leave)	30%	27%	36%	42%	36%
Absence (days/yr)	6–8	7	6	2	2
Musculo-skeletal pain	40–50%	27%	32%	29%	46%

6.3.6   Evaluating the Impact of Specific Interventions

Not all interventions were designed to affect all staff, and the unpredictable nature of organisational reality may mean that even those staff for whom an intervention was designed, may not necessarily 'receive' it in the way it was intended, if at all. A step-wise strategy to evaluate the impact of interventions was employed: (i) identifying numbers of staff aware of, or involved in, each intervention *(awareness/involvement rating)*, (ii) exploring amongst the latter group whether working conditions or well-being had improved in comparison with those not aware of or involved in that intervention *(success rating)*, and (iii) estimating whether the impact was considered to have been modest or considerable *(impact rating)*. The working conditions and well-being of staff who were aware of or involved in interventions could then be compared to the working conditions and well-being of those who were not.

Outlined below are brief summaries about the implementation and outcomes of five interventions: (i) changes to staff administration time, (ii) installation of computer facilities on wards, (iii) appointment of housekeeping staff, (iv) improved access to study leave and training, and (v) staff open forums. The first three were designed specifically to provide nurses with more time for administrative work and thus improve the reported conflict between the fulfilment of their managerial and clinical roles. Differences between Time 1 and Time 2 in outcome measures such as job satisfaction or well-being are given in the following tables where noteworthy.

(i) Changes to Allocated 'Office Hours'

A review of 'office hours' was undertaken, resulting in several changes. The G and H grade nurses were aware of this review and of the changes that had occurred as a result of it, since they had been given a specific allocation of office hours for administration. F grade nurses were not directly affected by this change (and had not been allocated office hours). As expected therefore, awareness among F grade nurses was less pronounced but those who were aware of the change generally reported that it had an appreciable, positive impact for them.

The impact of this intervention proved challenging to interpret. First there seemed to be a number of positive effects. Those involved in the intervention were less likely to report the following problems (Table 6.4): infrequent office days, lack of information about ward performance, lack of advice and support on day-to-day issues, lack of time to tackle short managerial tasks and lack of time to deal with clinical workloads. At interview, many staff reported that 'office days were now recognised as vital', or that there was 'time to do office work at work, not at home'. However, the intervention also appeared to be associated with improved working relationships with management. Praise, recognition and support from management were all less likely to be problems for those aware of the review. For example, F grade staff indicated that their manager (G grade) was more available and could do more to support and advise them since the introduction of regular office days.

Table 6.4 Likely impact of office hours review and resultant actions

Working Conditions		% Staff Reporting the Problem	
	Time 1	Time 2	
		Aware of Review	Not Aware of Review
Infrequent office days/time for managerial tasks	71%	43%	71%
Interruptions during time set aside for management and administration	76%	71%	48%
Lack of information about ward performance	79%	39%	62%
Conflicts between clinical and managerial workload	57%	71%	43%
Lack of advice and support on day-to-day issues	52%	18%	48%
Lack of time for dealing with short managerial tasks	38%	18%	52%
Lack of time to deal with clinical workload	38%	29%	52%

However, there did appear to be some problems associated with having extra office hours. Those with office time tended to report problems with interruptions, and continuing tensions between their managerial and clinical workload. Not all the time allocated could be well used. And while this intervention was regarded as positive by many, staff in some wards were clearly not feeling its full benefits. For example, where there were staff shortages, the take-up of office days was limited. Several staff commented that there were 'insufficient staff to manage office days', or that 'office days are always cancelled for ward work'.

(ii) Installation of Computer Facilities on all Wards

At the time of the evaluation, about half of the wards had received new computers. Staff in those wards where computers were fully operational reported that they had made a positive difference to their work. Data from the questionnaire survey (Table 6.5) suggested that these staff also reported improved well-being (lower worn-out scores) at Time 2 in comparison with Time 1. In contrast, staff in wards without computers reported slightly decreased well-being.

Table 6.5 Likely impact of the introduction of new computer facilities

	Time 1	Time 2	
		Wards With New Computer Facilities	Wards Without New Computer Facilities
Worn-out score	19.6	17.2	20.4
Working conditions		% Staff Reporting the Problem	
Lack of time for administrative tasks	71%	42%	68%
Lack of time available to deal with clinical workload	38%	29%	48%
Knowledge of whom to contact to solve problems	26%	17%	40%

Access to computers was thought to be easing communication problems and helping this group of nurses to manage their time better. Those with new computers were less likely to report that they had problems finding time to undertake administrative work, or that they had insufficient time to deal with their clinical workload – both problems that were significant risks for poor well-being (using odds ratios) in the risk assessment. Before the intervention, many staff spent a considerable amount of time finding computers on other wards that they could use. In interviews, many said that it was important that they were able to do 'ward work at work, not on the PC at home' and, that e-mail (where available) had eased the flow of communication considerably and was a useful means of communicating within a large service. They knew more about whom to contact elsewhere in the hospital when attempting to solve problems and noted that 'access to memos, news and information is much better'.

However, these benefits were not available to all. Staff indicated that there were a number of issues limiting the potentially positive impact of the new technology. Many indicated that computers had not yet been fully installed, did not have e-mail software, were not yet connected to networks, or did not otherwise function properly. Many staff had not been trained how to use the new facilities. These issues were recognised by management and were being addressed at the time of the evaluation.

(iii) Appointment of Housekeeping Staff

Just over a third of staff worked in wards that had been allocated a new housekeeper. There was unanimous agreement that the introduction of this post resulted in an improvement and there were plans to introduce them into other wards.

While it was noted that the housekeeping role was most relevant to the work carried out by more junior staff, the senior grades (F, G and H grades) nonetheless reported numerous benefits (see the first four items in Table 6.6).

Table 6.6 Likely impact of the appointment of housekeeping staff

Working conditions	% Staff Reporting the Problem		
	Time 1	Time 2	
		Wards with a housekeeper	Wards without a housekeeper
Inadequate advice and support on long-term planning issues	59%	20%	52%
Lack of support staff in the wards	52%	35%	57%
Lack of say over which staff worked on the ward	45%	26%	65%
Low staff morale within the ward	40%	40%	61%
Inadequate cover for sickness absence	64%	80%	48%
Lack of control over the way staff manage and use their own time at work	42%	45%	22%
Lack of time to deal with clinical workload	38%	60%	18%
Lack of control over the way staff manage the time of other staff on the ward	38%	45%	13%

Most agreed that housekeepers helped to keep the ward tidier and to maintain a high standard of hygiene. The most significant impact reported by staff was on reducing their peripheral workload (strictly non-'nursing' tasks) such as making beds, fetching drinks for patients and their relatives, and maintaining stock levels. Questionnaire data also indicated that morale was higher in wards with a housekeeper. An additional benefit was that because senior staff had been involved in the recruitment of housekeeping staff they reported being more satisfied with their level of input into staff recruitment.

This intervention was among those that were challenging to evaluate. The benefits felt by staff at interview were clear. However, it was clear from the questionnaire data (as represented in the final four items in Table 6.6) that those wards with housekeepers still felt more hard-pressed than those without. They had less adequate cover for sickness absence, less control over their own time, less time to deal with their clinical workload and less control over how they managed other staff. Interview data helped to clarify these initially puzzling results. All wards had originally made a 'case' for being awarded a housekeeper; those wards that were most hard-pressed were those that were given a housekeeper first. Therefore, although they perceived matters to have improved, staff from wards with housekeepers still reported being busier than those without such additional support.

(iv) Improved Access to Study Leave and Training

Policies on study leave were clarified and updated, ensuring that feedback on training courses was given by staff who had attended those courses, that approval for attendance on training courses was speeded up, and that allocation of training was more equitable. The view of most of those involved was that this intervention had a positive impact. They reported that study leave was indeed more readily available, that the cancellation of training courses at short notice was less of a problem, and that staff were now giving more feedback and information about training courses.

Table 6.7 Likely impact of changes to study leave and training policies

	Time 1	Time 2	
		Group involved	Group not involved
Job satisfaction	60%	65%	50%
		Satisfied	Satisfied
Working conditions		% Staff Reporting the Problem	
Lack of time for study leave and training	66%	38%	58%
Cancellation of training courses at short notice	38%	30%	50%
Lack of praise and recognition from manager	60%	27%	50%
Poor quality support and advice from senior staff	52%	21%	58%
Lack of time available to deal with short managerial tasks	38%	30%	92%

Table 6.7 compares data from the initial risk assessment (Time 1) and the intervention evaluation questionnaire (Time 2). The group involved in this intervention were more likely to be satisfied with their jobs at Time 2 than were those not involved.

There also appeared to be some indirect effects from these interventions to improve access to study leave and training. For example, staff affected by the change were more likely to report being adequately supported by management, and that their efforts were more adequately recognised. This may partly be explained by the fact that at interview, nurses clearly saw the provision of training as a sign of recognition of effort. And finally, those staff who were involved in changes to study leave and training policies were more likely to report being able to tackle short, simple managerial tasks. Once again, interview data revealed that organising and approving training was a significant component of 'short and simple managerial tasks'. The clarification and updating of policies had presumably made this aspect of their work more straightforward.

(v) Staff Open Forums

Staff open forums were introduced as a means of increasing communication and co-operation and to improve the sharing of new ideas and good practice. G and H grade staff were more likely be involved in the open forums than F grade staff. The majority of those involved were more satisfied with their jobs than before and viewed the impact of the open forums positively. Of course, one cannot rule out the explanation that only those already satisfied with their job attended the forums.

Table 6.8 Likely impact of Open Forums

	Time 1	Time 2	
		Group Involved	Group Not Involved
Job satisfaction	60% Satisfied	75% Satisfied	50% Satisfied
Working conditions	% Staff Reporting the Problem		
New ideas not permitted to be developed and implemented	66%	45%	64%
Lack of support for staff after distressing events	50%	25%	50%
Infrequent meetings with colleagues of same grade	37%	15%	45%

Table 6.8 reveals that those involved were less likely to report that new ideas were not permitted to be developed and implemented, that staff had inadequate support after distressing events, that there were infrequent meetings with colleagues of the same grade or that they did not know whom to contact elsewhere in the hospital when attempting to solve problems. Information from the interviews supported these findings from the questionnaire. Many staff indicated

that the forums had 'opened up communication channels' and had given them the 'opportunity to meet other staff' and be 'more informed'. Nonetheless, staff felt that the real impact of the forums had been relatively modest. A number of staff commented that they had yet to see any actions taken as a result of discussions at the forum. Further, not all staff were able to attend forums since they often took place at busy times or (especially for part-time staff) on days when they were not working.

6.4 DISCUSSION

There has been increased concern in recent years about occupational stress as a major cause of work-related illness, notably in the service professions. As such it is a foreseeable risk. At the same time, it has become clear that in many countries, employers are required in law to assess the risks arising from work and to prevent or otherwise manage them as far as is reasonably practicable. This chapter has presented a brief account of one such attempt, using a risk management framework, to assist employees to assess their own work and to introduce changes for the better. It focuses on the organisation as the generator of risk, and on prevention.

In summary, an initial risk assessment for stress in a group of senior nurses was used as a basis for the design of a collection of interventions, most of which were suggested, developed and implemented by staff from within the group. Researchers played an active role in the initial assessment and in the evaluation of subsequent interventions. The allocation of administration time, the introduction of computer facilities on wards and the appointment of housekeeping staff were introduced to address the major problem identified by this group, namely, a conflict between fulfilling both their clinical and managerial roles. The impact of these interventions varied; where well implemented, they were very positive, freeing up time for staff to to deal more satisfactoriy with the managerial aspects of their work. But in wards where they had not been fully implemented or where there were low staffing levels, their success was limited. The interventions designed to improve access to study leave and training were thought to be having the desired impact. Various interventions designed to improve communications (such as a staff newsletter, open forums, problem-solving workshops, the development of a new service development strategy) were introduced. There was some evidence to suggest that these latter interventions had speeded up problem-solving, improved the sharing of good practice, and improved communications both within and between wards. It is notable that overall, the interventions were viewed positively by most staff. This is not always the case; in other studies, with the best intentions, interventions do not always work out as planned.

It is naturally not possible to conclude definitively from the (quantitative) data presented in the tables above that only these specific interventions were responsible for subsequent reported changes in employee working conditions or well-being. However, as we have shown, the supplementary (qualitative) information provided by staff at interview often allowed otherwise apparently contradictory results to be explained in a plausible fashion, and elucidated any particular implementation problems. The quantitative data, on the other hand, provided a harder picture of the situation, and allowed the specific weaknesses of broadly well-received interventions to be exposed, and barriers identified. In this way, qualitative and quantitative data were complementary, allowing a richer and more satisfactory understanding of the changes in this group's working conditions to be reached.

Overall, the entire exercise, both in terms of its processes and outcomes, was well received by these senior nurses and by their managers. They felt that the quantitative, 'evidence-based', aspects to the process strengthened their understanding of the job and its possible effects on them, and gave them confidence to design and implement interventions. Given the straightforward nature of this evidence, it is a relatively simple matter for such work groups, once shown the system, to undertake future assessments themselves, without the support of researchers or consultants. In other groups studied by the authors, this has already been undertaken and has been successful.

6.5 ACKNOWLEDGEMENTS

The authors would like to thank the nurses involved in this project and to acknowledge financial support from the British Health and Safety Executive, UNISON and the Royal College of Nursing.

6.6 REFERENCES

Aiken, L.H., Clarke, S.P., Sloane, D.M. and Sochalski, J.A., 2001, Nurses' reports on hospital care in five countries. *Health Affairs*, **20**, pp. 45–53.

Bourbonnais, R., Comeau, M., Vezina, M. and Dion, G., 1998, Job strain, psychological distress and burnout in nurses. *American Journal of Industrial Medicine*, **34**, pp. 20–28.

Butterworth, T., Carson, J., Jeacock, J., White, E. and Clements, A., 1999, Stress, coping, burnout and job satisfaction in British nurses: findings from the clinical supervision evaluation project. *Stress Medicine*, **15**, pp. 27–33.

Callaghan, P., Tak-Ying, S.A. and Wyatt, P.A., 2000, Factors related to stress and coping among Chinese nurses in Hong Kong. *Journal of Advanced Nursing*, **31**, pp. 1518–1527.

Castledine, G., 1998, The role of the clinical nurse consultant. *Journal of Nursing*, **7**, p. 1054.

Chapman, C., 1998, Is there a correlation between change and progress in nursing education. *Journal of Advanced Nursing*, **28**, pp. 459–460.

Clegg, A., 2001, Occupational stress in nursing: a review of the literature. *Journal of Nursing Management*, **9**, pp. 101–106.

Cox, T., 1993, *Stress research and stress management: Putting theory to work* (Sudbury: HSE Books).

Cox, T. and Griffiths, A., 1995, The nature and measurement of work stress: theory and practice. In *The evaluation of human work: A practical ergonomics methodology*, edited by J. Wilson, and N. Corlett (London: Taylor and Francis).

Cox, T., Griffiths, A.J, Barlow, C., Randall, R. Thomson, T. and Rial González, E., 2000, *Organisational interventions for work stress: A risk management approach* (Sudbury: HSE Books).

Cox, T., Griffiths, A. and Randall, R., 2002, A risk management approach to the prevention of work stress. *Handbook of work and health psychology*, Second Edition, edited by M.J. Schabracq, J.A.M. Winnubst and C.L. Cooper (London: Wiley & Sons).

Cox, T., Randall, R. and Griffiths, A., 2002, *Interventions to control stress at work in hospital staff* (Sudbury: HSE Books).

Cox, T., Griffiths, A. and Rial González, E., 2000, *Research on work-related stress* (Luxembourg: Office for Official Publications of the European Communities).

Cox, T., Thirlaway, M., Gotts, G. and Cox, S., 1983, The nature and assessment of general well-being. *Journal of Psychosomatic Research*, **47**, pp. 353–359.

Dunn, L., 1997, A literature review of advanced clinical nursing practice in the United States of America. *Journal of Advanced Nursing*, **17**, pp. 814–819.

Edwards, D., Burnard, P., Coyle, D., Fothergill, A. and Hannigan, B., 2000, Stress and burnout in community mental health nursing: a review of the literature. *Journal of Psychiatric and Mental Health Nursing*, **7**, pp. 7–14.

European Commission, 1989, Council Framework Directive on the Introduction of Measures to Encourage Improvements in the Safety and Health of Workers at Work. 89/391/EEC. *Official Journal of the European Communities*, **32**, No L183, pp. 1–8.

European Commission, 1996, *Guidance on risk assessment at work* (Brussels: European Commission).

Griffiths, A.J., Cox, T. and Stokes, A., 1995, Work-related stress and the law: the current position. *Journal of Employment Law and Practice*, **2**, pp. 93–96.

Griffiths, A., 1999, Organizational interventions: facing the limits of the natural science paradigm. *Scandinavian Journal of Work, Environment and Health*, **25,** pp. 589–596.

Guppy, A. and Gutteridge, T., 1991, Job satisfaction and occupational stress in UK general hospital nursing staff. *Work and Stress*, **5**, pp. 315–323.

Health and Safety Commission, 1999, *Management of Health and Safety at Work Regulations: Approved Code of Practice and Guidance* (London: HMSO).

Health and Safety Commission, 1992, *Management of Health and Safety at Work Regulations* (London: HMSO).

Health and Safety Executive, 1990, *A Guide to the Health and Safety at Work etc. Act 1974* (Sudbury: HSE Books).

Health and Safety Executive, 1995, *Stress at Work: A Guide for Employers,* (Sudbury: HSE Books).

Hemingway, M.A. and Smith, C.S., 1999, Organizational climate and occupational stressors as predictors of withdrawal behaviours and injuries in nurses. *Journal of Occupational and Organizational Psychology*, **72**, pp. 285–299.

Hillhouse, J.J. and Adler, C.M., 1997, Investigating stress effect patterns in hospital staff nurses: results of a cluster analysis. *Social Science and Medicine*, **45**, pp. 1781–1788.

Hodgson, J.T., Jones, J.R., Elliott, R.C. and Osman, J., 1993, *Self-reported work-related illness* (Sudbury: HSE Books).

Jones, J.R., Hodgson, J.T., Clegg, T.A. and Elliott, R.C., 1998, *Self-reported work-related illness in 1995* (Sudbury: HSE Books).

Kunkler, J. and Whittick, J., 1991, Stress management groups for nurses. *Journal of Advanced Nursing*, **16**, pp. 172–176.

MacNeil, J.M. and Weisz, G.M., 1987, Critical care nursing stress: Another look. *Heart and Lung*, **16**, pp. 274–277.

Muncer, S., Taylor, S., Green, D.W. and McManus, I.C., 2001, Nurses representations of the perceived causes of work-related stress: a network drawing approach. *Work and Stress*, **15**, pp. 40–52.

Newton, T.J. and Keenan, A., 1987, Role stress re-examined: an investigation of role stress predictors. *Organizational Behavior and Human Decision Processes*, **40**, pp. 346–368.

Stordeur, S., D'hoore, W. and Vandenberghe, C., 2001, Leadership, organizational stress, and emotional exhaustion among hospital nursing staff. *Journal of Advanced Nursing*, **35**, pp. 533–542.

Wang, M., Eddy, J.M. and Fitzhugh, E.C., 1995, Application of odds ratio and logistic models in epidemiology and health research. *Health Values,* **19**, pp. 59–62.

CHAPTER SEVEN

Work Stress and its Effects in General Practitioners

Helen R. Winefield

7.1 THE NATURE OF WORK DONE BY GENERAL PRACTITIONERS (PRIMARY HEALTH CARE PHYSICIANS)

The medically trained professionals who seek to maintain an overview of the help-seeker's complete state of health, who try to distinguish which bodily system is malfunctioning and if necessary, to decide on a referral to a doctor with narrower but deeper special knowledge, and who seek to provide whole-person cradle-to-grave care, are known as General Practitioners (GPs) in Britain and Australia, and as primary care physicians in US and elsewhere. It is these first-contact medical practitioners who are the subject of this chapter. Although more numerous than the specialist doctors, GPs have remained relatively outside the spotlight of attention from occupational psychology research. The reasons for this are connected with their community locations (compared with the specialists' concentration in large hospitals) and with their traditional independence. Perhaps too because the status hierarchy amongst medical professions has until recently at least, assigned lower status to the generalists than the specialists, the GP community tends to resist intrusion by outsiders. Sometimes this suspicion is extended even to the academic GPs employed in university medical schools, and their efforts to stimulate research by practitioners.

Because of their vital role in community health care, GPs' work effectiveness needs to be understood. This chapter therefore applies the concepts and models of occupational/organisational psychology, to describe the sources and effects of work stress for this crucial group of human service workers. Although interventions to reduce GP work stress or improve their productivity have been few and rarely evaluated scientifically, the wider literature can suggest some testable guidelines.

In Britain GPs have a 'list' of enrolled patients for whom they care, and are salaried through the National Health Service according to the size of this population. By contrast in Australia any patient can in principle consult any GP, and payments come to the doctors via the national health insurance system Medicare. Primary health care research is in some ways easier in Britain where the size and demographics of the population of patients is known, as well as the comprehensive picture of their health which case notes provide. However, Australian health care consumers frequently attend several different general practices, sometimes choosing different doctors based on convenience of access and sometimes for different health problems (e.g. gynaecological examinations vs. children's ear infections vs. workers' compensation paperwork). This 'doctor-shopping' complicates any research seeking patient outcome measures (including research on quality of medical work). Apart from these differences the systems of primary health care are similar in Britain and Australia, more similar than either is to the diverse American systems. Nevertheless it is reasonable to draw upon the international literature in seeking to describe the work of general medical practice, how to assess its quality, and how to protect the well-being of the workers.

A large proportion of the Australian population, in the region of 82%, consults a GP in any one year. Britt *et al.* (2001) report that the most common 'reasons for encounter' in other words the patient's explanation for the consultation, are requests for check-ups (in 13.2 per 100 doctor–patient encounters), seeking a prescription (9.2 per 100), and cough (7 per 100). In the same large survey the medical view of these consultations was that the most common problems being managed were hypertension (in 8.6 per 100 encounters), acute upper respiratory tract infection (6.9), immunisation/vaccination (4.6) and depression (3.7). The most common management activities engaged in by GPs were prescribing, advising or supplying medication (at a rate of 108.2 times per 100 encounters), clinical treatments such as giving advice, information or counselling (37.2 per 100), physical procedures (e.g. excision of tissue, Pap smears, peak flow tests, in 12.2 per 100 encounters), referrals elsewhere (10.4 per 100), and ordering pathology tests or imaging. As can be seen clearly from these data, the GP is presented with the patient's undifferentiated illness and has to decide the meaning of presenting symptoms which are often expressed as vaguely as 'feeling tired all the time' or 'needing a check-up'. Many of the presentations are routine updates of prescriptions for the many chronic illnesses (e.g. hypertension, respiratory problems, diabetes) experienced by the ageing population of westernised countries. At the same time the GP must remain alert to the possibility of a low-frequency but life-threatening serious condition, for which early detection and intervention may be crucial. The cognitive demands of the GP's work are thus considerable.

7.1.1 Psychosocial aspects of the work

The face-to-face interaction with the patient which constitutes the medical consultation, in addition to being a problem-solving occasion, is a socio-emotional event with considerable affective significance for both parties. The doctor–patient relationship is well-known to be a powerful vehicle for nonspecific benefits to the patients. These benefits of contact with a caring and respectful expert include the reduction of anxiety, depression and isolation and all their somatic correlates such as pain. The consultation in fact represents a very commonly-occurring helpgiving occasion which contains many elements of more specialised psychotherapy (Barker and Pistrang, 2002; Winefield, 1992; Winefield, 1996). Doctors hesitate to reduce these nonspecific benefits by claiming ignorance of what the patient's symptoms mean or what can be done about them. Sometimes the nonspecific benefits constitute most of what the doctor can offer to the patient. Often a definitive diagnosis let alone cure of the underlying cause is an unrealistic goal, and amelioration of symptoms, minimizing the rate of physical decline, and providing social support, become the goals which can be realistically adopted. Patients very much value the doctor's communication of care and respect, including efforts to give explanations in simple language. They like to feel that the doctor is interested in them as an individual rather than as a medical problem, and they enjoy consultations where there is some nonmedical, social chat (Winefield and Murrell, 1991). Mead and Bower (2000) have recently reviewed this literature on 'patient-centredness' in medical care.

A trusting doctor–patient relationship with excellent communication is not only essential for the delivery of effective health care, it is also potentially a powerful source of job satisfaction for the GP. Many report that their long-term relationships with patients are extremely rewarding to them and maintain their commitment to the work. Conversely, patients who do not respond to the doctor's best efforts even though they attend for consultation with very high frequency, are worrying and frustrating. Nonetheless, the professional ethic requires service delivery without discrimination on the basis of personal liking or otherwise for the patients. 'Emotional work' such as that involved in facilitating positive emotional states in others, if necessary by controlling the expression of one's own feelings, is inherent in service occupations (Hochschild, 1983; James, 1989). Yet only recently has its potential contribution to burnout been recognised (Zapf *et al.*, 2001). Emotion work tends to remain invisible and is consequently unrecognised and under-valued. Explicit training for it also tends to be lacking. Though in recent decades medical professionals have all received some undergraduate training in the rudiments of effective medical communication skills, this is insufficient to prepare them for patients whose emotional demands are greater than average.

7.1.2 Workload and sick leave habits

The workload in primary health care is both complex and highly responsible. Outside of the consultation per se, the work of the general practitioner in medicine includes being the culturally designated gate-keeper to benefits associated with the sick role, such as insurance and other third-party claims and sickness certificates. The bureaucratic paperwork can amount to a significant demand on the GP's time which is however, not directly reimbursed. There are also all the usual demands in any workplace, of subordinate, peer and supervisor relationships. Increasingly GPs work in groups sharing reception and record-keeping staff, perhaps a practice manager and/or accountant, perhaps a practice nurse or visiting specialist consultants and allied health workers such as physiotherapists, psychologists, podiatrists and so on. In addition they belong to professional organisations which need local representatives and committee workers, they all have a responsibility to continually update their medical knowledge, and in some cases, they have teaching and research duties too. So the work is very diverse and the load often heavy. Yet it is unusual for doctors to have a personal doctor of their own, and most show extreme reluctance to take sick leave (Aasland *et al.*, 2001; Lens and van der Wal, 1997; Nuffield Provincial Hospitals Trust, 1996). Baldwin *et al.* (1997a) reported that their cohort of junior doctors working in Scottish hospitals, at an average age of 25, suffered from fairly frequent minor illnesses but usually continued to go to work. They tended to self-medicate or to ask a friend or colleague to treat them rather than going for an independent consultation, a particular concern for the psychiatric symptoms (including depression and alcohol problems) which they reported at above average rates. These concerns remain salient throughout practice careers, and there is some evidence of greater risks for solo practitioners and those who qualified in a different country, who may similarly lack local supports. The cultural climate within medicine of practitioners being reluctant to admit to illness or seek medical treatment, is causing concern as one possible reason for recruitment problems in the profession (Cupples *et al.*, 2002).

Workloads in terms of patients presenting for consultation tend to be high, and the breadth of necessary expertise continues to grow. For example in Australia the first national survey of mental health (Andrews *et al.*, 1999) recently revealed that, of the one in five adult citizens with a significant level of psychological distress, most of those who were receiving any help at all were receiving it from their GP rather than from a specialist mental health professional. Referrals for psychological problems by primary care providers (according to Britt *et al.*, 2001, currently occur at 0.3 per 100 encounters to psychiatrists, 0.2 per 100 to psychologists and 0.1 per 100 for counsellors, though formal referrals to the latter two need not be made through a GP). This finding resulted in calls for 'upskilling' of GPs in mental health care, an area where many feel under-prepared and some feel uninterested. Parenthetically, psychologists are pressing the case, through their professional association and elsewhere, that state funding should be available for more psychologists to provide mental health care in general practice settings. If such initiatives are successful GPs will need to develop a further range of multi-disciplinary relationships.

Similarly, it has been noted that the physical and mental health of people living in rural and remote areas of Australia (and other countries with large underpopulated areas) is suffering from the unwillingness of medical practitioners to live in such locations, far from collegial supports, professional development opportunities, employment for their partners and top-quality schooling for their children. Doctors who do practise in rural locations certainly have a diverse workload which many enjoy, but also face the role conflicts and confidentiality difficulties of any service professional living in close proximity to the clients, yet isolated from professional supports.

In the terms of Jahoda's (1982) useful model of the psychological benefits of work, GPs seem to have time structure, activity, and opportunities for social contact, all at levels which are at least adequate and may even be excessive for their own equanimity. Their social status is usually high and they can feel pride in the constructive and purposeful nature of their work. Finally considering the financial rewards of their work, although GPs accurately complain of being paid at a lower consultation rate than is the plumber who comes to fix the washing machine, GP incomes after tax and expenses are usually sufficient to enjoy a comfortable lifestyle.

7.2   THE SOURCES OF WORK STRESS

The Editor of the British Medical Journal Richard Smith was asked to write a Preface to Lens and van der Wal's (1997) book *Problem Doctors*, and neatly summarised the prevailing views of what can make medical work stressful (p. ix): 'medical students are put through a gruelling course and exposed much younger than their nonmedical friends to pain, sickness, death and the perplexity of the soul. And all this within an environment where 'real doctors' get on with the job and only the weak weep or feel distressed. After qualification, doctors work absurdly hard, are encouraged to tackle horrible problems with inadequate support, and then face a lifetime of pretending that they have more powers than they actually do. And all this within an environment where narcotics are readily available.'

Inherent in this quotation and the outline of the nature of GP work in the preceding section, is an account of the possible sources of work stress for this occupational group. In this section we shall be concerned not so much to establish whether or not GPs are a particularly stressed group of workers, but rather, to identify the causes and following that the consequences, of any work stressors to which they are vulnerable, with the aim of determining how to prevent or alleviate them.

First as helping professionals, with the work's high emotional demands and frequent lack of feedback about success, GPs seem to be prime candidates for burnout. Their everyday contact with pain, anguish, loss and death is unparalleled in our society. Malpractice litigation is an ever-present anxiety, workload demands are high, and supports (from peers as well as from supervisors) are often sparse. It is relevant to ask, not only how stressful work in general practice is, but also to what extent the stressful features of the work are counterbalanced by its considerable autonomy and social prestige.

As noted above many GPs report the long-term relationships with clients, in whose lives they represent a healthful influence, to be very emotionally sustaining. Other medical graduates are more drawn to the intellectual challenges of rare diagnoses and the exploration of new treatments – for some of them, the repetitive interactions with the chronically ill who make up the bulk of patients, are uninteresting to the point of being stressful. However job satisfaction is often high. Many GPs have continued to work past the conventional retirement age, and being self-employed in many important senses, do not need to comply with imposed regulations concerning the hours, places and duration of practice. Fairly recently, Australian GPs have been reimbursed at higher rebates if they have complied with continuing professional education recommendations – but there is latitude about which courses to attend and how actual learning outcome is assessed.

7.2.1   Patients as sources of stress

The acknowledged sources of stress for GPs can conveniently be grouped into three main categories (Allen, 1994; Arnetz, 2001; Calnan *et al.*, 2001; Hazell *et al.*, 1996; Kirwan and Armstrong, 1995; Schattner and Coman, 1998; Winefield *et al.*, 1994). The first category of stressors is interactions with patients, where doctors are faced continually with human suffering and distress. Considerable training is received by medical students, much by observation, in how to cope emotionally with sights and tasks which would be traumatic for the unprepared person, such as blood, injuries and disfigurements, death, medical emergencies, physical decay, sexual and excretory functions. Maintaining a balance between professional detachment and individualised care is a desirable and highly-refined skill, and it is hardly surprising then that patients sometimes complain of failures to display it. Although data are scarce, patient complaints of excessive detachment on the part of the doctor, perhaps reflecting the psychological process of depersonalisation, seem to be commoner than complaints of the doctor's emotional over-involvement.

One reason for emotional disengagement on the doctor's part is patient unresponsiveness to treatment, especially if medically unexplained. Patients who present their distress in somatic form and seek frequent medical help even when investigations fail to reveal any physical pathology, pose special challenges to doctors. Such patients are often suffering from undetected anxiety disorders or depression, and their numerous doctor and hospital visits cost the health care system very large amounts (Ninan, 2001; Roy-Byrne, 1996; Souetre *et al.*, 1994). A variety of pejorative and rejecting descriptors for such patients has been documented, for example 'the heartsink patient' (O'Dowd, 1988). Either the doctor is uncertain what the patient's medical problem is, or the doctor finds the patient personally abrasive, or both (Schwenk *et al.*, 1989). It can be difficult to distinguish between patients with the habit of attributing adverse sensations to physical illness, as in somatizing conditions and hypochondriasis, and those who intentionally produce or exaggerate symptoms in order to again admission to the sick role, or other benefits. Any suggestion that presenting symptoms have no medical basis, or attempted referral to a mental health professional, is likely to be met with patient anger and vehement denials that the problem is 'all in my mind', sometimes followed by new or exacerbated symptoms. The helplessness that doctors feel in this situation, combined with their real fear of having missed a crucial sign of illness, leads to many fruitless and expensive diagnostic procedures to be undergone by the patient, and much anxiety for the doctor. An equally aversive situation for the doctor is the feeling of being used by patients for unwarranted legitimation of their insurance or sickness claims.

Some patients may be perceived as actually dangerous, and GPs have indeed been murdered while on home visits, assaulted by psychotic or violently angry patients, and subjected to vexatious complaints or litigation (Marcovitch, 2002). Physical risks do not usually figure largely in lists of occupational stressors. Work overload on the other hand, especially lack of sleep during training, has consistently been associated with increased risks of feeling overwhelmed and ill (Baldwin *et al.*, 1997b). The risk of litigation is perceived as extremely stressful although in Australia to date, of low frequency (Schattner and Coman, 1998).

7.2.2 Nonpatients as sources of stress

The second category of GP stressors could be summarised as nonpatient interpersonal ones, and would include possibly difficult relations with co-workers of varying status relative to the GP, and also the juggling of emotional and time demands between work and family life. With increasing numbers of women doctors especially in general practice, and their determination to combine career and parental roles, the 'workaholic' culture traditional in medicine in the past may be modified. However during the transition women often report that they feel disapproval from male colleagues, for insisting on part-time work and/or on more vacations from work to suit their child care responsibilities (Swanson *et al.*, 1998). Belittling and intimidatory workplace behaviours by peers, senior staff or managers, have been reported by up to 40% of junior doctors (Quine, 2002). In this survey black, Asian and women doctors were at higher risks of being bullied at work.

7.2.3 Organisational sources of stress

Third at the organisational level, stressors can arise for GPs from government regulation of their work, their paperwork responsibilities, relative lack of career path, and decreasing professional autonomy (Calnan and Williams, 1995; Commonwealth Department of Health and Family Services, 1996; Edwards *et al.*, 2002). Their financial dependence on number of consultations encourages the maximum workload, and reimbursement for longer consultations is not as yet proportional, which means there are financial disincentives to offering the longer appointments needed for counselling. Not surprisingly perhaps, it is often women practitioners who are not the sole income-earners for their families, who tend to collect the practice clients with the more challenging psychological conditions (Turner *et al.*, 1994). They carry out more emotional work and are correspondingly unrewarded for it, at least in monetary forms.

The Australian national insurer the Health Insurance Commission monitors patterns of practice, and carries out investigations to determine why some practitioners exceed the average for example in rates of prescribing certain classes of medication or diagnostic tests. As their processes are not fully transparent to doctors, anxiety can arise about being the subject of such an investigation, sometimes involving unannounced confirmatory interviews with patients and other actions experienced as intrusive and threatening.

Another workstress hazard at the organisational level, which applies particularly to GPs due to their relatively scattered worksites, is the lack of any effective supervisory support. Often there is no time for collegial discussions with work peers and no opportunity to seek guidance over troubling cases or to receive constructive feedback on work performance. The stress surveillance which is usually recommended in high-risk occupations is not available, either formally or informally. Again women may be somewhat disadvantaged by lack of access to the 'old boy network' and the golf course or bar-room companionship of their male colleagues. There may be interactions between the personality tendency to self-criticism, a strong predictor of subsequent stress and depression, and the 'risky, competitive and occasionally humiliating medical culture' (Firth-Cozens, 2001, p. 217). The medical culture reinforces delay of gratification, suppression of feelings, and masochistic overwork, as neatly demonstrated when even an expert on medical stress admitted feeling dismayed at a young doctor's determination to work only '9 to 5' (Lamberg, 1999).

At the widest level of organisational structure, Arnetz (2001) has pointed to mergers, the commodification and corporatisation of health care, and the increased perception of physicians as advisors rather than experts, as possible stressors. In terms of the popular Demand-Control-Support model of occupational stress (Johnson and Hall, 1988), demands on medical practitioners especially GPs are increasing, control diminishing, and supports haphazard and frequently unavailable. Support for the Demand-Control-Support model has not however been universal (e.g. de Jonge *et al.*, 1996, with nurses). Calnan *et al.* (2001) concluded from their survey of 719 primary care workers, that Karasek's (1979) conceptual framework can usefully be supplemented by considerations of Siegrist's (1996) effort–reward imbalance model. A perceived inadequacy of rewards in relation to effort – such as may be experienced by a GP whose patients are rude or ungrateful, or who feels humiliated amongst peers by a malpractice litigation, or whose workplace lacks supportive collegial bonds – can be predicted to enhance that worker's perception of job strain and psychological distress.

7.3 OUTCOMES OR EFFECTS OF WORK STRESS IN THE PROFESSION

There are several reasons for concern about work stress in GPs, not least of which is the humanitarian one that chronic occupational stress is likely to reduce quality of life and increase risks of negative health and mental health outcomes, within that occupational group. In addition, if alcoholism, opioid abuse and suicidality were to increase, with consequent lower quality work performance, GPs' work stress would concern the public in more self-interested ways: nobody would wish to be the patient of a doctor disabled by substance abuse or severe depression. It has even been suggested that realistically, because physicians are seen as a socially privileged group, policy makers are only likely to act vigorously to reduce and prevent occupational stress in primary care physicians if there is clear evidence of danger to the public (Gerrity, 2001).

7.3.1 Psychological distress and early retirement

The health effects of work stress for doctors have usually been discussed in terms of increased risks for suicide and substance abuse problems rather than of physical illnesses. The evidence however is not very compelling when examined carefully. Perhaps because of the public's admiration for the medical profession, reports of doctors' frailties seem to attract undue publicity. Taking account of social class and age, female doctors may be more likely to commit suicide, especially early in their practice careers, and males at mid-life, when compared with other professionals (Aasland *et al.*, 2001). Rates of other forms of psychiatric morbidity and of alcohol abuse have also been reported as high, including over substantial follow-ups (Firth-Cozens, 2001). The difficulty has been to establish to what extent such adverse characteristics are a consequence of work stress, and to what extent they are the result of vulnerable or perfectionist people being placed in situations of ready access to the knowledge and the means for self-injury.

Any evidence of occupational stress and consequent unhappiness or distress is naturally, sufficient reason to search for remedies and seek to reduce the suffering of those affected. But in doctors there is the added impetus that any negative impact of work stress on the quality of their work may have very damaging consequences for the public. This could happen in two ways: by doctors leaving medical practice in favour of other careers, or by continuing to practise medicine but doing so badly.

The intention to take premature retirement or to leave medical practice has been explored in some studies (e.g. Baker *et al.*, 1995; Schattner and Coman, 1991), but there is little information about the predictors of number of years worked by the graduates from the long and publically-subsidised medical training. In 1994 36% of Australian GPs surveyed said they would like to retire, and cited reasons such as health, disillusionment with government interference, and the work being too demanding and stressful (Commonwealth Department of Health and Family Services, 1996, pp. 43–46). Schattner and Coman (1998) found that 53% of the GPs in their Australian sample had considered abandoning general practice because of occupational stress; the link between considering it and doing it is of unknown strength. In the United Kingdom Edwards *et al.* (2002) have called for doctors and their leaders to accept responsibility to work with managers and government to reverse the current deterioration in medical job satisfaction.

7.3.2 Poorer quality of work

The second possibility, that stressed doctors might perform their work more poorly, requires close analysis of how we could establish whether or not a given factor (such as overwork or contact with suffering, for example) was having an effect on the quality of work. It is a more complex question than establishing how to measure work quality of a manual performance such as assembling computer components. There, the speed and accuracy of the work can relatively easily be assessed. The same is not true for the GP working behind closed doors with patients who were already sick when they arranged the consultation. Seeing more patients per hour will not necessarily indicate better quality work performance – indeed patients prefer longer consultations (Howie *et al.*, 1991), and number of patients seen per day contributes directly to GP work stress (Orozco and Garcia, 1993).

The number of patients who survive their illnesses, or recover without expensive drugs, or who express gratitude to the doctor, are also insensitive measures and likely to be influenced by the demographics of the patient population. To adopt another strategy and look for indicators of *poor* quality work performance, we could ask what mistakes or unnecessary failures of treatment the doctor demonstrates. Immediately it is apparent that there has to be a benchmark against which the correctness of the doctor's diagnoses and treatment decisions can be assessed. Recent years have in fact seen an increase in 'protocols' or ideal treatment guidelines against which the treatment actually used for particular conditions can be compared.

The first stage of collective wisdom about what is the 'correct' treatment for example, is provided by the consensus (if one can discover it) of respected clinicians or expert committees. The next stage of development of solid evidence for which treatments are effective, is furnished by case studies and group comparisons before and after, preferably with some effort at control groups. Finally the 'gold standard' of scientific evidence on which to base treatment plans, is provided by the randomised control trial (RCT) with replication in independent research settings. Here sufferers of a very clearly-defined condition are randomly allocated to receive the new treatment X or the equally intensive, credible and enthusiastically delivered Treatment Y (the previous accepted method) or Treatment Z (an equally credible placebo or nonspecific treatment). After rigorous validated measurement of symptom severity and a realistic length of follow-up, the relative efficacy of Treatment X can be established. Meta-analyses of the RCTs will then show what is the average Effect Size of the treatment compared with controls. Then considerations of the treatment's effectiveness, including its acceptability to patients and suitability for less pure examples of the original disorder, and even its cost-effectiveness, can come into play. Collections of research studies bearing on the evidence for best practice in medicine are being made available widely to doctors including GPs, often electronically such as through the Cochrane Library (http://cochrane.hcn.net.au/clib).

There are two main problems however, with this method of establishing what is the correct method of diagnosis and treatment, and therefore, of deciding on the work quality of any given medical practitioner. The first, which in principle might prove temporary, is that these guidelines rarely include much detail about how to communicate the diagnosis to the patient, discuss the treatment options, persuade the patient to follow expert recommendations about the treatment regime, or console the patient for pain or incurability and maintain patient morale. An exception is the National Breast Cancer Centre (2000) psychosocial clinical guidelines which do include recommendations for doctors on how to break the bad diagnostic news to patients with empathy, and how to offer emotional support when the prognosis is poor. The second and perhaps more insurmountable problem with protocols as a measure of medical work quality, is that the more rigorous is the RCT evidence, the fewer patients it is likely to apply to. The need to exclude patients with comorbid conditions, language problems, cultural differences, and often with substance abuse or psychiatric problems, and the refusal of many eligible patients to consent to enter the research trial or to complete its requirements, mean that the conclusions can be seen as inflexible and inapplicable to most of the patients whom the doctor routinely treats. Not surprisingly then, doctors show considerable resistance to having their work evaluated in terms of its adherence to guidelines which they believe to be irrelevant to a majority of their own patients. Other reasons for resistance to evidence-based protocols have been suggested by Shekelle (2002) to include doctors' lack of time and training to participate in imposed quality assurance programs, and their fears of being blamed for any adverse effects of noncompliance with guidelines.

The damaging effects on work quality of fatigue in junior hospital doctors have been established (Baldwin *et al.*, 1997a, 1997b; Firth-Cozens and Morrison, 1989; Houston and Allt, 1997), and reforms are being implemented which cut their possible working hours (e.g. insisting on 11 hours of continuous rest in each 24 hour period, Pickersgill, 2001). Karasek (1992) noted that productivity losses may arise when stressed workers stay on the job but work less effectively, for example with poorer motivation, lowered quality of work, and less willingness to innovate. Empirical evidence for work stress causing poorer quality work in human service professionals is sparse, but Wright and Bonett (1997) found that one component of burnout (Emotional Exhaustion) predicted poor quality work three years later, in human service personnel. However work quality was rather unconvincingly assessed with one global rating made by a single top-ranking administrative officer. In hospitals, Jones *et al.* (1988) found that staff (5% of whom were physicians) in departments with high numbers of malpractice claims reported more stress than those with low malpractice claims. Stewart and Barling (1996) found that health workers caring for cancer patients rated their own performance more negatively after days when they had experienced more subjective work stress and negative mood.

7.3.3 Errors and mistakes

It is in primary care that statistically, most mistakes are likely to be made. For GPs as for other physicians there are no universally accepted measures of work quality, although patient satisfaction, complaints and malpractice claims have been used as indicators (Commonwealth Department of Health and Family Services, 1996,

Chapter 6). GPs may not have opportunities to amputate the wrong limb or administer the wrong anaesthetic during surgery, nonetheless of critical incidents reported by GPs, 27% had the potential for severe harm to the patient and 76% were preventable (Bhasale *et al.*, 1998; Britt *et al.*, 1997). The difficulties of identifying adverse events or incidents are great, given the confidentiality of medical consultations, the lack of technical expertise of the patients, and the disincentives for practitioners of admitting to mistakes (Kidd and Veale, 1998). Therefore we know little of what relationship may exist between work stress for GPs and the quality of their work performance in terms of their propensity to make errors in either diagnosis or treatment (Firth-Cozens, 2001; Gerrity, 2001).

In the medical literature the main focus has been the 'impaired physician' or 'problem doctor' (Lens and van der Wal, 1997), with the stigmatising implications of alcoholism, mental illness, and professional negligence. Willcock (2001) has drawn a useful distinction between *impairment*, which signifies risk to patients, and the more common *underperformance*, an emerging pattern of failure to attain standards. In providing routine patient care the line between these two conditions may be difficult to draw, and the prevalence of underperformance difficult to establish (though impairment is likely in 0.5 to 1% of all doctors at any one time). The most common complaints by patients refer to inadequate communication, doctors' failures to recognise the boundaries of their skills, and behaviours perceived as unacceptable. The recognised high-risk work conditions are geographical or professional isolation, inadequately managed psychiatric or substance abuse disorders, and very stressful client populations such as abused children (Willcock, 2001).

While doctors readily attribute their mistakes in patient care to tiredness, overwork and strain (Firth-Cozens and Greenhalgh, 1997), and while it seems logical from a theoretical perspective to predict that stressed practitioners should be less attentive, empathetic and effective in patient care, the empirical evidence is slender. Advances in understanding however require a cultural change towards systems thinking about medical mistakes (see below).

7.4   INTERVENTIONS AND OUTCOMES

It is accepted in public health that no epidemic was ever beaten by treating the sufferers. However stress reduction efforts for GPs have characteristically adopted an individual focus which can easily appear to constitute victim-blaming, in preference to tackling the really difficult but less obvious organisational issues. Although there is a considerable literature for doctors on how to deal with the difficult patient, and some on how to resolve work-home conflicts, organisational-level stressors such as lack of peer and supervisory support have been relatively ignored (Winefield, 1997; Winefield *et al.*, 1998). Firth-Cozens (2001) has summarised recommended interventions at both levels, that is for both individual medical practitioners and for health care organisations, and from both perspectives: preventive and reactive. The preventive organisational level which has been relatively ignored in the past includes a range of initiatives to monitor environmental risks including all accidents, improve the management skills of senior doctors and create a clear but supportive, no-blame learning environment, and to promote the health and competence of trainees.

Efforts are being made to reduce medication errors, which constitute a large proportion of the preventable errors in primary health care (Bhasale *et al.*, 1998), by introducing computer-assisted prescribing and prescription-generation (Coiera 2001). However, the impact of desktop computerisation on the mutually satisfying GP–patient relationship needs more study, as does dependence on online information about evidence-based guidelines and protocols. It is conceivable that in the future GPs not accepting such guidance will face higher insurance premiums.

Is better selection for the job likely to solve the problem of GP workstress? Improved selection is a commonly prescribed antidote to poor fit between worker and work environment, so worth exploring here. A major problem however is the familiar one that when competition to enter medical school is intense, and quality of work after graduation is difficult to measure and also many years in the future, to make connections between applicant characteristics and subsequent work performance is very problematic. What tends to happen in practice is that it is applicants within a relatively narrow band of academic achievement (high) and personality characteristics (engaging, sociable, able to make a favourable impression at interview) who enter training and eventually professional practice. The future success at practice of those who do not enter training can never be measured. As McManus and Vincent (1997) argue, trying to increase medical work quality through selection policies is likely to be much less efficient than trying to improve training. In particular they suggest that training should focus on the skills and knowledge relevant to good quality practice, such as decision-making, communication and interpersonal skills, and self-critical competence in the processes of problem-solving.

7.5 THEORETICAL CRITIQUE OF WORK STRESS RESEARCH IN THE PROFESSION

Because it is easier to collect data from hospitals than from the scattered community locations where GPs work, much of the advice on stress management for doctors refers to hospitals – often mixing the different occupational groups there – where the issues are different from those of primary care. Low return rates to surveys are typical for GPs and it must be admitted, naïve researchers who send out long and/or nonspecific questionnaires about sensitive topics without first establishing their bona fides, deserve no more. Time pressures are very real in primary health care and time spent providing data for somebody else's thesis is time not spent in gainful activity. Particularly when results promise only criticism and jargon-laden publications for other people, and any feedback to the GP participants contains no useful recommendations, disillusionment with the value of psychological research is understandable.

Tennant (2001) has identified common and serious methodological flaws in existing research on the psychological damage caused by work stress. There are many descriptive and cross-sectional studies and few well-controlled interventions. Measures of psychological distress, job reactions including burnout and satisfaction, and even job conditions such as workload and supervisor support, are all based on self-report, introducing risks of halo effects and the influence of unacknowledged third factors such as personality. The usual low consent rates to participate cause much doubt about the representativeness of those who do end up contributing to the knowledge bank. Demands for the anonymity of respondents makes longitudinal research difficult. Causality cannot be established and prediction of individual responses to interventions is accordingly risky.

It has proved difficult in nonmedical occupations too, to establish what may be the relationship between job stress, job satisfaction, and quality of work performance. A large recent meta-analysis of over 300 studies, including over 54,000 participants, has however concluded that the mean job satisfaction – job performance correlation is about .30 overall (Judge *et al.*, 2001). Job performance was mostly measured by supervisor ratings, less often by peer ratings or objective output, and on a few occasions, by subordinate or client ratings. Of greatest relevance for this chapter, the average true correlation between satisfaction and performance was higher (.52) in the case of high-complexity jobs. This is a strong relationship in terms of effect size, and certainly justifies further research on the stress/satisfaction relationship to performance in highly complex, highly autonomous medical practice.

7.5.1   Error reduction from a systems perspective

Another contribution to the understanding of the effects of GP job stress has recently come from outside medicine, in fact from the human factors approach to risk reduction as developed for dangerous workplaces such as airlines, oil rigs and the nuclear industry. This conceptual framework lends itself quite comfortably to errors and adverse events in medical services (Barach and Small, 2000; Leape *et al.*, 1998; Vincent *et al.*, 1998; Wilson *et al.*, 1999). The impetus of this approach is to look for system-level safety measures, taking for granted that human errors are inevitable in complex work environments. Some quite radical suggestions have been made for risk management in hospitals (Wilson and Fulton, 2000), including trying to settle the case quickly rather than establish the doctor's innocence, making the hospital financially responsible for litigation costs (as is not the case currently for public hospitals), and shifting the responsibility for safety procedures from doctors to hospital management. The principle of a blame-free, learning work environment makes sense if improved safety standards are the goal, but is contraindicated where the threat of litigation punishes disclosure of any errors. In addition the 'culture' of medicine which emphasises autonomy, collegiality and self-regulation, discourages the reporting of mistakes (Lawton and Parker, 2002).

The first step towards safer systems is the reporting and subsequent analysis of near misses and the apparently trivial incidents which nonetheless give valuable information about the patterns of events which also precede the much less frequent errors with serious outcomes. Schemes for reporting near misses, 'close calls' and warning events are in use in dangerous industries; to work well they require confidentiality, independent data-collection, timely feedback to reporters, and sustained leadership support (Barach and Small, 2000). Vincent *et al.* (1998) suggested that factors within the work team such as verbal and written communication, supervisory arrangements and the skills mix, are likely to determine the risk of errors in hospital medicine, and yet the relevant research has not been carried out. How these ideas might apply in primary health care has been similarly unstudied.

7.6   FINDINGS FOR AUSTRALIAN GPS

The considerable literature on sources and psychological concomitants of work stress for Australian GPs has been reviewed progressively above. With regard to the main focus of this chapter, namely whether job stress is associated for GPs with poor quality work, there is as little information available for Australia as there is for other countries. Indeed there is little reason to believe that the dynamics of work performance should differ according to nationality, even if some specific causes of work stress (such as those related to GPs' income determinants) do differ between countries.

We conducted a pilot study to explore the possibility of gathering first-hand information about GP work errors (Winefield and Veale, 2002). A trained interviewer visited randomly-chosen mid-career GPs and asked them to describe patient care incidents which they felt might have been related to work stress. Then these incidents were categorised according to seriousness, both actual and potential, using the system developed in Australia by Bhasale *et al.* (1998). Minor consequences were those that actually or might have resulted in another consultation but required no time off activities of daily living. Moderate consequences were those that actually or might have resulted in time off, or the need for assistance with, work or activities of daily living, or left the person with ongoing morbidity. Severe consequences did or might have resulted in hospitalisation, were or could have been life threatening or resulted in death.

The sample was small ($N = 30$), but that represented an 86% consent rate to participate, in those approached. Sheer workload including parental responsibilities did not predict work-related stress, in the form of intention to retire early, burnout or job dissatisfaction. Five participants (16.7%) did not describe any 'recent event where your patient care was or may have been affected by work stress', the rest did, one being merely expressing impatience. Of the other 24, 15 (62.5%) described mistakes in diagnosis and 9 (37.5%) mistakes in treatment. Three reported incidents had severe actual consequences but nine had severe potential consequences. Indicators of work stress (job satisfaction and burnout) did not show a clear relationship to the seriousness of the reported incident. On the other hand, results suggested a link between some practice characteristics (part-time, smaller groups of doctors) and incident severity, a finding congruent with the systemic perspective.

Participants also made clear how stressful it often is for them to be involved in such incidents, and the lack of an effective support structure for practitioners who feel guilty, worried or remorseful. Not one mentioned responding to work mistakes (either their own or by colleagues) by contacting the official bodies set up to provide help and counselling for doctors. Similarly, the lack of adequate support at work was associated with feeling less sense of accomplishment, reporting lower job satisfaction, and planning to retire at an earlier age. Overall conclusions of this preliminary investigation refer to the need for further system-level analyses of the connections between workplace characteristics, job satisfaction, and job performance in GPs, rather than more search for the impaired or imperfect practitioner. In agreement with the previous conclusions by Bhasale *et al.* (1998), the mechanisms to translate such findings into interventions which improve the safety of the primary health care system remain untested.

7.7   DISCUSSION AND FUTURE RESEARCH DIRECTIONS

This chapter has aimed to outline the nature of GP work, the possible sources of stress in that work, and the reasons for some individuals being more susceptible to experiencing the work as stressful. In fact there has been little focus on the personality or socio-demographics of the workers in this context; rather an organisational and systemic perspective has been adopted. This is a perspective which has been conspicuously absent in the majority of previous writings about GP work stress.

The work of the GP is demanding and highly complex both interpersonally and cognitively, and GPs often perform it in relative social isolation from peers, and almost always without supervision. Often the correct diagnostic or treatment decision is a matter of judgment which then needs to be adapted to the needs and wishes of the patient concerned. The responsibilities however are large, with death or disablement of clients a risk from wrong decisions by the GP. The conscientious and high-achieving people who enter this profession have to balance on the one hand the placebo benefits to their patients of seeming wise and powerful, with on the other hand the ambiguity of many situations and the uncontrollability of much pertaining to human health.

Thus a focus of this chapter has been the connection between the work environment and the work performance of GPs, with occupational stress as a mediating variable interacting with both. This conceptualisation seems to indicate some further investigations which may promote both GP psychological wellbeing and high-quality patient care. Clinical, health and organisational psychologists may be important resources for GPs as they seek these improvements.

The primary need seems to be to carry out longitudinal studies with multiple measures of work performance, of the effectiveness of interventions designed to provide more support for GPs. Helpful supports may be hypothesised to include the following kinds:

- informational support (about drug dosages and interactions, diagnostic criteria, treatment protocols, and public health issues). The timely availability of this information through the GPs' desktop computer is becoming a reality, but how will attention to the computer during the consultation affect the vital communication process with the patient? If GPs felt themselves to be relegated to keypad users rather than healers how would it affect their job satisfaction?
- emotional support to cope with the sadness of human suffering, the tough decisions, and the uncertainties of how to deliver the best quality care. At present this support seems to come mostly from spouses and the occasional social interaction with colleagues. Can a more supportive work culture be developed where peer interactions become a valued routine part of the work environment? The present practice reimbursement arrangements in Australia, with their piece-work fee schedules, seem to make this unlikely without some system-level changes.
- managerial support for work practices which pay attention to near misses and mistakes then use that information to reduce risks of adverse events and serious errors. Some of this may need to come from nonmedical administrators – though their acceptance by GPs will need to be fought for. Can the prized autonomy of general practice be maintained in such circumstances?

The public would probably prefer to continue receiving personalised health care delivered by respectful and familiar experts, rather than impersonal advice via computer. The GPs' role is likely to continue to be a central one in providing effective primary health care at community level. Psychologists have the research skills and the understanding of human service work, to be able to contribute to the design of health care systems. Taking account of sociopolitical imperatives such as control of health care budgets, and of sociodemographic changes such as an ageing increasingly chronically ill population, it seems appropriate for psychologists and GPs to collaborate in advancing scientific understanding of primary health care work.

7.8 REFERENCES

Aasland, O.G., Ekeberg, O. and Schweder, T., 2001, Suicide rates from 1960 to 1989 in Norwegian physicians compared with other educational groups. *Social Science and Medicine*, **52**, pp. 259–265.
Allen, I., 1994, *Doctors and their careers* (London: Policy Studies Unit).

Andrews, G., Hall, W., Teesson, M. and Henderson, S., 1999, *The Mental Health of Australians* (Canberra: Mental Health Branch, Commonwealth Department of Health and Aged Care).

Arnetz, B.B., 2001, Psychosocial challenges facing physicians of today. *Social Science and Medicine*, **52**, pp. 203–213.

Baker, M., Williams, J. and Petchey, R., 1995, GPs in principle but not in practice: A study of vocationally trained doctors not currently working as principals. *British Medical Journal*, **310**, pp. 1301–1304.

Baldwin, P.J., Dodd, M. and Wrate, R.W., 1997a, Young doctors making mistakes – II. Health and health behaviour. *Social Science and Medicine*, **45**, pp. 41–44.

Baldwin, P.J., Dodd, M. and Wrate, R.W., 1997b, Young doctors making mistakes – I. How do working conditions affect attitudes, health and performance? *Social Science & Medicine*, **45**, pp. 35–40.

Barach, P. and Small, S.D., 2000, Reporting and preventing medical mishaps: Lessons from non-medical near miss reporting systems. *British Medical Journal*, **320**, pp. 759–763.

Barker, C. and Pistrang, N., 2002, Psychotherapy and social support: Integrating research in psychological helping. *Clinical Psychology Review*, **22**, pp. 361–379.

Bhasale, A.L., Miller, G.C., Reid, S.E. and Britt, H.C., 1998, Analysing potential harm in Australian general practice: An incident-monitoring study. *Medical Journal of Australia*, **169**, pp. 73–76.

Britt, H., Miller, G.C., Knox, S., Charles, J., Valenti, L., Henderson, J., Kelly, Z. and Pan, Y., 2001, *General practice activity in Australia 2000–01* (Canberra: Australian Institute of Health and Welfare).

Britt, H.C., Miller, G.C., Steven, I.D., Howarth, G.C., Nicholson, P.A., Bhasale, A.L. and Norton, K.J., 1997, Collecting data on potentially harmful events: A method for monitoring incidents in general practice. *Family Practice*, **14**, pp. 101–106.

Calnan, M. and Williams, S., 1995, Challenges to professional autonomy in the United Kingdom? The perceptions of general practitioners. *International Journal of Health Services*, **25**, pp. 219–241.

Calnan, M., Wainwright, D., Forsythe, M., Wall, B. and Almond, S., 2001, Mental health and stress in the workplace: The case of general practice in the UK. *Social Science & Medicine*, **52**, pp. 499–507.

Coiera, E.W., 2002, Health informatics. *Medical Journal of Australia*, **176**, p. 20.

Commonwealth Department of Health and Family Services, 1996, *General practice in Australia: 1996* (Canberra: Commonwealth of Australia).

Cupples, M., Bradley, T., Sibbett, C. and Thompson, W., 2002, The sick general practitioner's dilemma – to work or not to work? *British Medical Journal*, **324**, p. 139S.

De Jonge, J., Janssen, P.P.M. and van Breukelen, G.J.P., 1996, Testing the Demand-Control-Support Model among health care professionals: A structural equation model. *Work & Stress*, **10**, pp. 209–224.

Edwards, N., Kornacki, M.J. and Silversin, J., 2002, Unhappy doctors: What are the causes and what can be done? *British Medical Journal*, **324**, pp. 835–838.

Firth-Cozens, J., 2001, Interventions to improve physicians' well-being and patient care. *Social Science & Medicine*, **52**, pp. 215–222.

Firth-Cozens, J. and Greenhalgh, J., 1997, Doctors' perceptions of the links between stress and lowered clinical care. *Social Science & Medicine*, **44**, pp. 1017–1022.

Firth-Cozens, J. and Morrison, L.A., 1989, Sources of stress and ways of coping in junior house officers. *Stress Medicine*, **5**, pp. 121–126.

Gerrity, M.S., 2001, Interventions to improve physicians' well-being and patient care: A commentary. *Social Science & Medicine*, **52**, pp. 223–225.

Hazell, P., Pearson, S.A. and Rolfe, I., 1996, Influences on the quality of life of general practitioners in New South Wales, Australia. *Education for Health*, **9**, pp. 227–235.

Hochschild, A.R., 1983, *The Managed Heart* (Berkeley, University of California Press).

Houston, D.M. and Allt, S.K., 1997, Psychological distress and error making among junior house officers. *British Journal of Health Psychology*, **2**, pp. 141–151.

Howie, J.G.R., Porter, A.M.D., Heaney, D.J. and Hopton, J.L., 1991, Long to short consultation ratio: A proxy measure for quality of care in general practice. *British Journal of General Practice*, **41**, pp. 48–54.

Jahoda, M., 1982, *Employment and unemployment: A social-psychological analysis* (London: Cambridge University Press).

James, N., 1989, Emotional labour: Skill and work in the social regulation of feelings. *Sociological Review*, **37**, pp. 15–42.

Johnson, J.V. and Hall, E. M., 1988, Job strain, workplace social support and cardiovascular disease: A cross-sectional study of a random sample of Swedish working population. *American Journal of Public Health*, **78**, pp. 1336–1342.

Jones, J.W., Barge, B.N., Steffy, B.D., Fay, L.M., Kunz, L.K. and Wuebker, L.J., 1988, Stress and medical malpractice: Organizational risk assessment and intervention. *Journal of Applied Psychology*, **73**, pp. 727–735.

Judge, T.A., Thoresen, C.J., Bono, J.E. and Patton, G.K., 2001, The job satisfaction–job performance relationship: A qualitative and quantitative review. *Psychological Bulletin*, **127**, pp. 376–407.

Karasek, R., 1992, Stress prevention through work reorganization: A summary of 19 international case studies. *Conditions of Work Digest*, **11**, pp. 23–41.

Karasek, R.A., 1979, Job demands, job decision latitude and mental strain: Implications for job redesign. *Administrative Science Quarterly*, **24**, pp. 285–308.

Kidd, M.R. and Veale, B.M., 1998, How safe is Australian general practice and can it be made safer? *Medical Journal of Australia*, **169**, pp. 67–68.

Kirwan, M. and Armstrong, D., 1995, Investigation of burnout in a sample of British general practitioners. *British Journal of General Practice*, **45**, pp. 259– 260.

Lamberg, L., 1999, 'If I work hard(er) I will be loved.' Roots of physician stress explored. *Journal of the American Medical Association*, **282**, pp. 13–15.

Lawton, R. and Parker, D., 2002, Barriers to incident reporting in a healthcare system. *Quality and Safety in Health Care*, **11**, pp. 15–18.

Leape, L.L., Woods, D.D., Hatlie, M.J., Kizer, K.W., Schroeder, S.A. and Lundberg, G.D., 1998, Promoting patient safety by preventing medical error. (Editorial) *Journal of the American Medical Association*, **280**, p. 1444.

Lens, P. and Van Der Wal, G. (eds), 1997, *Problem doctors: A conspiracy of silence* (Amsterdam: IOS Press).

Marcovitch, H., 2002, GMC must recognise and deal with vexatious complaints fast. *British Medical Journal*, **324**, p. 127.

McManus, I.C. and Vincent, C.A., 1997, Can future poor performance be identified during selection? Chapter 16 in P. Lens and G. van der Wal (eds) *Problem Doctors* (Amsterdam: IOS Press), pp. 213–236.

Mead, N. and Bower, P., 2000, Patient-centredness: A conceptual framework and review of the empirical literature. *Social Science & Medicine*, **51**, pp. 1087–1110.

National Breast Cancer Centre Psychosocial Working Group, 2000, *Psychosocial clinical practice guidelines: Information, support and counselling for women with breast cancer* (Canberra: National Health and Medical Research Council).

Ninan, P.T., 2001, Dissolving the burden of generalized anxiety disorder. *Journal of Clinical Psychiatry*, **62**, pp. 5–10.

Nuffield Provincial Hospitals Trust, 1996, *Taking care of doctors' health* (London: Nuffield Provincial Hospitals Trust).

O'Dowd, T.C., 1988, Five years of heartsink patients in general practice. *British Medical Journal*, **297**, pp. 528–530.

Orozco, P. and Garcia, E., 1993, The influence of workload on the mental state of the primary health care physician. *Family Practice*, **10**, pp. 277–282.

Pickersgill, T., 2001, The European working time directive for doctors in training. *British Medical Journal*, **323**, p. 1266.

Quine, L., 2002, Workplace bullying in junior doctors: Questionnaire survey. *British Medical Journal*, **324**, pp. 878–879.

Roy-Byrne, P.P., 1996, Generalized anxiety and mixed anxiety–depression: association with disability and health care utilization. *Journal of Clinical Psychiatry*, **57**, pp. 86–91.

Schattner, P.L. and Coman, G.J., 1998, The stress of metropolitan general practice. *Medical Journal of Australia*, **169**, pp. 133–137.

Schwenk, T.L., Marquez, J.T., Lefever, D. and Cohen, M., 1989, Physician and patient determinants of difficult physician-patient relationships. *Journal of Family Practice*, **28**, pp. 59–63.

Seigrist, J., 1996, Adverse health effects of high-effort/low-reward conditions. *Journal of Occupational Psychology*, **1**, pp. 27–41.

Shekelle, P.G., 2002, Why don't physicians enthusiastically support quality improvement programmes? *Quality and Safety in Health Care*, **11**, p. 6.

Souetre, E., Lozet, H., Cimarosti, I., Martin, P., Chignon, J.M., Ades, J., Tignol, J. and Darcourt, G., 1994, Cost of anxiety disorders: impact of comorbidity. *Journal of Psychosomatic Research*, **38**, pp. 151–160.

Stewart, W. and Barling, J., 1996, Daily work stress, mood and interpersonal effectiveness: a mediational model. *Work & Stress*, **10**, pp. 336–351.

Swanson, V., Power, K.G. and Simpson, R.J., 1998, Occupational stress and family life: A comparison of male and female doctors. *Journal of Occupational and Organizational Psychology*, **71**, pp. 237–260.

Tennant, C., 2001, Work-related stress and depressive disorders. *Journal of Psychosomatic Research*, **51**, pp. 697–704.

Turner, J., Tippett, V. and Raphael, B., 1994, Women in medicine: Socialization, stereotypes and self-perceptions. *Australian and New Zealand Journal of Psychiatry*, **28**, pp. 129–135.

Vincent, C., Taylor-Adams, S. and Stanhope, N., 1998, Framework for analysing risk and safety in clinical medicine. *British Medical Journal*, **316**, pp. 1154–1157.

Willcock, S., 2001, The underperfoming GP. *Medicine Today*, **2**, pp. 128–133.

Wilson, L.L. and Fulton, M., 2000, Risk management: How doctors, hospitals and MDOs can limit the costs of malpractice litigation. *Medical Journal of Australia*, **172**, pp. 77–80.

Wilson, R.M., Harrison, B.T., Gibberd, R.W. and Hamilton, J.D., 1999, An analysis of the causes of adverse events from the Quality in Australian Health Care Study. *Medical Journal of Australia*, **170**, pp. 411–415.

Winefield, H.R., 1992, Doctor–patient communication: An interpersonal helping process. In S. Maes, H. Leventhal, and M. Johnston (eds) *International Review of Health Psychology*, Vol. 2. (Chichester: Wiley), pp. 167–187.

Winefield, H.R., 1996, Counselling work in normal general practice: An analysis of Australian consultations. *Journal of Health Psychology*, **1**, pp. 223–234.

Winefield, H.R., 1997, Sources and prevention of occupational stress in medical professionals. *Australian Journal of Social Research*, **3**, pp. 95–108.

Winefield, H., Farmer, E. and Denson, L., 1998, Work stress management for women general practitioners: An evaluation. *Psychology, Health & Medicine*, **3**, pp. 163–170.

Winefield, H.R. and Murrell, T.G.C., 1991, Speech patterns and satisfaction in diagnostic and prescriptive stages of general practice consultations. *British Journal of Medical Psychology*, **64**, pp. 103–115.

Winefield, H.R., Murrell, T.G.C. and Clifford, J.V., 1994, Sources of occupational stress for Australian general practitioners, and their implications for training. *Family Practice*, **11**, pp. 213–217.

Winefield, H.R. and Veale, B.M., 2002, Work stress and quality of work performance in Australian General Practitioners. *Australian Journal of Primary Health*, in press.

Wright, T.A. and Bonett, D.G., 1997, The contribution of burnout to work performance. *Journal of Organizational Behavior*, **18**, pp. 491–499.

Zapf, D., Seifert, C., Schmutte, B., Mertine, H. and Holz, M., 2001, Emotion work and job stressors and their effects on burnout. *Psychology and Health*, **16**, pp. 527–545.

CHAPTER EIGHT

Teacher Stress

Esther R. Greenglass and Ronald J. Burke

8.1 INTRODUCTION

Teaching is a time-honoured profession. Traditionally, teachers were seen as the providers of on-site education. In the past, there was a standard curriculum; subjects that were deemed important to all students were taught in routine and didactic ways. Teaching involved imparting information and knowledge to students whose role was to memorise this information and feed it back to their teachers during exams. Today, however, the role of teaching has considerably changed in the light of advanced technology and increased globalisation. The classroom as we knew it, bound in time and space, has now been enlarged in cyberspace. Teachers are confronted by the rapid development of knowledge and technology. They must take into account the development of other sources of information outside the school (e.g. television, the Internet, etc.) as well as young people's changing relationship to information, culture and knowledge. With widespread use of computers, the Internet, and e-mail, the classroom can exist wherever technology can be extended. With the increase in knowledge in so many fields, the emphasis today is more on learning skills to handle advanced information, rather than memorisation of facts. In fact, due to increased technology, the classroom is no longer necessary for teaching and learning, as we know it. According to Oakley and Stevens (2000), learners can access a wide range of information and attend virtual classes. Increased flexibility has led increasingly to education for individual rather than classes. Teachers now have to consider technological, organisational, and pedagogical changes and move from being providers of on-site education to providers of lifelong learning. This involves teaching new skills to assist students in the management and organisation of knowledge. The new technology provides greater flexibility in pedagogy and greater access to educational resources. At the same time, there is the need for teachers to learn the technology and become facilitators for their students.

The increasingly important role played by knowledge and skills, as well as innovation and research, in economic growth and development, the emergence of the information society, and the need for lifelong education are all resulting in increased social pressure on the education system and teachers in particular. Raising education standards continuously and ensuring universal access to a recognised professional or vocational qualification, i.e. to a quality certificate or diploma, have become social imperatives. Some state that the role of educators should go beyond specific subjects or technology and help prepare students for a global education. For example, Taylor (1998) argues that, by providing curricula and instruction that emphasise a global perspective, teachers not only prepare youths for successful adulthood but work toward creating a more humane society. For some, teaching is a social practice in which there is greater emphasis on instruction in social responsibility, democracy, social justice, and civility (Smyth, 2000). At the same time, there is a need for teachers to be mentors to adolescents so that they can grow up to be responsible adults, accepting of cultural differences needed to function successfully in an increasingly pluralistic society.

8.2 SOURCES OF WORK STRESS

It is well documented that teaching is a highly stressful occupation. This chapter is about stress in primary and secondary teachers rather than tertiary teachers. Several writers have noted recurring demands in the teaching environment that lead to significant levels of stress and burnout in teachers (Winzelberg and Luskin, 1999). There are several sources of stress for teachers. One of the most frequently acknowledged stressors for teachers centers around students. Problems with students include disruptive behaviour, such as verbal and physical abuse, emotional demands of students, their special needs, and heterogeneity in abilities, all of which tax the time and energy of an already busy teacher (Boyle *et al.*, 1995; Hodge *et al.*, 1994). Another major stressor for teachers involves conflicting demands made by supervisors, colleagues, students, and their parents. A third set of stressors in teaching relates to a teacher's workload itself, mainly their work overload. Quantitative work overload involves too many demands on teachers with too little time in which to meet these demands adequately. Often it is accompanied by feelings of being rushed and impatience. The importance of work overload as a stressor is underlined by the fact that teachers themselves in countries all over the world have consistently cited work overload as a major stressor in their job. Important dimensions associated with work overload are excessive paper work, lack of time for preparation for class, making reports, and submitting grades for deadlines. Having to meet constant deadlines throughout the year including those associated with grading, tests, exams and written reports by students is cited as another major stressor by teachers. Oversized classes with either inadequate or no teaching assistance also function as stressors in teachers. Other demands on teachers that contribute to their stress levels include role conflict and role ambiguity (Chen and Miller, 1997).

    Additional stressors associated with teaching are found at the administrative level, particularly with regard to problems associated with teachers' schools, i.e. poor organisation of schools and lack of technical and administrative support (Brown and Ralph, 1992; Smith and Bourke, 1992, Travers and Cooper, 1993). A frequently cited administrative stressor is the exclusion by the administration of teachers from decision-making that bears directly on their workload and the quality of their work life. Research has shown that when teachers do not have the opportunity to participate in the school's decision-making processes, their morale, motivation, self-esteem, and job satisfaction are more likely to decline. Lack of recognition of teachers' professional accomplishments by supervisors and school administrators alike contributes to an increase in stress levels in teachers. When teachers experience high role conflict in their jobs, role ambiguity, and low autonomy, they are more likely to experience stress (Chen and Miller, 1997). At the administrative level, additional stressors include inflexible employment policies on staffing and leave, and poor communication with teachers.

Lack of recognition of teachers' professional accomplishments by supervisors and school administrators alike contributes to an increase in stress levels in teachers. Feelings of diminished personal accomplishment occur when teachers perceive themselves as ineffective in helping students to learn and in fulfilling other school responsibilities. Low feelings of accomplishment have been identified as a significant component of stress and burnout in teachers. In a study of 208 elementary and middle school teachers in Japan, Ito (2000) found that lack of personal accomplishment was negatively associated with self-evaluated teaching ability and human relations. A comparison of new and experienced teachers showed that the newer ones had lower feelings of personal accomplishment, and evaluated themselves more poorly on their ability to guide their classes. In teachers, burnout can stem from an individual's perception of a significant gap between expectations of successful professional performance and an observed, far less satisfying reality (Friedman, 2000).

8.3   OUTCOMES OF WORK STRESS IN TEACHING

Given the stress and burnout that can accompany teaching, there has been considerable research on the effects of work stress on teachers. Perhaps more than any other public service professionals, teachers are affected by job burnout, resulting in negative attitudes toward students and loss of energy, idealism and purpose (Schamer and Jackson, 1996). With increasing burnout, there is a continual eroding of the person's ability to cope effectively with the continual bombardment of perceived stressors.

In early studies of burnout in the teaching profession, burnout was perceived as a general concept which included almost any negative reaction of teachers to pressure related to their work such as becoming frustrated, mentally exhausted, excessively worried, feeling depression and anxious (Perlman and Hartman, 1982). Burnout is a concept that is used to characterise a reaction to long-term stress, which is specifically linked to the emotional strain of working frequently and intensively with other people. The most widely accepted conceptualisation of burnout is found in the work of Maslach and Jackson (1986) who first applied it to the study of human service professionals. Moreover, research shows that human service professionals including teachers, are particularly vulnerable to burnout.

Freudenberger's (1975) original idea of burnout was that workers who find themselves under increasing pressure to help others, would demand more of themselves than they were able to give, and would ultimately exhaust themselves.

Maslach and Jackson considered burnout as consisting of three components: emotional exhaustion, depersonalisation and reduced personal accomplishment. Of these dimensions, emotional exhaustion most resembles the prototype of stress responses, as described in the literature (Greenglass and Burke, 1990; Lee and Ashforth, 1990). Emotional exhaustion refers to a depletion of one's emotional resources and the feeling that one has nothing left to give to others psychologically. Such feelings are more likely to occur when teachers show strong involvement with the needs of their students.

Depersonalisation is described in terms of the development of indifferent, callous and negative attitudes towards students in particular. It is assumed that teachers will distance themselves from their students in order to cope with their feelings of emotional exhaustion. The third dimension of burnout involves a negative evaluation of one's personal accomplishment in working with other people. When teachers suffer from burnout, their feelings of accomplishment are low.

In later studies, the Maslach Burnout Inventory (MBI) (Maslach and Jackson, 1986) was adapted for measuring burnout in teachers by researchers in the United States and in other countries (Farber, 1982; Friedman, 1991). Friedman's (1993) reports findings based on 1592 teachers who filled out an adapted form of the MBI in which they reported their feelings related to burnout. Results indicate the burnout is a complex construct composed of several components each expressing feelings in different degrees of burnout severity. Listed in descending order of severity of burnout, these components are:

1. A sense of frustration and a desire to sever all contact with one's job or profession comprise the peak of burnout.
2. Negative feelings towards the recipients of one's services; depersonalisation, appears closest to the peak of burnout.
3. A sense of emotional exhaustion is close to depersonalisation.
4. A sense of professional non accomplishment is less severe than depersonalisation or emotional exhaustion as far as burnout is concerned.
5. It is notable that the desire to leave work and depersonalisation, together with emotional exhaustion are the core meaning of the notion of burnout.

Teachers show emotional exhaustion when they perceive they are unable to give of themselves to students as perhaps they did earlier in their careers (Byrne, 1994). Depersonalisation occurs when teachers develop negative, cynical and sometimes callous attitudes toward students, parents and/or colleagues. Feelings of diminished personal accomplishment occur when teachers perceive themselves as ineffective in helping students to learn and in fulfilling other school responsibilities. For individuals who experience burnout, one of the first reactions is distancing themselves from the job and from students. Overall, burned out teachers are likely to be less sympathetic toward students, have a lower tolerance for classroom disruption, are less apt to prepare adequately for class, and feel less committed to their work. According to teachers exhibit signs of emotional exhaustion when they perceive that they are unable to give of themselves to their students, as they did earlier in their career.

A great deal of research has been devoted to the understanding of factors contributing to burnout and its consequences for individuals and their health. Research indicates that stress and burnout are significant factors in the development of both physical and psychological illness (McGrath *et al.*, 1989). Further research findings show that burnout is correlated with numerous self-reported measures of personal distress (Belcastro and Gold, 1983; Greenglass, 1991; Schaufeli and Enzmann, 1998). These findings parallel those reported for teachers that burnout also correlates positively with depression, anxiety and somatisation (Bakker *et al.*, 2000; Burke and Ondrack, 1990; Greenglass *et al.*, 1990).

Considerable research has documented other effects of teacher stress. As a result of stress and burnout, teachers might develop physical symptoms, i.e. headaches or peptic ulcers (Belcastro *et al.*, 1982), psychological symptoms, depression and anger (Greenglass *et al.*, 1990) as well as behavioural symptoms, e.g. lowered commitment to teaching and absenteeism (Blasé, 1986).

8.4 THEORETICAL CRITIQUES

There have been several theoretical treatises offered to interpret the development and experience of teacher stress and burnout. These have focused on the process of the development of burnout over time, the factors contributing to stress and burnout and in some instances have integrated these factors into a conceptual framework. Increasingly research has focused on an examination of burnout as a process that develops over time. According to Cherniss (1980) and other researchers, burnout is a process that develops over time. As such, researchers have been interested in discovering the paths through which burnout develops. Some have demonstrated the existence of two separate paths to burnout: first, a cognitive one, manifested in personal and professional feelings of accomplishment, and second, an emotional path reflected in a sense of overload and emotional exhaustion. It is generally held that the process of burnout begins to develop with external stressors, i.e. disruptive students, excessive paperwork, conflicting demands, that lead to emotional exhaustion. Emotional exhaustion is considered the prototype of stress that is seen as most responsive to the nature and intensity of work demands. Various job conditions are more strongly related to emotional exhaustion than to the other two components. Emotional exhaustion is predicted mainly by occupational stressors, such as role overload, role conflict, inadequate use of skills and interpersonal conflicts. Emotional exhaustion is considered the affective component of burnout that leads to depersonalisation (Leiter, 1988). Through depersonalisation, teachers attempt to cease the depletion of emotional energy by treating their students as objects rather than human beings (Lee and Ashforth, 1990). Emotional exhaustion may also lead to reduced feelings of personal accomplishment in one's teaching role which may be seen as indicative of an impoverishment of people's perceptions of themselves in their teaching role (Greenglass *et al.*, 1997).

In a study of factors relating to burnout using time-lagged data, Greenglass and Burke (1990) report results of multiple regressions in which sources of teacher stress and total role conflict at time 1 were significant predictors of emotional exhaustion one year later in a sample of teachers. In this study, stress consisted of doubts about competence, problems with students, lack of stimulation, bureaucratic interference and lack of collegiality. Thus, while teachers are expected to be emotionally involved to some extent with their students, the pressures associated with these stresses create a frustrating work environment. As the teacher strives to do his/her job, these pressures take increasing amounts of energy. Continued attempts to meet one's demands in the face of such obstacles result in emotional exhaustion over time. Research findings suggest that depersonalisation of one's students contributes to decreased feelings of accomplishment (Greenglass *et al.*, 1997). Since effective teaching is contingent on teachers' communicating with students on a personal level, teachers may be particularly vulnerable to low feelings of accomplishment when perceiving that they are treating their students impersonally.

One of the advantages of viewing burnout as a process over time is that it allows for investigation of the antecedents of burnout, particularly those associated with the school, the teacher's interpersonal relationships and his/her workload. Being able to specify antecedents of burnout in teachers has both theoretical and applied value. By specifying variables that contribute to burnout, theoretical knowledge of the process of the development of stress and burnout is advanced. On a practical level, it is valuable to the school's administration to be able to specify just what causes burnout. Examining the development of stress and burnout over time offers the possibility of being able to specify causes of stress as well as enabling investigators to understand more fully the complex causal relationships involved. This presumably can lead to the development of better and more effective intervention techniques that can prevent harmful consequences due to burnout, and improve teacher morale.

Other conceptual models of the development of burnout have been put forth. For some, burnout has been placed within the framework of action theory, defining burnout in terms of disturbed action process (Burisch, 1993). For others, burnout has been viewed within the framework of organisational theory where it is seen as the result of organisational structure and culture (Winnubst, 1993). Considerable empirical research has been generated by the theory on burnout by Buunk and Schaufeli (1993) and Schaufeli and Buunk (1996) which is based on Equity Theory (Adams, 1965). To the extent that the psychological contract is violated when employees perceive their job security is threatened due to downsizing by their employers, workers are more likely to feel betrayed and, as a result, experience anger, cynicism and hostility. According to Equity Theory, people pursue reciprocity in their interpersonal and organisational relationships. What they invest and gain from a relationship should be proportional to the investments and gains of the other party in the relationship. When they perceive relationships are inequitable, they feel distressed and are motivated to restore equity (Buunk and Schaufeli, 1999). For these researchers, in order to understand the development of burnout, it is important to examine how individuals interpret others' behaviour at work.

This approach is based on two major assumptions. First, in all social relationships, social exchange processes and expectations of equity and reciprocity play a role. People expect a balance of investments and gains in their relationships. However, in the area of teaching as is the case in many other areas of human services, the chronic disequilibrium of teachers investing more in relationships with students than they receive, may lead to depletion of emotional resources, and eventually to a lowering of the teacher's investment in the relationship with students (depersonalisation). Second, teachers who experience stress, will try to cope with their problems by comparing their experiences with those of others. Teachers who have high emotional demands, will compare themselves with other teachers in order to assess the appropriateness of their own emotional reactions. This approach to burnout has generated a great deal of research in recent years. For example, research has found significant relationships between perceptions of imbalance, emotional exhaustion and depersonalisation in several different occupations (Schaufeli *et al.*, 1996; Van Dierendonck *et al.*, 1994).

Burnout has also been conceptualised as a work-related syndrome stemming from the individual's perception of a significant gap between expectations of successful professional performance and an observed, far less satisfying reality (Friedman, 2000). Like the majority of professionals, teachers enter their job with expectations about what the job will be like and what can be accomplished on the job. Schwab *et al.* (1986) point out that teachers enter the profession with a commitment to help their students, with expectations of making a difference. When these expectations are not met, the individual will have a sense of defeat and failure. The match between initial job expectations and actual job role experiences partly determines teachers' reactions to their jobs. Schwab *et al.* (1986) provide research data suggesting that unmet job expectations are associated with burnout. Conflicting job expectations contributed significantly to emotional exhaustion but not to depersonalisation or to lack of personal accomplishment. Friedman (2000) cites self-reports of novice teachers' experiences in their first year of teaching, reflecting a world of shattered dreams of idealistic performance.

Similar ideas have been put forth by Byrne (1994), as mentioned in Section 8.3, who argues that teachers exhibit signs of emotional exhaustion when they perceive that they are unable to give of themselves to their students, as they did earlier in their career. They develop signs of depersonalisation when they develop negative, cynical and callous attitudes toward students, parents, and colleagues, and feelings of diminished personal accomplishment when they perceive themselves as ineffective in helping students to learn and in fulfilling other school responsibilities. Thus, Byrne (1994) links unmet professional expectations of the teacher to the later development of each of the burnout components.

8.5 INTERVENTIONS

Over the years, there have been some investigations of interventions where stress and burnout in teachers can be reduced. Many of these interventions have led to lower stress and burnout levels in teachers by changing the work environment, or reducing stress at its source, i.e. in the workplace. Specific examples include staff development and counselling (Cherniss, 1980), increased worker involvement and participation in decision making (Schwab *et al.*, 1986), improving supervision through clarification of work goals (Maslach and Jackson, 1984), and facilitating the development of social support resources. Stress and burnout in teachers can be reduced by establishing clear lines of authority and responsibility in helping to reduce ambiguity and conflict. With teachers, it may be valuable to solicit their input in areas that affect their work life in order to gain greater involvement and commitment.

Most of the research on interventions has been conducted within the area of social support and its ameliorative effects on teacher stress. Development of social support among teachers can be achieved by providing adequate time and location interaction. Encouraging the development of mentoring relationships between older and younger teachers can also be useful in increasing commitment among teachers. Considerable research has demonstrated the effectiveness of social support in lowering burnout and stress levels in teachers. Social support is seen as the exchange of information leading a person to believe that he/she is cared for. It can also involve provision of information, tangible, practical, and/or emotional help. Research suggests that social support may be effective in reducing burnout directly or as a buffering agent. When there is a direct effect, social support is positively related to physical and psychological health, regardless of the presence or absence of work stressors. Social support may also moderate the impact of stressors on burnout so as to assist people with high stress to cope better. The buffering argument suggests that stressors may affect some people adversely, but that those with social support resources are relatively resistant to the deleterious effects of stressful events.

The role of social support in alleviating stress is well known (Cohen and Wills, 1985; LaRocco *et al.*, 1980). Social support can have direct effects through reducing levels of job-related stress, a direct effect through improving emotional and physical well-being, and an indirect effect on the job-stressor–strain relationship. Research shows that lower levels of burnout are found in teachers receiving high levels of social support. In particular, support from one's supervisor and co-workers may be particularly important when examining the social support/burnout relationship (Himle *et al.*, 1989; Himle *et al.*, 1986). It may be that the availability of support lessens the impact of exhaustion on the other two components of burnout. Thus, individuals with strong social networks and therefore with greater opportunity to enhance their professional effectiveness, may be less likely to experience the other aspects of burnout (Leiter, 1991).

Research has been conducted on several modes of intervention to reduce stress and burnout in teachers, including educational programs and workshops, provision of social support and the introduction of stress management programs. Steel (2001) discusses staff support for teachers through supervision. Teacher stress is a consequence of working in an environment where human relationships are critical for the successful management of young people with social, emotional, and behavioural difficulties. Six primary ways in which organisations demoralise and demotivate employees are work overload, lack of autonomy, poor rewards, loss of belonging, unfairness, and value conflicts. According to Steel (2001), one way of alleviating this stress is through introducing supervision in the field of education. Supervision in other helping professions has been successful in providing support, changing perceptions, managing emotions, and coping with stressful situations, and in so doing has improved relationships with others and work performance. The most recognised models of supervision are managerial, non-managerial, individual, and group.

Lapp and Attridge (2000) examined the impact of a year-long educational intervention program designed to reduce the stress of secondary teachers and school staff. Health promotion services were provided at the work site for public inner city high school teachers (aged 20–75 years). Two control schools with similar student and staff characteristics also participated in this study. Three kinds of interventions were featured during the 9-month study period. Educational training was delivered at work in a group format. On-site workshops were offered to build stress management skills. In addition, personal resources consisting of tools and materials for reducing stress were provided. Survey data were collected from staff at 3 schools at 3 time points during the 1997–1998 school year. Results show that the majority of teachers were stressed, with more than 1/3 experiencing a high level of stress. The delivery of multiple, brief, educational interventions had a small but positive impact on reducing stress.

In another study, the effects of support and adverse work environments were examined in the context of other risk/protective factors in an extension of a short-term longitudinal study involving 184 newly appointed women teachers (Schonfeld, 2001). Regression analyses revealed that, adjusting for pre-employment levels of the outcomes and negative affectivity, social support and adversity in the fall work environment were among the factors that affected spring depressive symptoms, self-esteem, job satisfaction, and motivation to teach. Social support from non-work sources was directly related to future improved symptom levels and self-esteem; supervisor and colleague support were directly related to future job satisfaction. Effects of occupational coping, professional efficacy, locus of control, and school factors (e.g. special vs. regular education) were largely non significant (Schonfeld, 2001).

Individuals vary in their reactions to workplace distress. Research supports the idea that personal resources can affect an individual's reactions to stress and burnout. Individuals who are healthy, capable and optimistic, are resourceful and thus are less vulnerable to work stress. When confronting stress, perceived competence, labelled as perceived self-efficacy, is crucial. Perceived self-efficacy and optimism are seen as prerequisites for coping effectively with stresses. Perceived self-efficacy, as a personal resource, reflects the person's optimistic self-beliefs about being able to deal with critical demands by means of adaptive actions. It can also be regarded as an optimistic view of one's capacity to deal with stress. For Leiter (1991), burnout is inconsistent with a sense of self-determination or self-efficacy; burnout diminishes the potential for subsequent effectiveness. Low self-efficacy is a central factor in the etiology of burnout (Cherniss, 1990). These data are consistent with others showing that high self-efficacy contributed to lower distress in nurses who were experiencing hospital restructuring. High self-efficacy was associated with lower emotional exhaustion, less cynicism, less depression and anxiety, and greater job satisfaction and professional efficacy (Greenglass and Burke, 2000). Additional findings show that, despite failure, appraisals of individuals high in self-efficacy remained positive (Jerusalem and Schwarzer, 1992).

An individual's coping strategies have been the subject of extensive research in the area of job stress and burnout. Research on coping has distinguished between problem-focused, or instrumental approaches, to coping and emotional-focused coping. While problem-focused coping is directed toward managing the source of stress, emotion-focused coping is aimed at regulating emotional responses elicited by the situation (Folkman and Lazarus, 1980). Many agree that problem-focused coping is an effective individual coping strategy given research findings that it is negatively related to distress symptoms (Billings and Moos, 1984; O'Neill and Zeichner, 1985). Greenglass (1988) reports negative relationships in managers between job anxiety and problem-focused coping, and in particular, internal control, a coping strategy that depends on one's own efforts to change a situation. The same research also found significantly negative correlations between job anxiety and preventive coping. Additional findings indicate that emotional or palliative coping, including wishful thinking and self-blame, are positively correlated with psychological distress such as job anxiety, depression and somatisation, and negatively associated with job satisfaction (Greenglass, 1993), thus suggesting that palliative coping itself may be a distress symptom.

Latack (1986) has developed a measure of coping that differentiates between control and escape oriented coping. As such, the scales encompass the dichotomy between emotion-oriented and problem-oriented coping discussed originally by Lazarus and Folkman (1984). Control coping involves discussions with supervisors, making a plan of action, and generally consists of actions and cognitive reappraisals that are proactive and take-charge in tone. Escape coping by definition tends not to be problem-focused, since it is designed to get the person away from the situation causing the stress. Escape coping involves denial and avoidance processes. Previous research indicates that control coping may be more effective in reducing distress than escape or more passive forms.

Coping strategies and behaviours at work involving mastery or problem-solving are associated with more positive outcomes and decreased distress than are escape or more passive forms of coping (Armstrong-Stassen, 1994; Leiter, 1991). Several studies have examined the role of individual coping efforts in dealing with organisationally generated stressors and the results are mixed. Data indicate that mental health workers who respond to problems at work with control oriented strategies are less likely to experience burnout (Leiter, 1990, 1991). Use of control coping strategies is related to lower levels of emotional exhaustion and to a more positive assessment of one's professional accomplishments (Leiter, 1991). Escape coping, on the other hand, was more likely to lead to greater levels of emotional exhaustion.

In a study of the relationship between coping and burnout in men and women teachers, Greenglass and Burke (1990) found that lower burnout levels were significantly related to higher levels of coping. In this study, two coping strategies in particular were examined. Internal control is a coping strategy based on one's own efforts to change a situation. A second strategy included for study was preventive coping which includes coping techniques aimed at promoting one's well-being and reducing the likelihood of anticipated or potential potential problems. Thus, the more teachers used strategies which were aimed at reducing potential problems and the more they used techniques which were based on their own efforts, the lower their levels of burnout. Additional data using LISREL analyses showed that control coping contributed positively and directly to personal accomplishment, one of the burnout components (Burke *et al.*, 1995). Implications of these findings are that, to the extent that a school administration can encourage and maintain autonomy in their teachers, they will at the same time be increasing their levels of personal accomplishment in their teaching role.

Griffith *et al.* (1999) assessed associations between teacher stress, psychological coping responses and social support. A questionnaire survey of 780 primary and secondary school teachers (53.5% response rate; mean age 38.1 years) was used. In stepwise multiple regression, social support at work and the coping responses, behavioural disengagement and suppression of competing activities, predicted job stress independently of age, gender, class size, occupational grade and negative affectivity. High job stress was associated with low social support at work and greater use of coping by disengagement and suppression of competing activities. Behavioural disengagement and suppression of competing activities are generally considered maladaptive responses in a teaching environment and may actually contribute to job stress. The authors conclude that coping and social support not only moderate the impact of stressors on well-being but influence the appraisal of environmental demands as stressful. While some argue that individual coping strategies can reduce teacher stress and burnout, others have pointed out the limitations of individual coping (i.e. Pines *et al.*, 1981).

8.6   TEACHER RESEARCH – GREENGLASS AND BURKE

Teacher stress and burnout were studied in a sample of teachers in a large Canadian city. Using a self-report anonymous questionnaire, the design of the study employed two waves of data collection, one year apart. In the first wave 833 teachers filled out the questionnaire, 51% were males and 49% were females. The second wave comprised 473 respondents, with 52% male and 48% female. The majority of the respondents were full-time teachers with approximately 11 years or more of teaching experience. Teachers came from three levels of school, including elementary, middle and secondary school.

In this research, five sources of stress were identified and assessed, including: Doubts about competence, problems with clients, lack of stimulation and fulfilment, bureaucratic interference and lack of collegiality (Burke *et al.*, 1984; Cherniss, 1980). In addition, teacher stress was measured using a modified version of the Stress Profile for Teachers (Wilson, 1979). In particular, the stressors that were measured included, Student behaviour, Parent/teacher relations, and Time management.

Burnout was assessed using the MBI – Maslach Burnout Inventory (Maslach and Jackson, 1981, 1986) that yields scores on three subscales – Emotional Exhaustion, Depersonalisation and Personal Accomplishment. Emotional exhaustion refers to feelings of being emotionally overextended and drained by others. Depersonalisation refers to a callous response toward people who are recipients of one's services. Reduced personal accomplishment refers to a decline in one's feelings of competence and successful achievement in one's work with people. The inventory provides for measures of frequency and intensity, however, only frequency was used in this study as in most other research.

Several questions were raised in this research and explored in a series of papers using univariate and multivariate analyses. First, relationships were established between various stressors and outcome measures including burnout, psychosomatics and job outcomes such as job satisfaction. Findings showed that burnout scores increased significantly with greater sources of stress and teacher stress. Additional results showed that there were significant positive correlations between total burnout scores and depression, anxiety, somatisation, conflict and ambiguity, and lowered work goals, and a negative relationship between burnout and job satisfaction (Greenglass and Burke, 1990; Greenglass *et al.*, 1996). Findings from this research further showed that external demands were significant contributors to emotional exhaustion, results that parallel models put forth by Leiter (1993) and Cordes and Dougherty (1993). The results were that bureaucratic interference in the work situation led to increased levels of emotional exhaustion in teachers. In turn, emotional exhaustion was associated with increased somatisation, findings also reported by Schaufeli and Van Dierendonck (1993) in a sample of Dutch nurses.

Further study indicated that effects of stressors on emotional exhaustion were evident when studied over the period of one year (Greenglass and Burke, 1990). According to Cherniss (1980) and others, burnout is a process that develops over time. It was hypothesised that there is a causal sequence of antecedent conditions, i.e. work demands, and psychological reactions. The model put forth here presumes that there are long term effects associated with stressors and with burnout itself.

Greenglass and Burke (1990) observed stressors at time 1 as predictors of burnout at time 2, one year later. Respondents for the study were 361 teachers who filled out a self-report questionnaire at times 1 and 2. Approximately one-half of the sample were women. A series of multiple regressions were done, separately for men and women. The criteria were time 2 emotional exhaustion, depersonalisation and lack of personal accomplishment, and the predictors were time 1 role conflict, teacher stress, supervisor support, sources of stress and marital satisfaction. With emotional exhaustion as the criterion, the model accounted for 21% of the variance in emotional exhaustion in women, and 29% of the variance in men, with sources of stress a significant predictor of emotional exhaustion in both. The data showed that certain stressors associated with one's job as a teacher were significant contributors one year later to emotional exhaustion in both women and men. The stresses referred to here include: doubts about competence, problems with students, interference on the part of the school bureaucracy, lack of fulfillment and lack of collegiality. When depersonalisation and lack of personal accomplishment were the criteria, the model did not predict as well since only approximately 12–13% of the variance in each of these criteria were explained by the model. In men only, sources of stress at time 1 predicted to depersonalisation and lack of personal accomplishment. The results showing strong and significant relationships between time 1 stressors and time 2 emotional exhaustion parallel the earlier reported findings within time 1, namely that work demands and stressors function mainly through emotional exhaustion and to a lesser degree with the other two burnout components.

A second issue investigated in our research pertains to social support and its effects on teacher stress and burnout. The measurement of social support involved asking teachers to respond to a series of questions that tapped different types and sources of social support. Social support was assessed in the present study using a modified version of those employed by Caplan *et al.* (1975). We examined the availability of three types of social support:Practical, e.g. 'How much can (people) be relied on to provide you with assistance when you really need it the most?'; Emotional, e.g. 'How much do (people) boost your spirits when you feel low?'; and, Informational, e.g. 'How much useful information does each of these (people) provide you with when you really need it most?' Three items assessing each of the above types of support were used to measure social support from the teacher's supervisor, co-workers and family/friends (Burke and Greenglass, 1989; Greenglass *et al.*, 1996).

Intercorrelations were computed among burnout components, sources of stress and social support components from the three sources. Significant negative correlations were found between burnout components and each type of social support (practical, emotional and informational) from supervisors, co-workers and family and friends. In addition, sources of support were found to be negatively correlated with each type of support from supervisors, co-workers and family and friends. The data showed that the higher the social support, the lower the teacher's emotional exhaustion, depersonalisation and lack of personal accomplishment. (Greenglass *et al.*, 1996).

Additional findings showed that, in addition to main effects, social support functioned as a buffer. In one study, a series of hierarchical multiple regressions were performed where the criteria were each of the three burnout components, emotional exhaustion, depersonalisation, lack of personal accomplishment and total burnout scores. The predictors were, sources of stress (mean value over the five sources of stress), components of social support (i.e. practical co-worker), and stress × support interactions (Greenglass *et al.*, 1996). Main effects were considered in the first step and multiplicative terms were entered in the second step, with each of the social support components separately entered in an interaction with sources of support. Results showed that, of the three source of social support, a teacher's co-workers were the most important buffers of burnout in teachers. When sources of support increased, the buffering effects of co-worker support were seen in emotional exhaustion, in depersonalisation and in total burnout scores. Information and emotional social support were important as buffers of burnout.

These findings point to the importance of examining specific components of social support as buffers, particularly from co-workers. They also underline the beneficial effects of companionship in the form of co-worker support in buffering high sources of stress in a teacher's workplace. While information support was the main buffer of emotional exhaustion, the findings showed that emotional support from co-workers and family and friends had a buffering effect on teacher depersonalisation. In the present context, emotional support refers to the provision of morale boosting, listening to one's work-related problems, and being a confidante. Thus, to the extent that teachers perceive emotional support from those close to them when sources of work stress are at high levels, they are less likely to depersonalise their students. It would appear then that when teachers are able to turn to others in times of high stress and receive emotional 'boosting', they are less likely themselves to act in a detached and impersonal manner towards their students. It is possible that the perception of emotional support in those close to them raises the salience for teachers of the importance of relating to students as individuals, to give the students the encouragement they themselves received in the form of emotional support.

A third question raised in our research program related to gender differences in burnout. In the past when researchers asked who suffered more from burnout, women or men, data supported the view that women were more susceptible to burnout since they often took primary responsibility for children, in addition to their employment. However, in research examining sex differences in burnout, often men and women occupy different occupational roles, which would result in a confounding of sex and occupation. For example, Maslach and Jackson (1985) examined sex differences in a wide range of human service occupations. Women were higher on emotional exhaustion and lower on personal accomplishment than men. In others words, women were more likely to feel emotionally drained by their work than men. But, in this study gender was confounded with type of occupation. Police officers and psychiatrists were usually men and nurses, social workers and counsellors were typically women. Therefore, the sex differences reported may in fact reflect differences in occupations.

Our data indicate that men experience higher scores on depersonalisation than women, when women and men in the same occupational role are compared (Anderson and Iwanicki, 1984; Greenglass *et al.*, 1988; Greenglass *et al.*, 1990;

Ogus *et al.*, 1990; Schwab and Iwanicki, 1982). In particular, work by Greenglass and Burke consistently suggests that male teachers score significantly higher than their female counterparts on depersonalisation (Greenglass and Burke, 1988; Ogus *et al.*, 1990; Burke and Greenglass, 1989). Why should men be more prone to depersonalisation, an attitude characterised by callousness and being impersonal towards one's students or clients? One explanation is found in accepted norms associated with the masculine gender role, which emphasises strength, independence, separation and invulnerability (Greenglass, 1991). In this context, depersonalisation may be regarded as a reflection of men's repressed emotionality. Another explanation is found in accepted norms associated with the masculine gender role that emphasise strength, independence, separation, and invulnerability. In this context, depersonalisation may be regarded as a reflection of men's repressed emotionality. Described in this way, depersonalisation may be seen as an ineffective coping form which allows men to continue their work with people yet remain untouched in any significant way by others' suffering. Moreover, it is through the distance he has imposed between his students and himself that the man can remain uninvolved, thus appearing callous. It is also possible that depersonalisation is a result of an inability on the part of men to cope with work strain. Since it is socially unacceptable for men openly to express vulnerability, men have fewer options for emotional expression.

Another explanation derives from the emphasis on achievement, which is an integral part of the masculine gender role. If men are also competitive and their feelings of masculinity depend on successful achievement, their cynicism may derive from distrust of those with whom they are competing. This may lead to anti-social and hostile feelings, particularly when threatened under stress. Additional data indicate that men are significantly higher than women on cynical distrust, a measure of hostility and distrust in others (Greenglass and Julkunen, 1989, 1991; Greenglass, 1998). These findings parallel earlier reported findings that men are higher on depersonalisation. These results also coincide with observations by Solomon (1982) that feelings of anger, hostility and aggression are an expected part of the masculine gender role even though avoidance of expressiveness is encouraged. Moreover, Hobfoll *et al.* (1996), in their studies of the Multiaxial Model of Coping, report that men utilise more aggressive and antisocial action in their coping.

Additional findings in our research have shown that higher depersonalisation in men is associated with lack of collegiality, higher absenteeism, greater medication use and a lower quality of lifestyle, i.e. greater smoking levels and greater consumption of caffeine and alcohol (Ogus *et al.*, 1990). Given that a male teacher with high depersonalisation is more likely to take drugs and generally to neglect his health, it is reasonable to suggest that these kinds of behaviour may interfere with his effectiveness as a teacher.

A number of studies of psychological burnout have been conducted in several countries. For example, in studies of burnout from North America and The Netherlands, burnout rates are related to demographic variables such as sex and age (Anderson and Iwanicki, 1984; Greenglass *et al.*, 1990; Van Ginkel, 1987;

Van Poppel and Kamphuis, 1992) as well as to factors related to work such as experience in teaching and type of school (Anderson and Iwanicki, 1984; Russell *et al.*, 1987; Van Ginkel, 1987). Across countries, there has been some similarity in results. North American data indicate on average that male teachers report higher scores on depersonalisation, whereas female teachers report higher scores on emotional exhaustion and lower scores on personal accomplishment (Anderson and Iwanicki, 1984; Greenglass *et al.*, 1990). Among Dutch teachers, however, emotional exhaustion has been found to be significantly higher for men (Van Ginkel, 1987) and, similar to North America, men display higher scores on depersonalisation in The Netherlands than women. Men's higher depersonalisation scores are probably due to socialisation associated with the masculine gender role that prescribes less display of emotion. Accordingly, men are more likely to distance themselves from the people with whom they work. These gender role prescriptions would appear to override cultural differences and thus account for the men's higher depersonalisation scores in both cultures.

A fourth issue that we have investigated in our study of teacher stress relates to gender differences in non work spheres. Some researchers have found that women have higher scores than men on emotional exhaustion (Maslach and Jackson, 1985). Women may experience greater emotional exhaustion because, compared to men, their total workload (including paid and unpaid work) is higher. Although the number of men doing housework and childcare has increased in recent years, according to research findings, women still do most of the work in the home (Barnett and Shen, 1997). Research also shows that women experience greater role conflict (between work and family roles) than their male counterparts (Greenglass and Burke, 1988). At the same time, other investigators argue that the dual role of employment and raising a family places excessive demands on employed women (Haynes *et al.*, 1980).

Cherniss (1980) suggests that the level of burnout experienced is a function of an individual's demands and supports. Demands may include sources of stress on the job while support may be given by one's supervisor and spouse in terms of practical assistance. Marital satisfaction may function as well as a form of support for married people since it may represent satisfaction with support received from one's partner. Given uneven societal expectations regarding work and familial roles among women and men, factors contributing to burnout in teachers should vary with gender. With the emphasis on family roles for women in our society, regardless of whatever else they do, it is expected that variables related to familial roles will be more significant predictors of burnout in women than in men. Thus, it is expected that marital dissatisfaction and conflict between work and family roles will be significant predictors of burnout in women but not in men.

We tested these hypotheses with a group of teachers who had children since the conceptual focus here was on family and work issues and their relationship to burnout. The sample consisted of a sub-sample of men and women teachers with children ($n = 556$) within the teacher sample we surveyed in wave 1 of our research program (Greenglass and Burke, 1988). In addition to their burnout, we also included a measure of their sources of stress, teacher stress, extent of supervisor support, total role conflict and marital satisfaction. All variables were entered as predictors in a series of multiple regressions, separately in men and women, with total MBI burnout scores as the criterion. Results showed that in women, but not in men, marital dissatisfaction and role conflict were significant predictors of burnout scores. In men, predictors of burnout tended to be confined to the work sphere. The men in this study, like their female counterparts, held both occupational and familial roles, since all respondents had children. For women and men, sources of stress was a significant predictor of burnout, however, in men, beliefs such as doubts about competence, problems with clients and lack of fulfillment, all identified as work sources of stress, played a singularly important role in predicting burnout in men. While marital dissatisfaction and role conflict played a significant role in predicting burnout in women, they were not significant predictors of burnout in men. Since Canadian men generally invest less in their family roles than their professional female counterparts, their feelings of self-esteem hinge less on their family roles and more on occupational ones. Thus, it is not surprising that beliefs such as doubts about competence, problems with clients and lack of fulfillment, all identified as work sources of stress by Cherniss, should play such a singularly important role in predicting burnout in men. By taking a gender-role perspective in the examination of predictors of burnout, significantly different trends are observed in factors identified as predictors of burnout in men and women.

8.7   DISCUSSION AND FUTURE RESEARCH

Recent research indicates that teachers, like many other employees within the government supported sphere, are experiencing new sources of stress due to widespread fiscal cutbacks (Bolick and O'Keefe, 1991; Fisher *et al.*, 2000). Although more demands are being placed on public schools, less money is available to meet them. With more burdens being placed on schools and on teachers, their resources are being stretched to their limit. This may also result in demoralisation and demotivation in teachers in six main ways: Work overload, lack of autonomy, poor rewards, loss of belonging, unfairness, and value conflicts (Steel, 2001). School boards are implementing policy changes that result in increases in the number of students per class and reduction in classes for students with special needs, with the consequence being higher workloads, longer work days and widespread feelings of unfairness in teachers. As indicated above, workload and perceived unfairness are major stressors that can lead to increases in psychological and physical symptoms. Teachers' effectiveness may also be compromised by the effects of budget cuts that undermine their professional feelings. To the extent that teachers perceive they are not valued or respected as professionals, their effectiveness in the classroom will diminish. Given these considerations, there are steps that school boards can take to alleviate some of the fall-out from cutbacks. For example, Burke and Leiter (1999) argue that keeping employees engaged during major organisational change requires that management provides employees with information about changes and also involves them in active participation and planning. Thus, to the extent that school boards can provide information about changes that involve their teachers and include them in planning and developing policy, the impact of these changes should be considerably decreased. Having knowledge of the school's plans for the future is a form of information control or predictability. And, perceived control has been linked to decreased stress levels and improved worker health (Spector, 1986).

At the same time, it is important that schools show their appreciation for teachers' efforts and that they be rewarded for their work. Research indicates that opportunities for recognition, pay, and promotion are positively related to perceived organisational support (Eisenberger *et al.*, 1999). It is assumed that in order to meet socioemotional needs and to assess the organisation's readiness to reward increased efforts, employees form general beliefs concerning how much the organisation values their contributions and cares about their well-being. It is argued that employees exchange emotional attachment to the organisation for the benefits they receive from their organisation (Mottaz, 1988). According to organisational support theory (Eisenberger *et al.*, 1986, 1997), perceived organisational support is strengthened by favorable work experiences that employees believe reflect voluntary and purposeful decisions made by the organisation. To the extent that a school board is seen as caring about the teacher, the impact of the cuts should be less. Research is needed to determine the mediators of teachers' affective commitment to their schools and its impact on teacher stress and burnout, particularly during periods of organisational change.

A related issue deserving of further research pertains to the development of stress and burnout over time. As discussed earlier in the section on Theoretical Critiques, several investigators have offered theories on the development of teacher stress and burnout over time. Focusing on the factors that contribute to teacher stress, theories can specify causes and consequences of stressors within a process-oriented framework. However, to date there is a paucity of studies examining the development of stress over time in teachers. There is a gap between researchers in the area of teacher stress and practitioners who implement interventions to alleviate teacher stress. Future research can be undertaken to determine the effectiveness of various organisational interventions using process model theories and research design in which systematic examination can be made of the factors involved in these approaches. In these ways relationships can be established between causes and mediators of stressful outcomes identified in teachers.

8.8 ACKNOWLEDGEMENTS

Special thanks are due to Margaret Chuong for her assistance in the preparation of this chapter. Reprint requests may be directed to Dr Esther Greenglass, Department of Psychology, York University, North York, Ontario, Canada M3J 1P3. e-mail: estherg@yorku.ca

8.9 REFERENCES

Adams, J.S., 1965, Inequity in social exchange. In L. Berkowitz (ed.), *Advances in experimental social psychology* (Vol. 2) (New York: Academic Press), pp. 267–299.

Anderson, M.B.G. and Iwanicki, E.F., 1984, Teacher motivation and its relationship to burnout. *Educational Administration Quarterly*, **20**, pp. 109–132.

Armstrong-Stassen, M., 1994, Coping with transition: A study of layoff survivors. *Journal of Organizational Behaviour*, **15**, pp. 597–621.

Bakker, A., Schaufeli, W.B. and Van Dierendonck, D., 2000, Burnout: prevalentie, risicogroepen en risicofactoren. [Burnout: prevalence, risk-groups and risk factors] In I.L.D. Houtman, W.B. Schaufeli and T. Taris (eds), *Psychische vermoeidheid en werk: Cijfers, trends en analyses* (Alphen and Rijn: Samsom), pp. 65–82.

Barnett, R.C. and Shen, Y.C., 1997, Gender, high and low schedule control, housework tasks, and psychological distress: A study of dual-earner couples. *Journal of Family Issues*, **18**, pp. 403–428.

Belcastro, P.A., 1982, Burnout and its relationship to teachers' somatic complaints and illness. *Psychological Reports*, **50**, pp. 1045–1046.

Belcastro, P.A. and Gold, R.S., 1983, Teacher stress and burnout: Implications for school health personnel. *Journal of School Health*, **53**, pp. 404–407.

Billings, A.G. and Moos, R.H., 1984, Coping, stress and social resources among adults with unipolar depression. *Journal of Personality and Social Psychology*, **46**, pp. 877–891.

Blasé, J.J., 1986, A qualitative analysis of sources of teacher stress: Consequences for performance. *American Educational Research Journal*, **23**, pp. 13–40.

Bolick, N. and O'Keefe, 1991, School budget blues. *American School Board Journal*, **178**, pp. 34–36.

Boyle, G.J., Borg, M.G., Falzon, J.M. and Baglioni, A.J., 1995, A structural model of the dimensions of teacher stress. *British Journal of Educational Psychology*, **65**, pp. 49–67.

Brown, M. and Ralph, S., 1992, Towards the identification of stress in teachers. *Research in Education*, **48**, pp. 103–110.

Burisch, M., 1993, In search of theory: Some ruminations on the nature and etiology of burnout. In W.B. Schaufeli, C. Maslach and T. Marek (eds), *Professional burnout: Recent developments in theory and research* (London: Taylor and Francis).

Burke, R.J. and Leiter, M.P., 1999, Contemporary organizational realities and professional efficacy: Downsizing, reorganization and transition. In T. Cox, P. Dewe and M.P. Leiter (eds), *Coping and health in* organizations (London: Taylor and Francis).

Burke, R.J. and Greenglass, E.R., 1989, Sex differences in psychological burnout in teachers. *Psychological Reports*, **65**, pp. 55–63.

Burke, R.J., Greenglass, E.R. and Konarski, R., 1995, Coping, work demands, and psychological burnout among teachers. *Journal of Health and Human Resources Administration*, Summer, pp. 90–103.

Burke, R.J., Shearer, J. and Deszca, E., 1984, Burnout among men and women in police work: An examination of the Cherniss model. *Journal of Health and Human Resources Administration*, **7**, pp. 162–188.

Buunk, B.P. and Schaufeli, W.B., 1993, Burnout: A perspective from social comparison theory. In W.B. Schaufeli, C. Maslach and T. Marek (eds), *Professional burnout: Recent developments in theory and research* (London: Taylor and Francis).

Buunk, B.P. and Schaufeli, W.B., 1999, Reciprocity in Interpersonal Relationships: An Evolutionary Perspective on its Importance for Health and Well-being. In W. Stroebe and M. Hewstone (eds), *Europen Review of Social Psychology*, Volume 10, pp. 260–291.

Byrne, B.M., 1994, Testing for the validity, replication, and invariance of causal structure across elementary, intermediate, and secondary teachers. *American Educational Research Journal*, **31**, pp. 645–673.

Caplan, R.D., Cobb, S., French, J.R.P., Jr, Van Harrison, R. and Pinneau, S.P., 1975, *Job demands and worker health: Main effects and occupational differences* (Washington, DC: U.S. Government Printing Office).

Chen, M. and Miller, G., 1997, *Teacher Stress: A Review of the International Literature*. http://orders.edrs.com/members/sp.cfm.

Cherniss, C., 1980, *Professional burnout in human service organizations* (New York: Pareger).

Cherniss, C., 1990, The human side of corporate competitiveness. In D.B. Fishman and C. Cherniss (eds), *The human side of corporate competitiveness* (Newbury Park, CA: Sage Publications).

Cohen, S. and Wills, T., 1985, Stress, social support and the buffering hypothesis. *Psychological Bulletin*, **98**, pp. 310–357.

Cordes, C.L. and Dougherty, T.W., 1993, A review and integration of research on job burnout. *Academy of Management Review*, **18**, pp. 621–656.

Eisenberger, R., Huntington, R., Hutchison, S. and Sowa, D., 1986, Perceived organizational support. *Journal of Applied Psychology*, **71**, pp. 500–507.

Eisenberger, R., Cummings, J., Armeli, S. and Lynch, P., 1997, Perceived organizational support, discretionary treatment and job satisfaction. *Journal of Applied Psychology*, **82**, pp. 812–820.

Eisenberger, R., Rhoades, L. and Cameron, J., 1999, Does pay for performance increase or decrease perceived self-determination and intrinsic motivation? *Journal of Personality and Social Psychology*, **77**, pp. 1026–1040.

Farber, B.A., 1982, *Stress and burnout: Implications for teacher motivation*. Paper presented at the annual meeting of the AERA, New York.

Fisher, D., Grove, K.A. and Sax, C., 2000, The resilience of changes promoting inclusiveness in an urban elementary school. *Elementary School Journal*, **100**, pp. 213–227.

Folkman, S. and Lazarus, R.S., 1980, An analysis of coping in a middle-aged community sample. *Journal of Health and Social Behaviour*, **21**, pp. 219–239.

Freudenberger, H.J., 1975, The staff burn-out syndrome in alternative institutions. *Psychotherapy: Theory, Research and Practice*, **12**, pp. 73–82.

Friedman, I.A., 1991, High- and low-burnout schools: School culture aspects of teacher burnout. *Journal of Educational Research*, **84**, pp. 325–333.

Friedman, I.A., 1993, Burnout in teachers: The concept and its unique core meaning. *Educational and Psychological Measurement*, **53**, pp. 1035–1044.

Friedman, I.A., 2000, Burnout in teachers: Shattered dreams of impeccable professional performance. *Journal of Clinical Psychology*, **56**, pp. 595–606.

Greenglass, E.R., 1988, Type A behaviour and coping strategies in female and male supervisors. *Applied Psychology: An International Review*, **37**, pp. 271–288.

Greenglass, E.R., 1991, Burnout and gender: Theoretical and organizational implications. *Canadian Psychology*, **32**, pp. 562–572.

Greenglass, E.R., 1993, The contribution of social support to coping strategies. *Applied Psychology: An International Review*, **42**, pp. 323–340.

Greenglass, E., 1998, Anger, hostility and coping. Paper presented at the 12th conference of the European Health Psychology Society, Vienna, August 31– September 2, 1998.

Greenglass, E.R. and Burke, R.J., 1988, Work and Family precursors of burnout in teachers: Sex differences. *Sex Roles*, **18**, pp. 215–229.

Greenglass, E.R. and Burke, R.J., 1990, Burnout over time. *Journal of Health and Human Resources Administration*, **13**, pp. 192–204.

Greenglass, E.R. and Burke, R.J., 2000, Hospital downsizing, individual resources, and occupational stressors in nurses. *Anxiety, Stress and Coping*, **13**, pp. 371–390.

Greenglass, E.R., Burke, R.J. and Konarski, R., 1997, The impact of social support on development of burnout in teachers: Examination of a model. *Work and Stress*, **11**, pp. 267–278.

Greenglass, E.R., Burke, R.J. and Ondrack, M., 1990, A gender-role perspective of coping and burnout. *Applied Psychology: An International Review*, **39**, pp. 5–27.

Greenglass, E.R., Fiksenbaum, L. and Burke, R.J., 1996, Components of social support, buffering effects and burnout: Implications for psychological functioning. *Anxiety, Stress, and Coping*, **9**, pp. 185–197.

Greenglass, E.R. and Julkunen, J., 1989, Construct validity and sex differences in Cook-Medley hostility. *Personality and Individual Differences*, **10**, pp. 209–218.

Greenglass, E.R. and Julkunen, J., 1991, Cook-Medley Hostility, anger, and the Type A behaviour pattern in Finland. *Psychological Reports*, **68**, pp. 1059–1066.

Greenglass, E.R., Pantony, K.L. and Burke, R.J., 1988, A gender-role perspective on role conflict, work stress and social support. *Journal of Social Behaviour and Personality*, **3**, pp. 317–328.

Griffith, J., Steptoe, A. and Cropley, M., 1999, An investigation of coping strategies associated with job stress in teachers. *British Journal of Educational Psychology*, **69**(4), pp. 517–531.

Haynes, S.B., Feinleib, M. and Kannel, W.B., 1980, The relationship of psychosocial factors to coronary heart disease in the Framingham Study. III. Eight-year incidence of coronary heart disease. *American Journal of Epidemiology*, **III**, pp. 37–58.

Himle, D.P., Jayaratne, S.D. and Chess, W.A., 1986, Gender differences in work stress among clinical social workers. *Journal of Social Service Research*, **10**, pp. 41–56.

Himle, D.P., Jayaratne, S. and Thyness, P., 1989, The effects of emotional support on burnout, work stress and mental health among Norwegian and American social workers. *Journal of Social Service Research*, **13**, pp. 27–45.

Hobfoll, S.E., Schwarzer, R. and Kym-Koo, C., 1996, Disentangling the stress labyrinth: Interpreting the meaning of the term stress as it is studied. Position paper presented at the First Meeting of the International Society of Health Psychology, Montreal.

Hodge, G.M., Jupp, J.J. and Taylor, A.J., 1994, Work stress, distress and burnout in music and mathematics teachers. *British Journal of Educational Psychology*, **64**, pp. 65–76.

Ito, M., 2000, Burnout among teachers: Teaching experience and type of teacher. *Japanese Journal of Educational Psychology*, **48**, pp. 12–20.

Jerusalem, M. and Schwarzer, R, 1992, Self-efficacy as a resource factor in stress appraisal processes. In R. Schwarzer (ed.), *Self-efficacy: Thought control of action* (Washington, DC, USA: Hemisphere Publishing Corp), pp. 195–213.

Lapp, J. and Attridge, M., 2000, Worksite interventions reduce stress among high school teachers and staff. *International Journal of Stress Management*, **7**, pp. 229–232.

LaRocco, J.M., House, J.S. and French, J.R.P., Jr, 1980, Social support, occupational stress, and health. *Journal of Health and Social Behaviour*, **21**, pp. 202–218.

Latack, J.C., 1986, Coping with job stress: Measures and future directions for scale development. *Journal of Applied Psychology*, **71**, pp. 377–385.

Lazarus, R.S. and Folkman, S., 1984, *Stress, appraisal and coping* (New York: Springer).

Lee, R.T. and Ashforth, B.E., 1990, On the meaning of Maslach's dimensions of burnout. *Journal of Applied Psychology*, **75**, pp. 743–747.

Leiter, M.P., 1988, Burnout as a function of communication patterns: A study of multidisciplinary mental health teams. *Group & Organizational Studies*, **13**, pp. 111–128.

Leiter, M.P., 1990, The impact of family resources, control coping, and skill utilization on the development of burnout: A longitudinal study. *Human Relations* **43**, pp. 1076–1083.

Leiter, M.P., 1991, Coping patterns as predictors of burnout: The function of control and escapist coping patterns. *Journal of Organizational Behaviour*, **12**, pp. 123–144.

Leiter, M.P., 1993, Burnout as a developmental process. In W.B. Schaufeli, C. Maslach and T. Marek (eds), *Professional burnout: Recent developments in theory and research*, (Washington, DC: Taylor and Francis).

Maslach, C. and Jackson, S.E., 1981, The measurement of experienced burnout. *Journal of Occupational Behaviour*, **2**, pp. 99–113.

Maslach, C. and Jackson, S.E., 1984, Patterns of burnout among a national sample of public contact workers. *Journal of Health and Human Resources Administration*, **7**, pp. 189–212.

Maslach, C. and Jackson, S.E., 1985, The role of sex and family variables in burnout. *Sex Roles*, **12**, pp. 837–851.

Maslach, C. and Jackson, S.E., 1986, *Maslach Burnout Inventory Manual* (2nd edn) (Palo Alto, CA: Consulting Psychologists Press).

McGrath, A., Houghton, D. and Reid, N., 1989, Occupational stress, and teachers in Northern Ireland. *Work and Stress*, **3**, pp. 359–368.

Mottaz, C.J., 1988, Determinants of organizationl commitment. *Human Relations*, **41**, pp. 467–482.

Oakley, W. and Stevens, K., 2000, A Lifelong Opportunity for Canadian Students. *Education Canada*, **40**, pp. 32–33.

Ogus, D., Greenglass, E.R. and Burke, R.J., 1990, Gender role differences, work stress and depersonalization. *Journal of Social Behaviour and Personality*, **5**, pp. 387–398.

O'Neill, C.P. and Zeichner, A., 1985, Working women: A study of relationships between stress, coping and health. *Journal of Psychosomatic Obstetrics and Gynaecology*, **4**, pp. 105–116.

Perlman, B. and Hartman, E.A., 1982, Burnout: Summary and research. *Human Relations*, **35**, pp. 283–305.

Pines, A., Aronson, E. and Kafry, D., 1981, *Burnout: From Tedium to Personal Growth* (New York: Free Press).

Russell, D.W., Altmeier, E. and Van Velzen, D., 1987, Job related stress, social support, and burnout among classroom teachers. *Journal of Applied Psychology*, **72**, pp. 269–274.

Schamer, L.A. and Jackson, M., 1996, Coping with Stress: Common Sense about Teacher Burnout. *Education Canada*, **36**, pp. 28–31.

Schaufeli, W.B. and Buunk, B.P., 1996, Professional burnout. In M.J. Schabracq, J.A.M. Winnubst and C.L. Cooper (eds), *Handbook of Work and Health Psychology* (New York: Wiley), pp. 311–346.

Schaufeli, W.B. and Enzmann, D., 1998, *The burnout companion to study and practice: A critical analysis* (London: Taylor and Francis).

Schaufeli, W.B. and Van Dierendonck, D., 1993, The construct validity of two burnout measures. *Journal of Organizational Behaviour*, **14**, pp. 631–647.

Schaufeli, W.B., Van Dierendonck, D. and Van Gorp, K., 1996, Burnout and reciprocity: Toward a dual-level social exchange model. *Work and Stress*, **10**, pp. 225–237.

Schonfeld, I.S., 2001, Stress in 1st-year women teachers: The context of social support and coping. *Genetic, Social, and General Psychology Monographs*, **127**, pp. 133–168.

Schwab, R.L. and Iwanicki, E.F., 1982, Who are our burned out teachers? *Educational Research Quarterly*, **1**, pp. 5–16.

Schwab, R.L., Jackson, S.E. and Schuler, R.S., 1986, Educator burnout: Sources and consequences. *Educational Research Quarterly*, **10**, pp. 15–30.

Smith, M. and Bourke, S., 1992, Teacher stress: Examining a model based on context, work load and satisfaction. *Teaching & Teacher Education*, **8**, pp. 31–46.

Smyth, J., 2000, Reclaiming Social capital through critical teaching. *Elementary School Journal*, **100**, pp. 491–511.

Solomon, K., 1982, The masculine gender role description. In K. Solomon and N. B. Levy (eds), *Men in translation: Theory and therapy* (New York Plenum), pp. 45–76.

Spector, P.E., 1986, Perceived control by employees: A meta-analysis of studies concerning autonomy and participation at work. *Human Relations*, **39**, pp. 1005–1016.

Steel, L., 2001, Staff support through supervision. *Emotional and Behavioural Difficulties*, **6**, pp. 91–101.

Taylor, H.E., 1998, How in the World Does One Teach Global Education? *Momentum*, **29**, pp. 16–18.

Travers, C.J. and Cooper, C.L., 1993, Mental health, job satisfaction and occupational stress among UK teachers. *Work & Stress*, **7**, pp. 203–219.

Van Dierendonck, D., Schaufeli, W.B. and Sixma, H.A., 1994, Burnout among general practitioners: A perspective from equity theory. *Journal of Social and Clinical Psychology*, **13**, pp. 86–100.

Van Ginkel, A.J.H., 1987, *Demotivation among teachers: A study into burnout and demotivation among secondary school teachers*. Lisse: Swets and Zeitlinger.

Van Poppel, J. and Kamphuis, P., 1992, Health, work and working conditions in schools: 1990–1991 (Tilburg: IVA).

Wilson, C.F., 1979, *Stress Profile for Teachers*. San Diego (CA: Department of Education, San Diego County).

Winnubst, J., 1993, Organizational structure, social support, and burnout. In W.B. Schaufeli, C. Maslach and T. Marek (eds), *Professional burnout: Recent developments in theory and research*, (Washington, D.C.: Taylor and Francis).

Winzelberg, A.J. and Luskin, F.M., 1999, The effect of a meditation training in stress levels in secondary school teachers. *Stress Medicine*, **15**, pp. 69–77.

CHAPTER NINE

Stress in University Academics

Anthony H. Winefield

9.1 REVIEW OF LITERATURE

Universities play a vital role in the economic and social life of all developed nations. They train the nation's scientists, engineers, lawyers, doctors and other professionals and produce much of its cutting-edge research. In order to fulfil this role successfully they need to attract and retain high quality staff and provide a supportive working environment. Their ability to do so has been threatened over the past decade by deteriorating working conditions resulting from cuts to their operating grants. In Australia, for example, the average student to staff ratio increased steadily from around 13:1 in 1990 to around 19:1 in 2000 (Senate Committee Report, 2001).

University teaching has traditionally been regarded as a low stress occupation. Although not highly paid, academics have been envied because they enjoyed tenure, light workloads, flexibility, 'perks' such as overseas trips for study and/or conference purposes, and the freedom to pursue their own research interests.

During the past fifteen years most of these advantages have been eroded in many countries. Academic salaries have fallen in real terms in countries such as US, UK, Australia and New Zealand. Increasing numbers of academic positions are now untenured, workloads have increased, and academics are under increased pressures to attract external funds for their research and to 'publish or perish'. Universities and academic departments are being subjected to external 'quality' audits which scrutinise their research output in terms of quantity and quality as well as teaching. Future funding support is determined by the outcomes of such audits. As Shirley Fisher (1994) says in relation to British universities in her recent book *Stress in Academic Life*:

> The demands on academics have risen rapidly over the last ten years...there has been a steady erosion of job control. All the signs are that this will continue (p. 61).

Several UK studies have supported Fisher's contention, including those by Hind and Doyle (1996) and Daniels and Guppy (1994).

It is now well recognised that workplace stress in Universities world-wide is increasing and has a multitude of detrimental effects on individuals and organisations. Clearly, given the role that universities play in education and training, it is important that universities are able to obtain creative solutions from a well-educated workforce. One way in which this may be done is through assisting staff to overcome their stress.

Research from US indicates that the phenomenon of academic stress is alarmingly widespread. In a survey of almost 2000 faculty members, Melendez and de Guzman (1983) found that 62% acknowledged severe or moderate job stress. In his review of the literature, Seldin (1987) states that the academic environment of the 1980s has imposed surprisingly high levels of job stress on academics, and that the level of stress will continue to increase in future decades.

The impact of faculty stress is less well documented. In their study on faculty stress in US, Shuster and Bowen (1985) characterised faculty morale as 'very poor' at a quarter of the campuses they researched. They further reported that many of the senior faculty members they interviewed were angry, embittered and felt devalued and abandoned. High levels of faculty stress have also been associated with high academic turnover.

In the early 1990s, the Carnegie Foundation for the Advancement of Teaching sponsored an international survey of the academic profession in which 14 countries participated (Australia, Brazil, Chile, England, Germany, Hong Kong, Israel, Japan, Korea, Mexico, The Netherlands, Russia, Sweden, US). The data were collected from 1991–1993 (Altbach, 1996). According to Altbach:

> For a number of years, the professoriate has been undergoing change and has been under strain almost everywhere. Fiscal problems for higher education are now evident in all of these fourteen countries . . . In most of the nations, the somewhat unprecedented phenomenon of increasing enrollments has been allowed to supersede allocated resources . . . At the same time, professors in a number of countries are being asked to be more entrepreneurial – for example, in bringing research grants and contracts to their institutions (pp. 4–5).

Somewhat surprisingly, despite widespread complaints about their working conditions, most of the respondents said that their overall morale was high because of the intellectual pleasure provided by their work with 63% (England) to 85% (Israel) disagreeing with the proposition 'If I had it to do over again, I would not become an academic'. A major source of dissatisfaction was institutional leadership: 'An unusually large number express dissatisfaction with and doubts about the quality of the leadership provided by top-level administrators at their colleges and universities.' (Altbach, pp. 28–29). On the other hand, Armour (1987) found that a high percentage of academics planned to leave academia, which they attributed to the high levels of stress encountered in the profession. The research literature further indicates that faculty stress significantly affects the quality of both teaching and research. Research has highlighted that some of the effects of faculty stress, such as detachment, low job satisfaction and low job commitment can be contagious for students and colleagues (Armour, 1987). It is apparent that the consequences of academic stress may be far more wide ranging than the occasional stress illness.

In contrast to the volume of research conducted in America, there has been very little research on the job-related stress experienced by academic staff in Australian Universities. In the late 1980s, Australian universities underwent major restructuring in the late 1980s similar to what happened to universities in England in the early 1990s. In both countries, the binary system was abolished, so that former Teachers Colleges and Polytechnics/Institutes of Technology were granted university status. These changes took somewhat different forms. In Australia, for example, but not in England, tertiary institutions were encouraged to merge, which resulted in major disruptions and numerous multi-campus universities. In England, many of the new universities lacked a research tradition and were seen as inferior to the traditional universities. In both countries, the restructuring was inevitably disruptive and augmented the ongoing problems associated with reduced funding.

In a recent study, assessing the level of stress of both academic and general staff at an Australian University, Jarrett and Winefield (1995) surveyed all staff at the University of Adelaide. The survey attracted more than 2000 replies which represented an overall response rate of 72% of non-casual staff (77% for general staff and 65% for academic staff). The overall level of psychological distress was very high, particularly among academic staff, even though their overall level of job satisfaction was moderately high. Indeed, dissatisfaction was reported with only 2 (out of 15) aspects of work: 'Your chance of promotion/reclassification' and 'The way the University is managed.' Results of the study were published recently by Winefield and Jarrett (2001).

Similar studies have been reported at Monash (Sharpley *et al.*, 1995) and the University of New England (Dua, 1994). At Monash, stress was perceived as a major problem for about 25% of staff with lack of feedback on performance, lack of promotion opportunities, worries about amalgamations and lack of equipment and/or infrastructure support identified as frequent sources of stress. At New England, Dua found: (a) staff at more junior levels reported more stress than those at more senior levels; (b) stress was associated with poorer self-reported health; (c) staff who perceived high levels of control over their work environment experienced less stress than those who perceived low levels of control.

Finally, several key researchers have suggested that while stress is an inevitable part of academia, universities must bear the responsibility for assisting employees to manage job-related stress (Seldin, 1987). Certainly with the increasing frequency of stress-related claims and the resultant costs (Armour *et al.*, 1987), it is within a university's best interest to be proactive rather than reactive in managing faculty stress.

Recent research has highlighted the need to include both positive and negative work-related events in our understanding of the stress coping process

(Wearing and Hart, 1996). However few theoretical models of stress and well being have attempted to incorporate both these factors. Indeed, there is currently little research demonstrating the theoretical relationship between cognitive appraisal of positive and negative work events and other well known moderators of stress, namely personality, coping resources, and coping strategies.

Surprisingly, even less theoretical development has focused on the direct effect stressors and uplifts have on meaningful organisational (productivity, absenteeism, turnover, morale, commitment) and individual (psychological well being, physical health) outcome variables.

To assist organisations with assessing and managing workplace stress, researchers have devised a number of general stress scales (e.g. Holmes and Rahe, 1967; Kanner *et al.*, 1981; Nowack, 1990) and occupational stress scales (e.g. Cooper *et al.*, 1988). In recent years researchers have argued that occupation-specific and industry-specific stress scales are more reliable and valid predictors of stress and effects of stress than general occupational stress scales. As a result of this argument, a great deal of recent stress research on various occupational groups (e.g. teachers, nurses, and police officers) has used occupational stress scales specifically designed for these groups. Dua (1994), for example has used scales specifically designed for university staff.

Though stressors in the workplace cause stress which, in turn, leads to strains (negative effects on health and quality of work), researchers have identified a number of moderating variables which can reduce the experience of stress or reduce the negative effects of stress. Some examples of these variables are coping strategies, social support, and hardiness. The last of these variables comprises control, commitment, and challenge. Research has shown that social support, positive coping, problem-focused coping, and hardiness reduce the level of stress and the impact of stress on health, and negative coping and type A behaviours increase the level of stress and its impact on health (e.g. Bernard and Krupat, 1994; Brannon and Feist, 1992; Sharpley *et al.*, 1995). It is also possible that personality dimensions such as introversion-extraversion and neuroticism or negative affect (Moyle, 1996) may also act as a mediator/moderator of stress and stress–health relationship.

Given the role of universities in the social and economic well being of Australia, it is vital that there be a comprehensive study of occupational stress in university staff in Australian universities. Some recent studies from Australian and overseas have assessed the level of stress experienced by academic staff (Blix *et al.*, 1994; Gmelch *et al.*, 1986; Hind and Doyle, 1996; Stough and Gillespie, 1996; Taris *et al.*, 2001; Wilkinson and Joseph, 1995) while others have focused on administrative staff (e.g. Blix *et al.*, 1991). Although a few have sampled both types of employees (Boyd and Wylie, 1994; Daniels and Guppy, 1992; Jarrett and Winefield, 1995; Sharpley, 1994) none have assessed a large representative sample of both academic and general staff across many universities and none has used a longitudinal research design. The ongoing study by Winefield *et al.* (2001) is currently attempting to do this.

9.2 THE SOURCES OF WORK STRESS

Fisher (1994) has drawn attention to the problems facing universities in Britain in recent years and reports results from studies conducted over the period 1988–1993 showing high levels of stress experienced by academic staff following a decade of reductions in government funding. In Australia a similar decline in government funding has occurred, particularly between 1994 and the present. The current situation in Australia has been documented in a recently released Senate Committee Report *Universities in Crisis* (Senate Committee, 2001). Government statistics show that, despite increases in student enrolments, the Commonwealth government's contribution to University operating grants has declined from \$Aus4772 m in 1994 to \$Aus4461 m in 2000. Moreover, the student to staff ratio has gradually increased from 12.9 in 1990 to 18.8 in 2000.

The situation has recently been portrayed in a documentary film *Facing the Music* produced by the Australian Broadcasting Commission about the Music Department at the University of Sydney (Australian Broadcasting Commission, 2001) and shown in November on national television. According to the publicity:

> It's budget time at the University, and Professor Anne Boyd is in a panic. As gifted young students pursue the mysteries of musical creation, Boyd is fighting to preserve basic standards after nearly a decade of relentless funding cuts. But Boyd is an innocent when it comes to harsh economic realities and the very qualities that are her strengths as a composer and teacher – her passion, energy and emotion – leave her vulnerable. Forced to drop staff and courses and pick up the phone to plead for private sponsorship, the usually conservative Boyd is forced to fight for what she believes in. The problem is, how long can she continue?
>
> Facing the Music documents Anne Boyd's roller-coaster journey as she struggles to negotiate the most tumultuous year of her life. It's a compelling story of one person's choices and an eye-opening look at a pressing issue – the perilous state of higher education.

Reduced funding has also led most Australian universities effectively to abolish academic tenure since the mid-nineties. In many western countries academic freedom has been highly valued because the role (and responsibility) of the academic has been seen as the fearless pursuit and dissemination of knowledge and, where appropriate, as acting as social critic. Tenure has been regarded as the only guarantee of academic freedom. Consequently, although academic work has not been highly paid, academics have traditionally enjoyed high levels of autonomy, freedom to publish and to speak openly, even when their views are unpopular with authority, whether it be the university administration, the scientific establishment, or the government.

Critics of tenure have pointed out that it protects the lazy, incompetent and unproductive and denies opportunities to talented young scholars. During the past four or five years, in response to increasing financial pressures, many Australian universities have abandoned tenure (Coady, 2000; Molony, 2000). So-called tenured staff can (and have been) made 'involuntarily redundant' and there has been an increase in contract (as opposed to tenure track) appointments.

Gillespie *et al.* (2001) interviewed 178 academic and general staff in 22 focus groups at 15 Australian universities. They found that both academic and general staff reported a dramatic increase in stress over the past five years and academic staff reported higher levels of stress than general staff. The five most frequently cited sources of stress were insufficient funding and resources, work overload, poor management practice, job insecurity, and insufficient recognition and reward.

The higher levels of stress reported by academic staff were confirmed in a follow-up survey of 8732 staff at 17 Australian universities by Winefield *et al.* (2001). We found that psychological strain was higher, and job satisfaction lower, in academic staff than in general staff. We also found that academic staff in the older universities experienced less strain and greater job satisfaction than academic staff in the newer (post-World War II) universities. By contrast, strain and job satisfaction among general staff were less influenced by university age.

Higher stress levels among academic staff than general staff were also reported by Winefield and Jarrett (2001) in their study of staff at the University of Adelaide conducted in 1994, described in Section 4.6.2.

9.3 EFFECTS OF WORK STRESS IN THE PROFESSION

What are the consequences of the apparent widespread increase in academic stress? The impact of job-related stress on general employee performance (not specific to academia) has been well documented. Job-related stress has been found to increase: turnover of staff, absenteeism, non-productivity, inefficiency, frustration, tiredness and burnout (Melendez and de Guzman, 1983; Osipow and Spokane, 1991).

Employees experiencing high levels of job-related stress also report lower levels of job satisfaction, morale and general well-being, which in turn may negatively affect work performance (Nowack, 1989; Terry *et al.*, 1995). Thus workplace stress takes its toll upon the health of the organisation and the health of the workers (Matteson and Ivancevich, 1987).

Stress is often accompanied by negative feelings, anxiety, depression, sadness, hopelessness, helplessness, anger, and/or a sense of worthlessness. Stressed persons are more likely to be psychologically distressed (Dua, 1990; Dua, 1994; Dua and Price, 1992; Nowack, 1990) and stress has been linked to a number of physical illnesses, for example, heart disease and cancer (Hamburg *et al.*, 1985) gastrointestinal disease (Winter, 1983), and lowered immune status (Marriott *et al.*, 1994).

9.4   INTERVENTIONS TRIED OR KNOWN AND OUTCOMES

Organisational stress researchers generally distinguish three levels of intervention: primary, secondary and tertiary (Murphy, 1988). Primary and secondary interventions are generally seen as preventative, whereas tertiary interventions involve taking steps to help the individual who has experienced an adverse reaction to workplace stressors, either psychological or physical or both.

Primary interventions involve attempts to reorganise or restructure the work environment so as to make it less stressful. For example, they could involve improving the physical environment through introducing more efficient heating/cooling systems, or increasing space in order to reduce crowding, or purchasing new equipment, or more comfortable chairs, etc. Other kinds of primary intervention involve changes to work organisation aimed to reduce known sources of stress, such as role overload, role conflict and role ambiguity. They may be aimed at specific, identified sources of stress, arising from complaints by workers, or they may be theoretically inspired. For example, Karasek's theory assumes that stress is caused by a combination of high demand and low autonomy so a primary intervention might aim to reduce demand or increase autonomy.

Secondary interventions, on the other hand, are oriented to individual workers and designed to help them cope with potential stressors in their work environment. For example, time management and relaxation are widely used stress management techniques that can be taught to workers so as to help them combat a stressful work environment.

Other sorts of secondary interventions may be theoretically based. For example, the Person-Environment Fit theory (French *et al.*, 1984), assumes that job stress can be a consequence of two kinds of mismatch: a mismatch between the requirements of the job and the ability of the worker to meet those requirements; and a mismatch between the worker's expectation of what the job involves and what it actually involves.

The first kind of mismatch could be addressed by helping the worker to acquire or develop relevant skills. The second kind of mismatch could be more difficult to address and may require the worker to find different work.

Tertiary intervention is aimed at helping workers who are suffering as a consequence of work pressures, which may lead to other problems, such alcohol addiction, family problems etc. Employee Assistance Programmes (EAPs) were first introduced in US around 1970 largely to counsel employees suffering from alcoholism but more in recent years they have developed a wider orientation.

9.5   THEORETICAL CRITIQUE OF WORK STRESS RESEARCH IN THE PROFESSION

As with much of the reported research on occupational stress, the work stress research carried out on academics has relied largely on cross-sectional analyses of self-report data. As many critics have observed, such data are often difficult to interpret because they are vulnerable to the problem of method variance (leading to spurious correlations) and do not readily permit causal inference.

Improvements to such research designs include the collection of longitudinal data and the use of 'objective' data, in addition to 'subjective' self-report data. It is important to recognise however, that self-report data are often an invaluable source of information and in some situations, the only source of information. Critics of self-report data tend to overlook the fact that they are used by audiologists and optometrists for prescribing hearing and visual aids. Also, soft tissue injuries where there is no internal bleeding are not amenable to objective observation and may rely on self-report for their diagnoses.

In behavioural research, the most sensible approach to self-reporting is to assume that it is likely to be valid, unless the person has a good reason to mislead, or is distracted, or where there is reason to suspect that judgment may be impaired (say, through fatigue, drug or alcohol use). On the other hand self-report measures taken on a single occasion may be biased because of a temporary mood state leading to spurious correlations as referred to above. As Frese and Zapf (1988) have pointed out '... the use of subjective judgments of stressors can lead to an overestimate of the correlation between stressors and dysfunctioning... On the other hand, the use of objective (observers') judgments of stressors leads to an underestimate of the 'true' correlation' (p. 381).

Because of this, it is desirable to utilise data from a variety of sources. In organisational stress research it is often possible to assess the levels of strain experienced by workers by measuring absenteeism, turnover, sick leave and stress claims. These measures can be used to verify self-report measures indicating low levels of job satisfaction, strain etc. Similarly, the pressures within the work environment may also be assessed independently. In relation to academic stress, for example, the work load for academics involved in teaching can be estimated from the student to staff ratio. In general, most commentators agree that stress researchers should use multiple measures, including both 'objective' and 'subjective' measures and longitudinal designs in order to tease out the often complex relations between organisational stressors and their impact on individual workers.

9.6   INTRODUCTION TO DATA

The data reported here come from surveys conducted at an established Australian metropolitan university (The University of Adelaide) in 1994 and in 2000. The 1994 survey was conducted by Jarrett and Winefield (1995) as part of an organisational climate survey commissioned by the Vice Chancellor. The results are described in Winefield and Jarrett (2001).

The 2000 survey was conducted as part of a national survey of 17 Australian universities (Gillespie *et al.*, 2001; Gillespie *et al.*, 2001; Winefield *et al.*, 2001). The national survey was supported by a collaborative grant from the Australian Research Council, with the National Tertiary Education Union as the industry partner and cash contributions from Vice Chancellors of 17 participating universities. The support from the Vice Chancellors was obtained in response to a letter from the Vice Chancellor of the University of Adelaide, inviting them to join her in supporting the project.

The first survey was not intended to include longitudinal observations, therefore no attempt was made to identify responses. The second survey was part of a longitudinal study, therefore, although responses were anonymous, all participants were asked to provide a code identifier so that the researchers could match responses from Time 1 (2000) to Time 2 (2002) (Data from Time 2 had not been collected at the time of writing this chapter.).

9.6.1 Method

Participants

All participants were staff members from the University of Adelaide. The 1994 sample comprised 2040 employees, including casual staff. This represented an overall response rate of 57.1%, however this figure was depressed by the very low response rate among the casual staff, many of whom were either difficult to contact or, because they were students, no doubt saw the questionnaire as irrelevant. Among the non-casual staff, the overall response rate was 72.2%. The response rate was 77.2% among general staff and 65.4% among academic staff (faculty). Ages ranged from 17 to 69 and 51% of the respondents were men and 49% were women.

Seven categories of staff were distinguished, three categories of academic staff and four categories of general staff, ranging from full-time to casual (and including various categories of part-time staff). The highest response rates were for the full-time, non-casual general staff from whom 916 (out of 1186 sent out) survey forms were received (77.2%) and for the full-time, non-casual academic staff from whom 602 (out of 920 sent out) survey forms were received (65.4%). The lowest response rate was for the casual general staff from whom 216 (out of 725 sent out) survey forms were received (29.8%). Overall, 2040 (out of 3570) survey forms were received, yielding a rate of 57.1%. Because of the low response rate from casual staff, their results were excluded. Also, because of missing data the numbers reported in the following Tables were further reduced.

In 2000, only non-casual staff were surveyed. Of 2300 survey forms distributed, 661 were returned, yielding a response rate of 28.7%. Ages ranged from 18 to 67 and 51% of the respondents were men and 49% were women.

Because of missing data the numbers reported in the following Tables were usually less than this.

Measures

The outcome measures to be reported are a Job Satisfaction Scale (Warr *et al.*, 1979), a widely used measure of Strain (or Psychological Distress), the 12-item version of the General Health Questionnaire (Goldberg and Williams, 1988) – the GHQ-12, and a measure of 'negative affectivity'. The GHQ-12 has been recommended by Banks *et al.* (1980) as a valid indicator of mental ill-health (or psychological distress) in occupational studies.

Payne (1988) has argued that all studies of occupational stress should include a measure of 'negative affectivity' such as trait anxiety (or trait neuroticism). In the 1994 survey, the measure of negative affectivity used a short (10-item) version of the Spielberger trait anxiety scale (Spielberger *et al.*, 1983). In the 2000 survey, negative affectivity was measured using the 12-item neuroticism scale from the NEO ('big five') personality inventory (Costa, 1996).

The Job Satisfaction scale comprises 15 items exploring different facets of job satisfaction, as well as a 16th global item (not forming part of the scale) (All of the scales possessed satisfactory internal reliability with alpha coefficients ranging from 0.82 to 0.91.).

Procedure

At both times, the questionnaires were distributed to all staff through the university's internal mail system and were returned to an internal mail box. Strict anonymity was preserved.

9.6.2   Results

In reporting the results, groups with fewer than 20 respondents were arbitrarily removed. In 2000, the number of respondents from Architecture and Law were both well below 20, and there were fewer than 10 in the Academic – Teaching only staff and General service staff categories. A significance level of .05 was employed throughout. Unrelated samples *t*-tests were used to determine the significance of the change (All statistically significant differences are marked with a single asterisk *.).

(a) Job Satisfaction

The results for Job Satisfaction are shown in Table 9.1.

As Table 9.1 shows, there was a statistically significant decline in Job Satisfaction from 1994 to 2000 and this decline was evident on 14 of the 15 items on the scale. The only exception was item 3 'fellow workers'. The items showing the biggest declines were item 9 ('industrial relations'), item 13 ('hours of work') and item 11 ('the way the University is managed').

Table 9.2 shows mean job satisfaction scores for 7 different job classifications. All showed a significant decline in job satisfaction.

Table 9.1 Means the Responses to Individual Items on Job Satisfaction Scale[1]

Item		1994	2000	Diff
1.	Physical work conditions	4.8	4.5	−0.3*
2.	Freedom to choose method of working	5.6	5.3	−0.3*
3.	Fellow workers	5.4	5.3	−0.1ns
4.	Recognition for good work	4.5	4.2	−0.3*
5.	Immediate boss	5.2	5.0	−0.2*
6.	Amount of responsibility given	5.4	5.0	−0.4*
7.	Rate of pay	4.3	3.8	−0.5*
8.	Opportunity to use abilities	5.0	4.6	−0.4*
9.	Industrial relations	4.3	3.1	−1.2*
10.	Chance of promotion	3.4	3.1	−0.3*
11.	University management	3.4	2.7	−0.7*
12.	Attention to your suggestions	4.3	4.0	−0.3*
13.	Hours of work	5.1	4.2	−0.9*
14.	Amount of variety in job	5.5	5.1	−0.4*
15.	Job security	4.5	4.1	−0.4*
Total		70.7	64.3	−6.7*
**16.	Overall job satisfaction	5.0	4.5	−0.5*

[1] Scores ranged from 1 (=extremely dissatisfied) to 7 (=extremely satisfied)
* $p < .05$ ** This item does not form part of the scale

Table 9.2 All staff: Mean scores on Job satisfaction scale by job classification

Classification	$n1$	$n2$	Job Sat 1994	Job Sat 2000	Difference
Head	64	29	73.0	64.5	−8.5*
Acad T&R	445	198	68.0	62.3	−5.7*
Acad R only	109	50	71.1	66.6	−4.5*
Gen Cler/Ad	479	152	72.3	64.8	−7.5*
Gen Prof	208	68	70.7	67.8	−2.9
Gen Tech	317	87	73.4	65.5	−7.9*
Total			70.8	64.3	−6.5*

Table 9.3 shows mean job satisfaction scores for each of seven academic areas. There were significant declines in job satisfaction for all of the areas except for Engineering.

Tables 9.4 and 9.5 show mean job satisfaction for different levels of seniority for academic and general staff respectively. For the academic staff, only the most junior ranks (Levels A/Post Doctoral Fellows) did not show a significant decline.

For the general staff, on the other hand, all categories showed a significant decline. Overall, job satisfaction was somewhat higher for general staff at both times.

Table 9.3 All staff: Means scores on Job Satisfaction in 1994 and 2000 by academic area

Area	$n1$	$n2$	Job Sat 1994	Job Sat 2000	Difference
Ag Science	241	78	71.7	66.0	−5.7*
Hum/Soc Sci	199	42	66.9	59.9	−7.0*
Economics	72	21	72.5	65.6	−6.9*
Engineering	117	33	70.5	70.0	−0.5
Health Sci	310	122	73.6	66.0	−7.6*
Math Sci	84	26	70.2	60.8	−9.4*
Science	304	73	70.6	62.0	−8.6*
Total			70.8	64.3	−7.6*

Table 9.4 Academic staff: Mean scores on Job Satisfaction by seniority

Position	$n1$	$n2$	Job Sat 1994	Job Sat 2000	Difference
Levels D/E	127	73	70.7	61.9	−8.8*
Levels B/C	315	166	68.6	62.8	−5.8*
Level A/PostDoc	172	54	67.5	66.0	−1.5
Total			68.9	63.2	−5.7*

Table 9.5 General staff: Mean scores on Job Satisfaction by seniority

HEO Level	$n1$	$n2$	Job Sat 1994	Job Sat 2000	Difference
HEO 8 or 9	73	53	72.8	66.8	−6.0*
HEO 6 or 7	205	84	73.1	64.3	−8.8*
HEO 4 or 5	357	101	72.3	66.1	−6.2*
HEO 1, 2 or 3	399	70	70.1	64.3	−5.8*
Total			71.6	65.2	−6.4*

Table 9.6 shows that the decline in job satisfaction was similar for men and women.

Table 9.6 Mean scores on Job Satisfaction for men and women

Sex	$n1$	$n2$	Job Sat 1994	Job Sat 2000	Difference
Men	1006	310	70.7	63.1	−7.6*
Women	985	304	71.0	65.3	−5.7*

(b) Psychological Strain (GHQ)

Table 9.7 shows the mean psychological strain (GHQ) scores for different staff classifications. Although there was an overall increase in psychological strain, the increase was greater for the general staff than for the academic staff. Indeed, whereas in 1994, strain levels were significantly higher in the academic staff than in the general staff, t (1518) = 6.13, $p < .05$, in 2000 the situation was reversed.

The General Clerical and Administrative classification and the General Technical classifications showed the biggest increase. Indeed, whereas these two classifications were both below the overall mean in 1994, in 2000 they were the two highest.

Table 9.7 All staff: Mean scores on Psychological Strain (GHQ) by job classification

Classification	$n1$	$n2$	GHQ 1994	GHQ 2000	Difference
Head	64	29	13.8	12.8	−1.0
Acad T&R	445	198	13.9	13.6	−0.3
Acad R only	109	50	12.6	13.6	1.6
Gen Cler/Ad	479	152	11.6	13.9	2.3*
Gen Prof	208	68	12.1	11.5	−0.6
Gen Tech	317	87	11.8	14.1	2.3*
Total			12.2	13.6	1.4*

Table 9.8 shows the mean psychological strain (GHQ) scores in the nine different academic areas. In 1994, the Humanities/Social Sciences were clearly the highest, in 2000 this area was ranked third highest, and only marginally higher than Science. Only one area (Engineering) actually showed a decrease.

Table 9.8 All staff: Means scores on Psychological Strain (GHQ) in 1994 and 2000 by academic area

Area	$n1$	$n2$	GHQ 1994	GHQ 2000	Difference
Ag Science	241	78	12.6	13.5	0.9
Hum/Soc Sci	199	42	13.5	14.6	1.1
Economics	72	21	11.8	14.2	2.4
Engineering	117	33	12.2	10.7	−1.5
Health Sci	310	122	11.8	13.1	1.3*
Math Sci	84	26	12.1	14.0	1.9
Science	304	73	12.2	14.5	2.3*
Total			12.2	13.6	1.4*

Interestingly, as shown in Table 9.9, there was no overall increase in strain among the academic staff. On the other hand, as shown in Table 9.10, there was a marked increase in strain among the general staff, particularly in the junior levels.

Table 9.11 shows the percentages of possible 'cases' and possible 'severe cases' in 1994 and 2000 compared with national data published by Andrews *et al.* (1999). As can be seen, the percentage of possible 'cases' in 1994 was more than double the national average and by 2000 the percentage had increased by a further 20%, exceeding 50% altogether.

Table 9.9 Academic staff: Mean scores on Psychological Strain (GHQ) by seniority

Position	$n1$	$n2$	GHQ 1994	GHQ 2000	Difference
Levels D/E	127	73	13.0	13.4	0.4
Levels B/C	315	166	14.1	13.5	−0.6
Level A/PostDoc	172	54	12.5	13.7	1.2
Total			13.3	13.4	0.1

Table 9.10 General staff: Mean scores on Psychological Strain (GHQ) by seniority

HEO Level	$n1$	$n2$	GHQ 1994	GHQ 2000	Difference
HEO 8 or 9	73	53	12.3	14.2	1.9
HEO 6 or 7	205	84	11.9	12.9	1.0
HEO 4 or 5	357	101	12.3	14.1	1.8
HEO 1, 2 or 3	399	70	11.2	15.5	4.3*
Total			11.8	14.2	2.4

Table 9.11 Percentages of possible 'cases' (= 2) and possible 'severe cases' (= 4) in some Australian studies using the GHQ-12 with binary scoring

Source	Sample	N	% 0 or 1	% ≥ 2	% ≥ 4
Andrews *et al.* (1999)	Australian sample	10600	80.8	19.2	10.4
Adelaide (1994)	University staff	1917	56.3	43.7	27.6
Adelaide (2000)	University staff	661	47.1	52.9	34.9

(c) Negative Affectivity

Different measures of negative affectivity were used in 1994 and in 2000, therefore it is not possible to examine differences. On the other hand, negative affectivity is assumed to be a stable personality trait, whether measured as trait anxiety or (trait neuroticism).

Interestingly, there were no sex differences in either 1994 or in 2000, as shown on Table 9.12.

Table 9.12 Mean scores on Strain (GHQ) for men and women

Sex	$n1$	$n2$	GHQ 1994	GHQ 2000	Difference
Men	1006	310	12.1	13.5	1.4
Women	985	304	12.2	13.4	1.2

Table 9.13 shows the mean negative affectivity scores for 1994 (trait anxiety) and for 2000 (neuroticism) for the different staff categories and Table 9.14 shows similar data for the 10 different academic areas.

Table 9.13 All staff: Mean scores on Negative Affectivity by job classification

Classification	n1	n2	Trait anxiety 1994	Neuroticism 2000
Head	64	29	16.5	28.0
Acad T&R	445	198	17.5	30.2
Acad R only	109	50	18.4	32.1
Gen Cler/Ad	479	152	17.0	33.1
Gen Prof	208	68	17.0	29.1
Gen Tech	317	87	16.9	32.5
Total			17.5	31.2

Table 9.14 All staff: Means scores on Negative Affectivity in 1994 and 2000 by academic area

Area	n1	n2	Trait Anxiety 1994	Neuroticism 2000
Ag Science	241	77	17.3	31.6
Hum/Soc Sci	199	41	17.5	31.7
Economics	72	20	16.7	28.8
Engineering	117	33	16.4	26.2
Health Sci	310	122	16.8	31.2
Math Sci	84	24	18.3	30.6
Medicine	221	122	16.7	31.2
Science	304	71	17.6	30.6
Total			17.2	31.2

Finally, as Table 9.15 shows, the women showed greater negative affectivity than the men at both times.

Table 9.15 Mean scores on Negative Affectivity for men and women

Sex	n1	n2	Trait anxiety 1994	Neuroticism 2000
Men	1006	310	16.8	29.7
Women	985	304	17.5	32.7

9.6.3 Discussion

First, it needs to be pointed out that the response rate among non-casual staff was very much higher in 1994 than it was in 2000, even though the survey was supported at both times by both the University administration and by the Union. Perhaps the University was more willing to remind staff of the importance of completing the survey form in 1994 than it was in 2000 because the 1994 survey had been commissioned by the Vice Chancellor whereas the 2000 survey was part of a larger national study. Also, the 1994 survey form was considerably shorter than the 2000 survey form, which might also explain the higher response rate.

The results show a marked overall decline in job satisfaction from 1994 to 2000. This, in itself, is not surprising given the increasingly difficult financial situation experienced by all Australian universities during that period. The decline in job satisfaction was evident in both academic and general staff, although it was least evident in the most junior academic ranks.

Similarly, it was evident across all academic areas except for Engineering. The overall level of job satisfaction remained moderate to high for most facets of the job (11 out of 15), the exceptions being 'The way the University is managed', 'Industrial relations', 'Your chance of promotion' and 'Your rate of pay'. By comparison, in 1994 there was general dissatisfaction with only two: 'The way the University is managed' and 'Your chance of promotion'. The facets of the job showing the largest declines in satisfaction were (in order, as shown in Table 9.1): 'Industrial relations', 'Your hours of work', and 'The way the University is managed'.

These findings are consistent with those from an ongoing national study of stress in Australian university staff which is based on 17 Australian universities (Gillespie *et al.*, 2001; Winefield *et al.*, 2001).

Interestingly, the difference between academic and general staff at Adelaide did not correspond with the national university difference. At Adelaide, the average job satisfaction score for the academic staff (63.2) was marginally higher than that for academic staff at all 17 universities (62.7), whereas for the general staff the average at Adelaide (65.2) was below the average at all 17 universities (67.6).

A recently published meta-analytic study by Judge *et al.* (2001) based on 312 samples with a combined N of 54,417 found that the 'true' correlation between job satisfaction and job performance was estimated to be .30 (accounting for 9% of the variance) which is far higher than that of .17 (accounting for 2.9% of the variance) reported in the well-known meta-analytic study by Iaffaldano and Muchinsky (1985). They also reported that the correlation for high complexity jobs was even higher, .52 (accounting for 27% of the variance).

The authors criticise the earlier study on four grounds: (1) it excluded unpublished studies; (2) it included multiple correlations from single studies; (3) it used an inappropriate method for correcting for unreliability in ratings of job performance; and (4) it inappropriately combined specific facets of job satisfaction in estimating the average correlation between job satisfaction and job performance. Although the study by Judge *et al.* (2001) shed little light on nature of the relationship between job satisfaction and job performance (the authors discuss seven possible models), the size of the correlation which they report is likely to stimulate further research on the topic. After all, if job satisfaction accounts for 9% of the variance in job performance (and 27% of the variance in high complexity jobs), it is clearly going to be taken more seriously than if it only accounts for less than 3%.

This diverging pattern was far more pronounced in relation to the other main outcome measure, psychological strain. At Adelaide, the average GHQ score for the academic staff (13.4) was slightly lower than that for academic staff at all 17 universities (13.7), whereas for the general staff the average at Adelaide (13.7) was considerably higher than the average at all 17 universities (12.8). Moreover, Adelaide was one of the only two universities at which the average strain (GHQ) score was higher in the general than in the academic staff. A possible explanation for the slightly lower levels of strain among academic staff than among general staff at Adelaide is that the student: staff ratio was lower (i.e. more favourable) than at other Australian universities.

The overall decline in job satisfaction since 1994 is no doubt partly attributable to local difficulties affecting the University in recent years. For example, in August 2001 the Vice Chancellor who had been appointed in 1996, resigned before completing her initial term of office and was replaced by an acting Vice Chancellor who had recently retired from another Australian university.

The industrial problems experienced by the University are reflected in the detailed responses to the Job Satisfaction questionnaire shown in Table 9.1. In 2000, the facet of the job over which staff expressed greatest dissatisfaction was 'The way the University is managed' (2.7) followed by 'Your chance of promotion' (3.1) and 'Industrial relations between management and workers in the University' (3.1).

In most respects, the results from the academic staff at the University of Adelaide are similar to those from academic staff at other Australian universities (Winefield *et al.*, 2001b). For example, academics involved in both teaching and research show the lowest job satisfaction and the highest strain, both at the University of Adelaide (see Tables 9.2 and 9.7) and at the other Australian universities. Academic staff in Humanities/Social Sciences showed the lowest job satisfaction and the highest strain both at the University of Adelaide (see Tables 9.3 and 9.8) and at the other Australian universities.

An interesting difference between the academics at the University of Adelaide and the academics at the other Australian universities surveyed in 2000 was that at most of the universities the intermediate level academic staff (Levels B/C) had the lowest levels of job satisfaction and the highest levels of strain. As Tables 9.4 and 9.9 show, this was not the case at the University of Adelaide. As can be seen in Table 9.4, the more senior categories (Levels D/E) showed a greater drop in job satisfaction from 1994 to 2000 than the Levels B/C and were marginally lower in 2000. As Table 9.9 shows, although Levels B/C showed the highest strain in 1994, in 2000 they did not differ much from the more senior (Levels D/E) or more junior grades (Level A/PostDoc).

Overseas studies have reported conflicting findings in relation to seniority, with some showing that stress is higher in more junior than in more senior staff (Abouserie, 1996; Gmelch *et al.*, 1986) although others have found no difference (Richard and Krieshok, 1989).

The negative affectivity scores suggest that the observed differences on job satisfaction and strain can not be attributed to negative affectivity, even though the variables were positively correlated. For example, the women scored higher than the men on negative affectivity at both times (Table 9.15) yet they did not differ significantly from the men at either time on either job satisfaction (Table 9.6) or on strain (Table 9.12).

The lack of difference between men and women on the two main outcome measures is consistent with results reported by Abouserie (1996) and Richard and Krieshok (1989), although other studies have found that women reported higher stress levels than men (Blix *et al.*, 1994; Boyd and Wylie, 1994; Gmelch *et al.*, 1986; Sharpley, 1994).

9.7 FUTURE RESEARCH DIRECTIONS

As many critics of organisational stress studies have observed, cross-sectional studies relying solely on self-report measures are of limited value. Such data need to be augmented by data from other sources, such as longitudinal research designs and data obtained from independent sources. For example, high levels of psychological strain within an organisation, or a particular occupational group or profession are likely to lead to high levels of self-reported strain (using questionnaire inventories) which are maintained over a period of time (as revealed in longitudinal comparisons) as well as independent measures such as relatively frequent (and expensive) compensation claims, and high sickness, absenteeism, and turnover rates (Dollard *et al.*, 2001).

University academics are not traditionally an occupational group noted for making compensation claims or for showing high sickness, absenteeism or turnover rates. On the other hand, there is evidence that they are being subjected to an increasingly stressful work environment. Much of the increase is related to reduced government funding and increased demands for 'accountability'. In countries such as Australia, New Zealand and UK, salaries have been eroded, as has job security (tenure), workloads have increased, and additional demands have been imposed. At the same time, academics have experienced reduced control/ autonomy as 'collegiality' has been replaced by 'managerialism'.

Cooper (1998) contains a range of contemporary theories of organisational stress, each focussing on different aspects of the work environment. Four of the best known are Maslach's burnout theory, Karasek's demand-control theory, Siegrist's effort–reward imbalance theory, Hobfoll's conservation of resources theory, and person-environment fit theory. It is readily apparent how the changes to the work roles of academics in recent years would be conceptualised within each of these theories.

For example, burnout theory has been applied principally to members of the caring professions and to teachers, and teaching is one of the core activities of most university academics (the exceptions being those involved in research only or those who have moved into administration). Demand-control theory is clearly relevant to the situation in which the demands on academics are increasing (increased teaching loads arising from worsening student–staff ratios, increased demands to publish and attract external funding) and control is decreasing because of increased managerialism, and loss of tenure. Effort–reward imbalance theory refers to disequilibrium between the amount of effort involved in performing a job and the reward received. Again, the relevance is clear in a situation where academic work loads are becoming intolerable, yet academic salaries are falling relative to other groups. Conservation of resources theory assumes that people strive to maintain their resources and experience negative outcomes (e.g. strain, burnout) when these are threatened by excessive demands. (Taris *et al.*, 2001 have recently applied this theory in a study of job stress in Dutch university staff). Finally, person-environment fit predicts that strain will occur when there is a mismatch between the individual and the work environment. The mismatch can arise either because the worker is unable to meet the demands of the job or because the job fails to satisfy the needs of the worker. There are good reasons to suppose that both kinds of mismatch are increasing in relation to academics.

Future research on academic stress will be informed by the theories outlined above but needs also to explore the possibility and potential of intervention strategies. Unless academic stress can be reduced there will be significant costs both to the effectiveness of universities and to the psychological and physical well being of the academics responsible for their core activities of research and teaching. Universities themselves will find it increasingly difficult to attract and retain high quality academics. Bright young people are likely to be attracted to professions offering better, less stressful working conditions. Good, productive academics are likely to be attracted to other careers, or to academic careers in overseas countries offering better working conditions and remuneration (the 'brain drain'). There is already evidence of both these trends in Australian universities where many of the best students are choosing to study subjects such as information technology rather than physics or chemistry, and commerce and accountancy rather than economics, and where some of the best academics are going overseas.

The strategy adopted by several Australian universities of offering voluntary redundancy or early retirement packages is often counter-productive. Such packages are generally most attractive to those staff who are able to get other jobs – the very people whom the university least wishes to lose.

The costs to the university of maintaining a workforce of stressed academics are also obvious. Academics who are experiencing psychological strain are unlikely to perform at a high level. Consider their core activities, teaching and research. An outstanding university teacher is one who keeps abreast of current developments in the field, and who is able to communicate effectively with students, and inspire them by enthusiasm and excitement. Such qualities are unlikely to be found in people suffering from work overload and burnout.

The other core activity of academics is research. In order to perform research that is creative and original, researchers need time to be able to devote to reading, thinking and discussing their ideas. These fundamental requirements are unlikely to be achieved in an environment where there is never ending pressure to produce, as much and as quickly as possible.

Dollard and Winefield (1996) have drawn attention to the fact that Sweden and the United States have both recognised occupational stress as a national health issue. For example, Levi (1990) reported that the Swedish Government had established a Commission for Work Environment and Health to identify present, and predict future work-related illnesses and to propose recommendations and strategies to remove or reduce the risks.

In the United States, the National Institute for Occupational Safety and Health (NIOSH) recognises psychological disorders as one of the ten leading occupational diseases and generally refers to them under the rubric of 'job stress' (Sauter *et al.*, 1990).

By contrast, Australian mental health experts seem less willing to acknowledge the role of job stress as a determinant of mental health. This is revealed in a recent monograph *Promotion, prevention and early intervention for mental health* published by the Commonwealth (Australian) Government in 2000. In the section headed 'Psychosocial determinants of health and mental health' the authors refer to 'The benchmark Whitehall studies of British civil servants (Marmot *et al.*, 1984)' (p. 11) which identifies the following factors associated with ill health: 'low socioeconomic status, high stress levels, hardship or risk exposure in early life, social exclusion, high stress in the workplace, job insecurity, low social support, addictive behaviours, unhealthy food choices and unhealthy transport practices' (p. 11). However, the authors only list three (psychosocial) factors as contributing to health in the Australian context: poverty, ethnicity, and gender. These three no doubt subsume several of the ten listed by the Whitehall researchers. For example, 'poverty' and 'ethnicity' would no doubt subsume 'low socioeconomic status, hardship or risk exposure in early life, social exclusion, low social support, addictive behaviours, unhealthy food choices, and unhealthy transport practices', but two of the factors associated with job stress 'high stress in the workplace' and 'job insecurity' do not seem to be regarded as important by the Australian mental health experts who wrote the monograph.

Two of the main challenges for future researchers (in Australia at least) will be to persuade the Government first that job stress in general poses a threat to the psychological and physical health of workers, and second that academics in particular are experiencing increasingly high levels of job stress (even though the same may well be true for other workers). Perhaps the most effective means of persuasion will be to argue that job performance, as well as physical and psychological health, is likely to be adversely affected by job stress (increased strain and reduced job satisfaction). The recent meta-analytic finding of a .30 correlation between job satisfaction and job performance (accounting for 9% of the variance) and .52 for high complexity jobs (accounting for 27% of the variance) reported by Judge *et al.* (2001) is very much higher than the correlation of .17 (accounting for less than 3% of the variance) reported by Iaffaldano and Muchinsky (1985) and is likely to stimulate further research designed to explicate the relationship between them.

9.8 REFERENCES

Abouserie, R., 1996, Stress, coping strategies and job satisfaction in university academic staff. *Educational Psychology*, **16**, pp. 49–56.

Altbach, P.G. (ed.), 1996, *The International Academic Profession* (Princeton, N.J.: Carnegie Foundation for the Advancement of Teaching).

Andrews, G., Hall, W., Teeson, M. and Henderson, S., 1999, *Mental Health of Australians* (Canberra: Mental Health Branch, Commonwealth Department of Health and Aged Care).

Armour, R.A., 1987, Academic burnout: Faculty responsibility and institutional climate. *New Directions for Teaching and Learning*, **29**, pp. 3–11.

Australian Broadcasting Commission, 2001, *Facing the music* (Sydney: ABC).

Banks, M.H., Clegg, C.W., Jackson, P.R., Kemp, N.J., Stafford, E.M. and Wall, T.D., 1980, The use of the General Health Questionnaire as an indicator of mental health in occupational studies. *Journal of Occupational Psychology*, **53**, pp. 187–194.

Bernard, L.C. and Krupat, E., 1994, *Health psychology: Biopsychosocial factors in health and illness* (Orlando, FL: Harcourt Brace).

Blix, A.G., Cruise, R.J., Mitchell, B.M. and Blix, G.G., 1994, Occupational stress among university teachers. *Educational Research*, **36**, pp. 157–169.

Bowen, H.R. and Schuster, J.H., 1986, The professoriate needs our help-here's why. *AGB Reports*, **28**, pp. 18–23.

Boyd, S. and Wylie, C., 1994, *Workload and stress in New Zealand universities* (Auckland: New Zealand Council for Educational Research and the Association of University Staff of New Zealand).

Bradley, J. and Eachus, P., 1995, Occupational stress within a U.K. Higher Education Institution. *International Journal of Stress Management*, **2**, pp. 145–158.

Brannon, L. and Feist, J., 1992, *Health psychology: An introduction to behavior and health* (Belmont, CA: Wadsworth).

Coady, T., 2000, Universities and the ideals of inquiry. In T. Coady (ed.), *Why universities matter* (St Leonards, NSW: Allen & Unwin), pp. 3–25.

Cohen, J., 1988, *Statistical power analysis for the behavioral sciences* (New York: Erlbaum).

Cooper, C.L. (ed.), 1998, *Theories of organizational stress* (Oxford: Oxford University Press).

Cooper, C.L., Sloan, S.J. and Williams, S., 1988, *Occupational Stress Indicator: Management Guide* (Windsor, UK: NFER-Nelson).

Costa, P.T., 1996, Work and personality: Use of the NEO-PI-R in industrial/organizational psychology. *Applied psychology: an international review*, **45**, pp. 225–241.

Daniels, K. and Guppy, A., 1992, Control, information-seeking preferences, occupational stressors and psychological well-being. *Work and Stress*, **6**, pp. 347–353.

DEETYA, 1999, Higher Education: Report for the 1999 to 2001 triennium (Canberra: Commonwealth of Australia).

Dollard, M.F., Winefield, H.R. and Winefield, A.H., 2001, *Occupational strain and efficacy in human service workers* (The Netherlands: Kluwer).

Dollard, M.F. and Winefield, A.H., 1996, Managing occupational stress: A national and international perspective. *International Journal of Stress Management*, **3**, pp. 69–83.

Dua, J.K., 1994, Job stressors and their effects on physical health, emotional health, and job satisfaction in a university. *Journal of Educational Administration*, **32**, pp. 59–78.

Fisher, S., 1994, *Stress in academic life* (Buckingham, UK: The Society for Research into Higher Education and Open University Press).

French, J.R.P. Jr, Caplan, R.D. and van Harrison, R., 1984, *The Mechanisms of Job Stress and Strain* (New York: Wiley).

Frese, M. and Zapf, D., 1988, Methodological issues in the study of work stress: Objective versus subjective measurement of work stress and the question of longitudinal studies. In C.L. Cooper and R. Payne (eds), *Causes, coping and consequences of stress at work* (Chichester: Wiley).

Gillespie, N.A., Walsh, M., Winefield, A.H., Dua, J. and Stough, C., 2001, Occupational stress within Australian universities: Staff perceptions of the determinants, consequences and moderators of stress. *Work and Stress*, **15**, pp. 53–72.

Gmelch, W.H., Wilke, P.K. and Lovrich, N.P., 1986, Dimensions of stress among university faculty: Factor-analytic results from a national study. *Research in Higher Education*, **24**, pp. 266–286.

Goldberg, D.P. and Williams, P., 1988, *A user's guide to the GHQ* (London: NFER, Nelson).

Hind, P. and Doyle, C., 1996, A cross cultural comparison of perceived occupational stress in academics in higher education. Paper given at XXVI International Congress of Psychology: Montreal.

Holmes, T.H. and Rahe, R.H., 1967, The social readjustment rating scale. *Psychosomatic Medicine*, **11**, pp. 213–218.

Iaffaldano, M.T. and Muchinsky, P.M., 1985, Job satisfaction and job performance: A meta-analysis. *Psychological Bulletin*, **97**, pp. 251–273.

Jarrett, R.J. and Winefield, A.H., 1995, Report on climate survey. University of Adelaide (Unpublished report).

Judge, T.A., Thoreson, C.J., Bono, J.E. and Patton, G.K., 2001, The job satisfaction–job performance relationship: A qualitative and quantitative review. *Psychological Bulletin*, **127**, pp. 376–407.

Kanner, A.D., Coyne, J.C., Schaefer, C. and Lazarus, R.S., 1981, Comparison of two modes of stress measurement: Daily hassles and uplifts versusmaor life events. *Journal of Behavioral Medicine*, **4**, pp. 1–39.

Karasek, R.A., 1979, Job demands, job decision latitude, and mental strain: Implications for job redesign. *Administrative Science Quarterly*, **24**, pp. 285–308.

Lazarus, R.S., 1966, *Psychological stress and the coping process* (New York: McGraw-Hill).

Levi, L., 1990, Occupational stress: spice of life or kiss of death? *American Psychologist*, **45**, pp. 1142–1145.

Marmot M.G., Shipley M.J. and Rose G., 1984, Inequalities in death-specific explanations of a general pattern? *Lancet*, **1**, pp. 1003–1006.

Melendez, W. and de Guzman, R.M., 1983, Burnout: The new academic disease. *ASHE-ERIC Higher Education Research Report*, No 9. Washington, DC: Association for the Study of Higher Education.

Molony, J., 2000, Australian universities today. In T. Coady (ed.), *Why universities matte* (St Leonards, NSW: Allen and Unwin), pp. 72–84.

Murphy, L.R., 1988, Workplace interventions for stress reduction and prevention. In C.L. Cooper and R. Payne (eds), *Causes, coping and consequences of stress at work* (Chichester: Wiley).

Nowack, K.M., 1990, Initial development of an inventory to assess any health risk. *American Journal of Health Promotion*, **4**, pp. 173–180.

Payne, R.L., 1988, A longitudinal study of the psychological well-being of unemployed men and the mediating effect of neuroticism. *Human Relations*, **41**, pp. 119–138.

Richard, G.V. and Krieshok, T.S., 1989, Occupational stress, strain and coping strategies in university faculty. *Journal of Vocational Behaviour*, **34**, pp. 117–132.

Sauter, S.L., Murphy, L.R. and Hurrell, Jr, J.J., 1990, Prevention of work-related psychological disorders. *American Psychologist*, **45**, pp. 1146–1158.

Seldin, P., 1987, Research findings on causes of academic stress. *New directions for Teaching and Learning*, **29**, pp. 13–21.

Selye, H., 1976, *Stress in health and disease*. Reading, MA: Butterworth.

Senate Committee, 2001, *Universities in crisis*. Senate Committee Report (Canberra: Australian Government).

Sharpley, C.F., Dua, J.K., Reynolds, R. and Acosta, A., 1995, The direct and relative efficacy of cognitive hardiness, Type A behavior pattern, coping behaviour and social support as predictors of stress and ill-health. *Scandinavian Journal of Behaviour Therapy*, **24**, pp. 15–29.

Spielberger, C.D., Gorsuch, R.L., Lushene, R., Vagg, P.R. and Jacobs, G.A., 1983, *Manual for the state-trait anxiety inventory* (Palo Alto, CA: Consulting Psychologists Press).

Taris, T.W., Schreurs, P.J.G. and Van Iersal-Van Silfhout, I.J., 2001, Job stress, job strain, and psychological withdrawal among Dutch university staff: towards a dual-process model for the effects of occupational stress. *Work and Stress*, **15**, pp. 283–296.

Warr, P.B., Cook, J. and Wall, T.D., 1979, Scales for the measurement of some work attitudes and aspects of psychological well-being. *Journal of Occupational Psychology*, **52**, pp. 129–148.

Wearing, A.J. and Hart, P.M., 1996, Work and non-work coping strategies: Their relation to personality, appraisal and life domain. *Stress Medicine*, **12**, pp. 93–103.

Wilkinson, J. and Joseph, S., 1995, Burnout in university teaching staff. *The Occupational Psychologist*, **27**, pp. 4–7.

Winefield, A.H. and Jarrett, R.J., 2001, Occupational stress in university staff. *International Journal of Stress Management*, **8**, pp. 285–298.

Winefield, A.H., 2000, Stress in academe: Some recent research findings. In D.T. Kenny, J.G. Carlson, F.J. McGuigan and J.L. Sheppard (eds), *Stress and health* (Amsterdam: Harwood Academic Publishers).

CHAPTER TEN

Prostitution: An Illustration of Occupational Stress in 'Dirty Work'

Kara Anne Arnold and Julian Barling

10.1 INTRODUCTION

10.1.1 Prostitution

Prostitution is arguably one of the oldest human service professions (Cloud, 2000). Despite its often illegal nature and the social and moral taint associated with it, it is not likely to disappear soon (Belk *et al.*, 1998). Prostitutes provide a valuable societal service in that they offer sexual services for persons who are lonely, who perhaps cannot find companionship elsewhere, or who are 'physically repugnant to others' (Carr, 1995).

We define a prostitute as someone who exchanges sexual services for money (Alexander, 1987). The simplicity of this definition masks an important diversity within this occupation (Dank, 1999). Prostitutes range from 'drug addicted street hustlers, teenage Thai girls sold by their parents to brothel owners, [to] high-priced career call girls working as independent entrepreneurs or for escort agencies and mainstream corporations' (Dank, 1999). There is considerable debate among feminists as to whether prostitution should be conceived of as legitimate work (liberal feminism) or as violence against women (radical feminism) (Jaggar, 1997; Simmons, 1999). Without engaging the particulars of this debate, we acknowledge it and we take the position that prostitution is work, and that prostitutes experience numerous stressors, some unique to their profession, and many common across occupations. While the majority of research into prostitution has dealt with women (who make up the majority of the worlds' prostitutes), we also address male prostitution for the sake of a comprehensive picture.

10.1.2 Dirty Work

To understand the stress experienced by prostitutes, we will use the concept of 'dirty work', which was first introduced by Hughes (1951). He referred to dirty work as tasks within occupations that are perceived as physically disgusting or degrading. The social construction of dirty work was more fully explicated by

Ashforth and Kreiner (1999) to be work that was physically, socially or morally tainted. Work can be physically tainted if it is directly associated with garbage or death, or is performed under dangerous or harmful conditions. Social taint accompanies occupations that involve contact with stigmatised people or groups, or where the relationship of the worker to clients and others is submissive. An occupation is morally tainted when it is perceived of as sinful or of questionable nature (Ashforth and Kreiner, 1999). To truly be considered 'dirty' the work must be both 'necessary and polluting' (Ashforth and Kreiner, 1999). There are both high and low prestige dirty work occupations. Prostitution is an example of low prestige dirty work that carries all three forms of taint. It is also arguably perceived of as more 'evil' than 'necessary' by societies across the globe today. This has implications for the amount of effort that a prostitute will need to expend in order to maintain a positive self identity.

10.2   WORK STRESS FRAMEWORK

We conceptualise occupational stress within a traditional work stress framework (Pratt and Barling, 1988). In this framework, a stressor refers to an objective characteristic or event in the environment. Stress is the persons' subjective individual experience of the objective stressor. One individual may perceive role overload to be stressful – another may not. Strain refers to the psychological and/or physiological responses to stress.

10.2.1   Work Stress Framework Applied to Prostitution

From a review of the literatures on prostitution (and more broadly sex workers), a number of objective environmental stressors and subjective work stress that are present for prostitutes become apparent, as do strains of a psychological, behavioural and physical nature. We provide a conceptual model of our discussion in Figure 10.1.

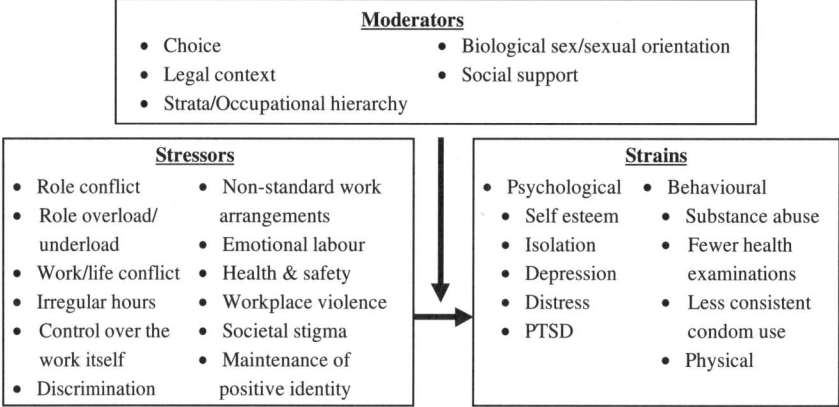

Figure 10.1 Conceptual Model of Occupational Stress in Prostitution.

Before embarking on this discussion, it is important to note that to our knowledge, there have been no studies directly investigating work stress in prostitution although some researchers have focused on post traumatic stress disorder in prostitutes (Farley and Barkan, 1998; Farley *et al.*, 1998) and some discuss psychological distress in prostitutes (Boyle *et al.*, 1997). Our model is best understood as a series of propositions, because many of the stressors and the relation to strains that we suggest have yet to be empirically tested with prostitutes.

10.3 TRADITIONAL WORK STRESSORS

10.3.1 Role Theory Stressors

Many of the work stressors that prostitutes face are also faced by other human service workers. Role conflict is frequently cited as a stressor at work (Beehr, 1995). Role conflict is prevalent in prostitution, as it applies to work/non-work roles. Many prostitutes do not let their families know what they do for a living (Daley, 2001) and a majority find it difficult to maintain personal relationships due to the nature of their profession (Pyett and Warr, 1999; Pyett *et al.*, 1996).

Role overload (Narayanan *et al.*, 1999) and underload (Fisher, 1993) are also apparent in prostitution. The encounters with clients can be unpredictable, potentially high risk and excessively demanding. At the same time there can be long periods of 'downtime' between clients which can be tedious (Brewis and Linstead, 2000; Scambler and Scambler, 1999). The extent to which these stressors are present may vary based on the context in which the prostitute is working. Street walkers may be exposed to more role overload as the number of clients per shift tends to be greater for these workers than for 'indoor' workers. A New Zealand study found that nearly half of all indoor workers surveyed reported only one or two clients per shift, while three quarters of street workers surveyed reported three or more clients per shift (Plumridge and Abel, 2001). However, this same study also found that significantly more street workers (85%) than indoor workers (55%) had refused a client in the past 12 months, even though the percentage of women who reported wanting to refuse a client was the same (53% and 59% respectively). This may suggest that while indoor workers enjoy fewer clients per shift, they may lack the same level of choice in selecting their clients.

10.3.2 Work/Life Conflicts and Hours of Work

The interface between work and family is a significant stressor for many people (Barling and Sorensen, 1997). Prostitutes are no different in this respect. Many have children, and for those who retain custody of their children, childcare issues can be a stressor. The irregular hours of work (mostly in the evening and at night) make finding childcare difficult for many of these women (Harcourt *et al.*, 2001). This juggling of work and family is a stressor that has been extensively investigated with women holding jobs that are not stigmatised (Ernst Kossek and Ozeki, 1998), and may be exacerbated for those doing 'dirty work'.

In terms of family and personal relationships, perhaps the greatest evidence of spillover of work to life occurs in the prostitute's own sex life (Brewis and Linstead, 2000). Many women working as prostitutes report that they do not feel like having sex with their partners when they get home after work, that their relationships are under pressure due to the nature of their work, and that their partners are jealous of their male clients (Brewis and Linstead, 2000).

Irregular hours of work can also be stressor (Martens *et al.*, 1999). People working irregular hours report significantly more health complaints, less well-being and lower quality sleep compared to those working non-flexible work schedules (Martens *et al.*, 1999). Working on the street generally leads to a higher income and also offers greater flexibility in terms of the hours of work for many prostitutes (Harcourt *et al.*, 2001). This flexibility may allow women to meet family obligations, but also increase the risk of negative outcomes for them in terms of health (physical, well being and sleep) and safety as there is ample evidence that 'streetwalkers' are at increased risk of facing serious violence (Harcourt *et al.*, 2001).

10.3.3   Control over the Work Itself

As with other occupations the amount of control prostitutes experience with respect to how the actual work is done is a critical determinant of subsequent strain, and the classic demand-control model proposed by Karasek (1979) applies to prostitution. The notion that the organisational context in which the work is performed is more important in determining strain than personal characteristics of workers seems to be especially true with respect to prostitution. The outcomes for prostitution in terms of strains depend on contextual factors rather than individual variables (Plumridge and Abel, 2001). Prostitutes who work in situations where they perceive they have little control over the acts they perform with clients are likely to experience greater strain than prostitutes who perceive they have control over this aspect of the work (Pyett and Warr, 1999).

10.3.4   Discrimination

Many workers encounter discrimination based on age, sex, race, nationality, class and other categories in occupational settings. Prostitutes can also be the victims of age and 'appearance'-based discrimination. Age discrimination may be apparent but it does not coincide with what gerontologists would label 'old age'. Evidence from a study of table dancers suggests that being old may be socially constructed in the sex industry, and that a dancer may be considered old in her early 20s (Rambo Ronai, 1992). The loss of physical attractiveness means the loss of credibility and influence in this profession, and aging equates to a loss of attractiveness for most dancers (Rambo Ronai, 1992). Some of the alternatives available to the 'aging' table dancer include finding a 'sugar daddy' (an older man to marry), carving out a niche in the bar scene for herself (becoming a manager or using social skills to gain repeat business for the bar), or 'make[ing] up for visual appeal with wholesale sexual activity' (Rambo Ronai, 1992). Aging dancers who cannot create a niche generally end up being employed in lower status clubs. As prostitutes age occupationally (which happens in young adulthood), they may find themselves with more limited occupational choices.

10.3.5 Non-Standard Employment Arrangements/Status Incongruence

Prostitutes work as contingent employees in the Netherlands, renting rooms from the brothel owner and hence being essentially self employed (Daley, 2001). The literature on non-standard work arrangements has found negative outcomes for people who are in these work situations but would rather be permanent, traditional full time employees (Ellingson *et al.*, 1998). Independent contractors such as these lack benefits that would be available if the person were an employee. In the Netherlands, brothel owners are fighting with the tax department about whether the prostitutes that rent their rooms are actually employees. At stake is who pays the social service costs for these prostitutes (Daley, 2001).

10.3.6 Emotional Labour

Emotional labour is defined as the 'act of displaying organizationally desired emotions during service transactions' (Morris and Feldman, 1996). Emotional labour is thought to be more prevalent in service occupations, as people who work in services generally are subjected to stronger norms about appropriate expression of emotion in certain situations (Schaubroeck and Jones, 2000). In particular what is disturbing to people is the disequilibrium or dissonance between how the worker feels and the emotions they must exhibit (Schaubroeck and Jones, 2000). This discrepancy between felt and expressed emotions has a negative effect on physical health (Schaubroeck and Jones, 2000).

Prostitutes are faced with unique demands in terms of emotional labour. Their work consists of acts that are intensely personal and intimate. They must feign affection and excitement to develop a regular clientele. One way that prostitutes cope with the emotional demands placed on them is 'categorization of different types of sexual encounters [as] relational, professional or recreational' (Brewis and Linstead, 2000). In this way they can maintain distance from the client encounter and maintain their self-identity (Brewis and Linstead, 2000). The literature also suggests that prostitutes maintain an emotional distance by the use of condoms in work or professional sex, and by refusing to kiss clients. Kissing 'is rejected because it is too similar to the kind of behaviour in which one would engage with a non-commercial sex partner; it smacks too much of genuine desire and love for the other person' (Brewis and Linstead, 2000).

10.3.7 Health and Safety Issues

Probably one of the most important stressors for prostitutes is health issues that arise due to the nature of the tasks that the job entails. Because prostitutes engage in more sexual relations than average (Karim *et al.*, 1995), they may be at greater risk for venereal disease. The risk of HIV infection would also be a major stressor. Due to concern about public health and safety, a large number of studies have investigated the incidence of HIV infection among prostitutes. Many of these studies find that the incidence of HIV positive prostitutes is relatively small in U.K., Australia and United States, and that most cases of HIV infection in these countries are related to sharing of needles rather than unsafe sexual activity (Brewis and Linstead, 2000). However in some areas such as Latin America, India and sub-Saharan Africa, as many as 80% of prostitutes are estimated to be HIV positive (Brewis and Linstead, 2000).

Unwanted pregnancy is also a significant stressor. The choice faced is to have the baby or have an abortion. In some parts of the world, abortion remains illegal, and even in areas where it is legal, strong moral condemnation against abortion exists in significant segments of society. Most women (even those working as prostitutes) faced with an unwanted pregnancy do not easily choose abortion (Korn, 1996). The alternative of having the baby brings other stressors into the life of a prostitute. The pregnancy disrupts the ability to engage in sexual services at some point. Depending on their social context, prostitutes may not have access to good prenatal health care. If they are drug addicted (as many streetwalkers have been found to be c.f. Plumridge and Abel, 2001), such choices and stressors are exacerbated.

Practicing safe sex is extremely important to the majority of prostitutes who have been studied (Albert *et al.*, 1995; Brewis and Linstead, 2000; Karim *et al.*, 1995; Pyett and Warr, 1999). In occupational safety terms, condoms can be considered personal protective equipment. Over-and-above condoms, prostitutes will tend to use additional forms of birth control (i.e. the pill, IUD) to protect against unwanted pregnancy. Most prostitutes appear to be well educated about the health risks of unsafe sex in terms of sexually transmitted diseases, even in areas of the world where the HIV infection rate is high (Browne and Minichiello, 1995; Karim *et al.*, 1995). The problem does not appear to be one of insufficient knowledge on the part of prostitutes. Part of the difficulty is *not* in getting the *worker* to use the safety equipment but in this case convincing the *client* to use it, highlighting the importance of lack of control as a central work stressor.

The ability of a female prostitute to insist upon safe sex and the use of condoms varies. For example, in Nevada where brothels are legal, condom use is legally mandated (Albert *et al.*, 1998). Public health signs informing clients of the legal requirement to wear condoms are posted throughout the brothel (Albert *et al.*, 1998). This is in sharp contrast to the situation a prostitute will be exposed to in an illegal context where there is no support for the use of condoms during sexual acts (Karim *et al.*, 1995). Even in the Nevada brothels, some prostitutes report that clients are still reluctant to wear a condom, and workers must use techniques to encourage them to do so (Albert *et al.*, 1998). Studies in South Africa (Karim *et al.*, 1995), Australia (Pyett and Warr, 1999), and New Zealand (Plumridge and Abel, 2001) show that prostitutes have varying degrees of control over whether or not the client wears a condom, and many times the insistence on condom use can result in assault and rape. Physical assault and difficulties enforcing condom use tend to be more frequently reported by street workers than brothel workers. Similar to laws in Nevada, in Australian states where brothels are licensed condom use is legally mandated – although management may or may not strictly adhere to regulations (Pyett and Warr, 1999). As in other industries, conditions that may improve the health or safety of workers are not always pursued by management or workers (Sells, 1994). Where management is perceived by prostitutes to be unsupportive regarding health and safety, a negative safety climate would ensue, with adverse health and safety implications (cf. Zohar, 2000).

It may appear that in environments where prostitution is legal condom use is supported and in turn the stressor of contracting HIV or becoming pregnant is reduced. In Nevada prostitutes are required to undergo weekly medical exams for gonorrhoea, herpes and venereal warts, and monthly blood tests for syphilis. They also must test negative to an initial HIV antibody test and test negative every month thereafter as a condition of employment (Albert *et al.*, 1995). If a prostitute tests positive for HIV in Nevada, her employment is immediately terminated. Arguably, this reflects a concern for public health rather than employee health and safety (Wilton, 1999). This illustrates the disdain which society holds for people who do dirty work, and highlights how dirty work exacerbates work stress.

10.3.8 Workplace Violence

As an occupational group, prostitutes face horrific workplace violence. A prostitute suffering from post traumatic stress disorder (PTSD) said of the failure of therapists to connect her history of violence with symptoms of PTSD: 'I wonder why I keep going to therapists and telling them I can't sleep, and I have nightmares. They pass right over the fact that I was a prostitute and I was beaten with 2 × 4 boards, I had my fingers and toes broken by a pimp, and I was raped more than 30 times. Why do they ignore that?' (Farley and Barkan, 1998). The risk of violence at work is increased in terms of frequency and/or specific kinds of violent acts for female prostitutes, compared to occupations that are less socially and morally tainted. Because prostitution is illegal in many jurisdictions, and is also subject to strong moral prohibition even when legal, the police can become a source of stress rather than a valuable source of support (Lewis and Maticka-Tyndale, 2000). Doctors and therapists may not provide adequate care to marginalised women as the above quote illustrates. Most violence is perpetrated by clients although some prostitutes report violence from voyeurs, residents and other sex workers (Harcourt *et al.*, 2001).

In terms of violence an Australian study found that 75% of the sample of prostitutes had experienced physical violence while at work. Physical violence included physical assault, rape at gun or knifepoint, robbery, threats, abduction and stalking (Harcourt *et al.*, 2001). A New Zealand study reported that 83% of respondents had experienced one or more violent events such as physical assault, threats, being held against their will, verbal abuse, rape, or being forced to have unprotected sex (Plumridge and Abel, 2001). There were significant differences between indoor and street workers in terms of the violence reported, with indoor workers experiencing significantly less than street workers. The percentages of female prostitutes reporting rape were approximately 33% in the Australian study, 27% of the street workers and 8% of the indoor workers in the New Zealand study. This is high in comparison to more socially acceptable occupations where even though approximately half of working women have experienced sexual harassment, only rarely are women raped at work (Barling *et al.*, 2001). A recent study of frontline in-home employees of Regional and Community Health Boards asked employees about sexual harassment that they have experienced at work. They found that 1% reported being threatened with a weapon such as a knife or gun, and 1% of the women surveyed reported being raped at work (Barling *et al.*, 2001). While the study of frontline health workers does not report percentages that have experienced physical assault per se they do ask about workplace violence and show that 1% report being threatened with a weapon other than a knife or gun, 3.9% report that someone has tried to hit them and 4.5% report being threatened with a gun (Barling *et al.*, 2001). In contrast, 44% of the Australian sample, 41% of the street workers and 21% of indoor workers in the New Zealand sample reported being physically assaulted. These data show that prostitutes face more severe and more frequent workplace violence and sexual harassment than do other in-home workers.

Why do prostitutes experience such severe and frequent workplace violence? When people employed in socially acceptable occupations work within their 'traditional organizational environments, they can be afforded necessary protection if they are threatened with, or experience, sexual harassment or workplace violence' (Barling *et al.*, 2001). This does not hold for workers inside clients' homes. Outside of the traditional organisation protection is limited at best. In much the same way, prostitutes are not afforded any 'organisational' protection if they work on their own. Many rely on pimps to protect them. However this arrangement can also become a stressor as the pimp may take a large percentage of their income and also coerce them into doing things they would not otherwise do.

Another reason for increased workplace violence is related to the characteristics of the job. A recent study that differentiated between co-worker initiated and public initiated violence has developed a scale to measure risk for workplace violence based on job characteristics (LeBlanc and Kelloway, in press). Prostitutes would score high on this scale in terms of their risk for violence from the public. People may feel less restricted in the violence they direct against prostitutes because of the social and moral taint attached to this occupation and because the legal system does not protect these workers in the same way as other socially acceptable workers. In Australia, for example, the rape of a prostitute 'has been frequently treated with less severity than the theft of a product' (Brewis and Linstead, 2000). In 1991 the Supreme Court of Victoria in Australia upheld a ruling that 'a prostitute as a result of her work, would be less psychologically damaged by sexual assault than a "chaste" woman' (Brewis and Linstead, 2000). Many prostitutes have been murdered while at work but these murders do not tend to get the same amount of police concern and investigation that others do.

The social and moral taint that becomes associated with women in this occupation is apparent by the attitude expressed in letters to the editor of a local Spokane, Washington paper about the money being spent on investigations into several prostitute murders in that location. 'Why waste time on prostitutes?' was the theme of the comments. As the chief of the serial homicide squad comments 'If they had been teachers, the dollars would have flowed' (Hornblower, 2000). A similar situation occurred in Vancouver, Canada where 31 prostitutes had gone missing but it was not until their families and friends demanded that the police take action that anything was done (Wood, 1999). Similarly, two people admitted setting fire to a brothel in Israel 'out of religious fervour against prostitution' (Lowe and Goldberg, 2000). This fire killed four women. The perpetrators pled guilty in order to receive a four year sentence.

The illegality of the profession, and the nature and consequence of dirty work, means that prostitutes 'find themselves in dark streets and defenceless situations. There are no eyes there. But there are predatory misogynists, serial killers, men who get off on violence. They see the women's vulnerability' (Wood, 1999).

10.4 STRESSORS UNIQUE TO PROSTITUTION AND DIRTY WORK

10.4.1 Societal Stigma

The concept of 'dirty work' suggests that prostitution is a low status occupation, and it is physically, socially and morally tainted (Ashforth and Kreiner, 1999). Prostitutes are well aware of the stigma attached to what they do for a living. Stigma can be a significant stressor for prostitutes and for others engaged in 'dirty work', and can render social validation problematic (Ashforth and Kreiner, 1999). The negative impact on self esteem is a stressor that people engaged in occupations that are less tainted do not have to deal with. When employed in a high status, untainted occupation, work can be central to a positive self-definition.

10.4.2 Effort to Maintain a Positive Identity

One of the fundamental tenets of social identity theory is that 'individuals seek to enhance their self-esteem through their social identities', and one major component of self-definition is occupational identity (Ashforth and Kreiner, 1999). Someone employed as a prostitute will be constantly engaging in various ideological techniques to neutralise the negative connotations associated with the work they do. Ashforth and Kreiner (1999) have delineated three such techniques, namely reframing, recalibrating and refocusing that are used at the group level to transform the meaning of the stigmatised work. Reframing, for example, allows prostitutes to transform the stigma into a badge of honour by claiming they are providing an educational and therapeutic service rather than selling their bodies (Miller, 1978). These protective techniques enhance the self-definition of prostitutes, and may be considered coping mechanisms. At the same time, the necessity for engaging in such techniques take mental and emotional energy and can simultaneously become a stressor. As well the requirement of a strong group culture in order to support these ideological techniques may not always be present for prostitutes. Even with efforts to engage in these techniques, most members of dirty work occupations will retain some ambivalence about their jobs as they are still members of the larger society that is constructing their work as 'dirty' and they do have ongoing contact with persons outside their occupation (Ashforth and Kreiner, 1999).

Prostitutes must also construct their self-identity in circumstances that put pressure on the relationship between their professional and personal lives (Brewis and Linstead, 2000). The prostitutes' body (Brewis and Linstead, 2000) and potentially their psyche is what is consumed by the client in the act of commercial sex, and this creates additional pressure to somehow create a division between the professional and the personal. Many of the techniques that prostitutes use to maintain this divide will be discussed as coping mechanisms.

10.5 MODERATORS

Not all stressors, however, lead to strain, and under some conditions, some stressors may be more closely linked with strain. An understanding of how these relationships are moderated is critical for a comprehensive understanding of work stress.

10.5.1 Choice

Until this point we have assumed that our discussion applies to people who freely choose to engage in prostitution as an occupation. Many workers (not just prostitutes) are limited in their career choices by social forces over which they have little control. Two examples in the sex industry where there is clearly no free choice involved are child prostitution (Anonymous, 1996) and illegal trafficking of human beings – generally women who are sold into sexual slavery (Hughes, 2000; Michelle, 2000). Children arguably do not have the capacity and understanding to make such a decision freely. There are also clearly situations where women are forced by brutal violence to remain in this occupation. In these cases, the strains resulting from the stressors of this work will be severe.

For others who, within the confines of social forces, choose to pursue this occupation, there are varying degrees of volition. Many are 'neither entirely free nor completely enslaved' (Jaggar, 1997). Some prostitutes will describe enjoying the work while others describe the work as 'disgusting and degrading' but perceive that they have no other options available to pursue (Brewis and Linstead, 2000). Thus, choice is perhaps best conceived of as a continuum which ranges from no choice whatsoever (children and slaves) to completely free choice and enjoyment. The less volition in choosing the occupation, the more negative the outcomes of the various stressors will be.

10.5.2 Legal Context

The laws governing prostitution are varied and currently changing rapidly. There are three legal frameworks that can be distinguished when discussing the regulation of prostitution (West, 2000). Prohibition outlaws all prostitution. Abolition is a modified form of prohibition, and 'allows the sale of sex but bans all related activities such as soliciting, brothel–keeping and procurement' (West, 2000). The second framework is legalisation, in which the state regulates the industry through licensing or registration and mandatory health checks. Within this framework, outlets or workers not granted permits are subject to criminal prosecution (West, 2000). Legalisation currently exists in Nevada, USA (Nieves, 2001). The final framework is decriminalisation which aims to normalise prostitution and remove the social taint which makes these workers vulnerable to exploitation (West, 2000). These frameworks are not mutually exclusive and in some countries (for example the Netherlands), a mix of all three may be evident (West, 2000).

There is some evidence that legalisation and decriminalisation may lessen the stressors associated with the various strains experienced by prostitutes (Wilton, 1999). For example, research shows a direct link between the amount of violence against prostitutes and the level of illegality of prostitution (Michelle, 2000). Hence, the legal context in which prostitution takes place will moderate the impact of the stressors on the strains with greater negative strains being experienced in situations where prostitution is regulated under a prohibitionist framework. In contrast, others suggest that in fact trafficking of women is greater in areas where prostitution has been legalised (Hughes, 2000). This argument would suggest a higher percentage of women forced into the occupation where it is legalised – hence higher strain in legalised frameworks. Clearly, therefore, research is needed to unravel just how the legal context moderates the effects of prostitution on well-being.

10.5.3 Occupational Hierarchy – Strata within the Occupation

Prostitution is an occupation within which many strata exist (West, 2000). What may benefit one group within this occupation may well penalise another. As an example, the legalisation of brothels can sometimes accompany increased harassment of women who work on the street (West, 2000). Our review of the literature suggests that street workers are most vulnerable to distress because they seem to be faced with more extreme stressors, and a lack of social support.

10.5.4 Biological Sex and Sexual Orientation

We have focused this discussion on female heterosexual prostitutes. Men also offer heterosexual and homosexual services as prostitutes, but much less research has investigated the experience of male prostitutes (Browne and Minichiello, 1996). The research that has been conducted with respect to male prostitutes finds that this group is also heterogeneous, and there is a hierarchical division according to place of work and type of service provided (Browne and Minichiello, 1996; Minichiello *et al.*, 1999). Most men working as prostitutes would be involved in providing sexual services to other men, as women are seldom the clients of commercial sex (Browne and Minichiello, 1996). Whether or not these men identify themselves as homosexuals, they are potentially subjected to even greater marginalisation than women providing services to men. They both engage in an occupation that is tainted physically, socially and morally and hence seen to involve dirty work, and adopt a sexual orientation that many in society would disdain. On the other hand male prostitutes would experience less violence at work than female prostitutes (Minichiello *et al.*, 1999). Transgender workers may be exposed to the most extreme forms of intolerance from the community and violence from clients (Harcourt *et al.*, 2001). One study found that transgender workers were subject to the same degree of physical assaults and rapes as women prostitutes (Farley and Barkan, 1998). The authors concluded that to be female, or to appear to be female, meant that the prostitute was at increased risk of being a target for violence. Biological sex and sexual orientation are important moderators that need to be empirically examined in this context.

10.5.5   Social Support

Most prostitutes do not have adequate social support. A study conducted in Australia gives a sense of this lack of support: '...Daily life was characterised by profound social isolation which was most evident in the absence of family, friends or any personal support network around them' (Pyett and Warr, 1999). Lack of social support will have negative implications for the outcomes experienced by prostitutes, as 'social support systems buffer people against the negative effects of stress' (Geller and Hobfoll, 1994). Different sources of social support are related to strains experienced following disasters (Barling *et al.*, 1987). The stigma attached to dirty work and prostitution means that any social support that prostitutes receive will usually come from peers. It is less likely to come from friends, family or society. In addition to family, friends and co-workers, trade unions can act as important social supports and have been found to be a source of emotional support that can mitigate the psychological distress experienced by workers (Bluen and Edelstein, 1993).

While prostitution may be one of the oldest professions, organisations that campaign for the rights of prostitutes and may provide social support are a relatively recent development (Jenness, 1990). The Rode Draad (Red Thread) reports on an initiative to form a trade union to represent Dutch prostitutes (Anonymous, 2001). In New Zealand in 1987 health department officials recruited sex workers to form an organisation, the New Zealand Prostitutes Collective (NZPC), devoted to HIV prevention. This organisation engages in political activism to push the agenda of legislative reform in New Zealand (Lichtenstein, 1999). The Prostitutes Collective of Victoria Inc. (P.C.V.) is a community-based organisation in Australia that has been developed to involve and represent sex industry workers' concerns as a part of a world-wide prostitutes rights movement. The P.C.V. works for basic human rights, occupational health and safety rights for prostitutes, and breaking down the myths of prostitution which characterise women as good or bad (Prostitutes Collective of Victoria, n.d.). One study found that the only support available to the prostitutes in their sample was the outreach worker of P.C.V. (Pyett and Warr, 1999).

In the United States the San Francisco Coalition on Prostitution began as a network of prostitutes, dancers and performers in the sex industry. Currently, this Coalition emphasises community building and public education (San Francisco Coalition on Prostitution, n.d.). Also in the United States, COYOTE ('Call Off Your Tired Ethics') was founded by Margo St James in 1973 (Jenness, 1990). COYOTE works for the rights of all sex workers: strippers, phone operators, prostitutes, porn actresses and so on of all genders and sexual persuasions. COYOTE supports programmes to assist sex workers in their choice to change their occupation, works to prevent the scapegoating of sex workers, and to educate sex workers, their clients and the general public about safe sex. COYOTE is a member of The National Task Force on Prostitution, and The International Committee for Prostitutes' Rights (COYOTE San Francisco, n.d.). In Canada the Sex Workers Alliance of Vancouver (SWAV) was founded in 1994 to fight for sex workers' right to fair wages and to working conditions that are safe, clean and healthy. Its members are people who work or have worked in the sex industry, and their friends (The Sex Workers Alliance of Vancouver, n.d.). Whether these and other organisations and initiatives are successful remains a question that requires empirical validation.

10.6 STRAINS, COPING MECHANISMS AND INTERVENTIONS

10.6.1 Strains

Psychological

There generally has been 'little consideration... given to the psychological problems suffered by the prostitutes themselves' (Carr, 1995), even though it appears that these are the most common and severe of all the strains this occupational group experiences. Studies that have been conducted with prostitutes show that the more common psychological strains experienced by prostitutes are low self esteem (Harcourt *et al.*, 2001), a profound sense of isolation (Pyett and Warr, 1999), depression (Boyle *et al.*, 1997), distress (El-Bassel *et al.*, 1997), and post traumatic stress disorder (Farley *et al.*, 1998; Farley and Barkan, 1998).

Sex traders (defined as women who had exchanged sex for money or drugs during the 30 days prior to the interview) have been found to score significantly higher than non-sex traders who also used drugs on two measures of psychological distress (El-Bassel *et al.*, 1997). This study controlled for age, ethnicity, pregnancy, recent rape, perceived risk for acquired immunodeficiency syndrome, current and regular crack use and current and regular alcohol use. One difficulty with research examining the levels of psychological distress experienced by prostitutes is determining if distress is related to factors associated with the work itself or with factors that pre-date entry into the occupation (Boyle *et al.*, 1997). A study that attempted to address this found that for female sex workers, involvement in the sex industry itself was not a source of psychological distress (Boyle *et al.*, 1997). Yet for women who scored from mild to moderate on the GHQ-28 (a general self report measure of psychological state including somatic symptoms, anxiety, insomnia, social dysfunction and severe depression), injecting drug use, leaving home prior to age 16, and wanting to leave the sex industry emerged as predictors of poor mental health. The authors of this study suggest that the risks to well-being associated with prostitution are magnified for persons with high levels of psychological distress. Similar conclusions were also drawn in another study (Romans *et al.*, 2001).

One significant outcome of prostitution may be post traumatic stress disorder (PTSD) (Farley and Barkan, 1998). Two studies that have investigated this strain have been conducted across cultures (in the United States, South Africa,

Thailand, Turkey, and Zambia). Overall, large percentages of prostitutes in both studies (68% in the San Francisco sample and 67% of the mixed country sample) met the DSM-III-R criteria for a diagnosis of PTSD. These samples reported a high percentage of prostitutes that want to leave the profession (88% and 92% respectively) and this could be one potential reason for the rates of elevated stress. These high percentages suggest that for the lowest on the occupational hierarchy and those who feel trapped, the strains experienced are severe. In terms of self esteem, these authors suggest that 'the hatred and contempt aimed at those in prostitution is ultimately internalised. The resulting self-hatred and lack of self-respect are extremely long-lasting' (Farley *et al.*, 1998).

Behavioural

By far the most prevalent stereotype of the prostitute is the down and out drug addicted street walker. Some studies suggest that women working on the street are also more likely to be hard drug users (Pyett and Warr, 1999). In places where brothels are legal, it is generally illegal to use drugs on the premises and women who are proven drug users are not employed by management (Pyett and Warr, 1999). There is no direct evidence of a causal relationship between prostitution and drug use and this relationship could be bidirectional. It has been suggested that some prostitutes use soft drugs in order to cope with the job (Brewis and Linstead, 2000). Whether this is a negative outcome (strain) or a technique that prostitutes use to cope with work stress is difficult to determine. What remains clear is that drug use and prostitution often go hand in hand (Scambler and Scambler, 1999).

Other potential behavioural strains include less consistent use of condoms (Carr, 1995), and fewer regular health examinations (Boyle *et al.*, 1997). These are behaviours that have been found to be more common in prostitutes who are experiencing more extreme psychological distress, and who work in less protected environments (Pyett and Warr, 1999). Whether these are a result of work stress or just a function of the circumstances surrounding particular work environments is another question that currently remains unanswered.

Physical

We found no studies in the work stress literature that investigate some of the more traditional physical strains (such as headaches or gastrointestinal disorders) that prostitutes face. This could be because more severe illnesses are typically associated with prostitutes, or it could be due to the marginalisation of this group. Prostitutes have many of the same health concerns as others, but generally do not get the same attention within healthcare systems and are often not comfortable within the healthcare system (Carr, 1995).

10.6.2 Coping Mechanisms and Interventions

There are very few studies that investigate how prostitutes cope with the work stress associated with their profession. One study that looked at this question found that female prostitutes do rely on many of the ideological techniques described by

Ashforth and Kreiner (1999) to manage their day to day lives (Brewis and Linstead, 2000). Certain ideologies behind sex work, such as the argument that prostitutes are service workers who just happen to be selling their sexuality instead of their social skills, is one such technique (Brewis and Linstead, 2000). Other techniques are used to reframe what a prostitute does in terms of actually providing potential education to clients in terms of safe sex, and many will describe their work as 'healing' which implies that the work is about more than just 'satisfying the pent up sexual desires of lonely or frustrated men' (Brewis and Linstead, 2000). There are distancing techniques that many use to maintain boundaries between their personal and professional identities. Some of these have been discussed previously and include the use of condoms with clients but not in personal relationships, refusing to kiss clients, as well as rituals to transition from work to home (Brewis and Linstead, 2000).

Drug use is another way that prostitutes cope with the work stress associated with their occupation. As discussed above many use soft drugs to manage the stress. Indeed, Brewis and Linstead (2000) go so far as to suggest that because '. . . prostitution may by turns be boring, terrifying, unpredictable, disgusting and risky in potentially equal measure, how do . . . prostitutes make it through the working day psychologically speaking *without* resorting to the use of drugs?'.

There have been very few, if any, interventions for work stress for prostitutes. We know of no studies looking at this issue and have not, in our review of the literature, come across case studies or examples of any programmes that have been initiated to relieve work stress or improve coping skills in this line of work. While there are programmes and treatment centres to enable prostitutes to exit from the occupation (Cole, 1998), there are no avenues of stress relief except for outreach programmes that provide condoms and some support to working women. Outreach programmes, however, have been criticised for their narrow focus on sexual health and HIV prevention (Rickard and Growney, 2001).

10.7 METHODOLOGICAL CHALLENGES

One of the main methodological challenges involved in studying this occupational group is the relative inaccessibility of this population (Browne and Minichiello, 1996). Most studies of both men and women involved in prostitution rely on convenience samples (Browne and Minichiello, 1996). In some instances researchers report that pimps or brothel management discourage participation in studies (Farley *et al.*, 1998), and the workers themselves may be hesitant due to distrust of authority, the intensely personal nature of the occupational tasks and the desire to protect their anonymity (Karim *et al.*, 1995). Another potential difficulty is the nature of the profession which means that any data used in a study is most often self report. Most organisations in this industry do not keep any useful records that could be used to answer research questions. As an additional methodological caveat, it must be noted that the majority of studies on which this discussion has been based are drawn from non-random samples within single countries. Finally, there is a lack of funding to support research into marginalised workers such as prostitutes (Browne and Minichiello, 1996), and this may well be true for research involving any type of dirty work.

10.8   IMPLICATIONS AND CONCLUSIONS

This chapter has reviewed the literature on prostitution and work stress in order to understand work stress experienced by prostitutes. The stressors and resulting strains and some potential moderators were discussed. Coping mechanisms used by prostitutes as well as interventions tried were also examined.

Prostitutes are arguably subjected to many stressors at work. Prostitution is atypical in terms of most human service work in that the work in many cases is illegal and carries with it forms of physical, social, and moral taint, factors which both are potentially significant sources of work stress. Some of the stressors that we have identified are also pertinent to other human service workers as well as to dirty workers more generally. Many of these are aggravated for prostitutes.

We conclude by considering the possible consequences of continuing to choose not to study the stress and strain associated with prostitution. The most obvious consequence would be continuing ignorance. There is, however, a larger issue. In his early writing on dirty work, Hughes (1951) pointed to the benefits of studying those social phenomena with the least prestige and we suggest his comments apply to the understanding of stress associated with 'dirty work'. If researchers paid more attention to workers in both low wage and low prestige 'dirty jobs', a lot more could be learned about the full range of work stress that workers must deal with. The lack of research attention to some of the potentially most stressful occupations may result in theories that do not reflect the full range of the phenomenon of interest – in this case the stress that human service workers face every day at work.

10.9   REFERENCES

Albert, A.E., Warner, D.L. and Hatcher, R.A., 1998, Facilitating condom use with clients during commercial sex in Nevada's legal brothels. *American Journal of Public Health*, **88**, pp. 643–646.

Albert, A.E., Warner, D.L., Hatcher, R.A., Trussell, J. and Bennett, C., 1995, Condom use among female commercial sex workers in Nevada's legal brothels. *American Journal of Public Health*, **85**, pp. 1514–1523.

Alexander, P., 1987, Why this book? In *Sex work: Writings by women in the sex industry*, edited by F. Delacoste and P. Alexander (San Francisco, California: Cleis Press), pp. 14–19.

Anonymous, 1996, Saving children from sex. *The Economist*, **340** (7981), p. 17.

Anonymous, 2001, Oldest profession solicits recruits for trade union. *The Globe & Mail* (Oct. 4).

Ashforth, B.E. and Kreiner, G.E., 1999, How can you do it?: Dirty work and the challenge of constructing a positive identity. *Academy of Management Review*, **24**, pp. 413–434.

Barling, J. and Sorensen, D., 1997, Work and family: In search of a relevant research agenda. In *Creating tomorrow's organizations*, edited by C.L. Cooper, and S.E. Jackson (NY: John Wiley & Sons Ltd.), pp. 157–169.

Barling, J., Bluen, S.D. and Fain, R., 1987, Psychological functioning following an acute disaster. *Journal of Applied Psychology*, **72**, pp. 683–690.

Barling, J., Rogers, A.G. and Kelloway, E.K., 2001, Behind closed doors: In-home workers' experience of sexual harassment and workplace violence. *Journal of Occupational Health Psychology*, **6**, pp. 255–269.

Beehr, T.A., 1995, Role ambiguity and role conflict in the workplace. In *Psychological stress in the workplace*, edited by T.A. Beehr (NY: Routledge), pp. 55–82.

Belk, R.W., Ostergaard, P. and Groves, R., 1998, Sexual consumption in the time of AIDS: A study of prostitute patronage in Thailand. *Journal of Public Policy and Marketing*, **17**, pp. 197–214.

Bluen, S.D. and Edelstein, I., 1993, Trade union support following an underground explosion. *Journal of Organizational Behavior*, **14**, pp. 473–480.

Boyle, F.M., Dunne, M.P., Najman, J.M., Western, J.S., Turrell, G., Wood, C. and Glennon, S., 1997, Psychological distress among female sex workers. *Australian and New Zealand Journal of Public Health*, **21**, pp. 643–646.

Brewis, J. and Linstead, S., 2000, 'The worst thing is the screwing' (1): Consumption and the management of identity in sex work. *Gender, Work & Organization*, **7**, pp. 84–97.

Browne, J. and Minichiello, V., 1995, The social meanings behind male sex work: Implications for sexual interactions. *The British Journal of Sociology*, **46**, pp. 598–622.

Browne, J. and Minichiello, V., 1996, Research directions in male sex work. *Journal of Homosexuality*, **31**, pp. 29–56.

Carr, S.V., 1995, The health of women working in the sex industry: A moral and ethical perspective. *Sexual & Marital Therapy. Special Issue: Women's sexual health*, **10**, pp. 201–213.

Cloud, J., 2000, The oldest profession gets a new museum. *Time*, **156**, p. 4.

Cole, W., 1998, Life off the streets. *Time*, **152** (20), p. 8.

Coyote San Francisco, n.d. Retrieved Oct. 18, 2001, from the World Wide Web: http://www.bayswan.org/COYOTE.html

Daley, S., 2001, New rights for Dutch prostitutes, but no gain. *New York Times* (August 12).

Dank, B.M., 1999, Sex work, sex workers, and beyond. In *Sex work & sex workers*. Vol. 2, edited by B.M. Dank and R. Refinetti (New Brunswick, New Jersey: Transaction Publishers), pp. 1–6.

El-Bassel, N., Schilling, R.F., Irwin, K.L. and Faruque, S., 1997, Sex trading and psychological distress among women recruited from the streets of Harlem. *American Journal of Public Health*, **87**, pp. 66–70.

Ellingson, J.E., Gruys, M.L. and Sackett, P.R., 1998, Factors related to the satisfaction and performance of temporary employees. *Journal of Applied Psychology*, **83**, pp. 913–921.

Ernst Kossek, E. and Ozeki, C., 1998, Work-family conflict, policies, and the job-life satisfaction relationship: A review and directions for organizational behavior-human resources research. *Journal of Applied Psychology*, **83**, pp. 139–149.

Farley, M. and Barkan, H., 1998, Prostitution, violence, and posttraumatic stress disorder. *Women & Health*, **27**, pp. 37–49.

Farley, M., Baral, I., Kiremire, M. and Sezgin, U., 1998, Prostitution in five countries: Violence and post-traumatic stress disorder. *Feminism & Psychology*, **8**, pp. 405–426.

Fisher, C.D., 1993, Boredom at work: A neglected concept. *Human Relations*, **46**, pp. 395–417.

Geller, P.A. and Hobfoll, S.E., 1994, Gender differences in job stress, tedium and social support in the workplace. *Journal of Social and Personal Relationships*, **11**, pp. 555–572.

Harcourt, C., Van Beek, I., Heslop, J., Mcmahon, M. and Donovan, B., 2001, The health and welfare needs of female and transgender street sex workers in New South Wales. *Australian & New Zealand Journal of Public Health*, **25**, pp. 84–89.

Hornblower, M., 2000, The Spokane murders. *Time*, **156**, pp. 42–44.

Hughes, D.M., 2000, The 'Natasha' trade: The transnational shadow market of trafficking in women. *Journal of International Affairs*, **53**, pp. 625–651.

Hughes, E.C., 1951, Work and the self. In *Social psychology at the crossroads: The university of Oklahoma lectures in social psychology,* edited by J.H. Rohrer, and M. Sherif, (New York: Harper & Brothers Publishers), pp. 313–323.

Jaggar, A.M., 1997, Contemporary western feminist perspectives on prostitution. *Asian Journal of Womens Studies*, **3**, pp. 8–29.

Jenness, V., 1990, From sex as sin to sex as work: Coyote and the reorganization of prostitution as a social problem. *Social Problems*, **37**, pp. 403–420.

Karasek, R.A., 1979, Job demands, job decision latitude, and mental strain: Implications for job redesign. *Administrative Science Quarterly*, **24** (June), pp. 285–308.

Karim, Q.A., Karim, S.A., Soldan, K. and Zondi, M., 1995, Reducing the risk of HIV infection among South African sex workers: Socioeconomic and gender barriers. *American Journal of Public Health*, **85**, pp. 1521–1528.

Korn, P., 1996, *Lovejoy: A year in the life of an abortion clinic*. (New York: The Atlantic Monthly Press).

Leblanc, M.M. and Kelloway, E.K. (in press), Predictors and outcomes of workplace violence and aggression. *Journal of Applied Psychology*.

Lewis, J. and Maticka-Tyndale, E., 2000, Licensing sex work: Public policy and women's lives. *Canadian Public Policy/Analyse de Politiques*, **26**, pp. 437–449.

Lichtenstein, B., 1999, Reframing 'eve' in the aids era: The pursuit of legitimacy by New Zealand sex workers. In *Sex work & sex workers*. Vol. 2, edited by B.M. Dank and R. Refinetti (New Brunswick, U.S.A.: Transaction Publishers), pp. 37–59.

Lowe, C. and Goldberg, B., 2000, Arson bargain. *The Jerusalem Post* (Oct. 22).

Martens, M.F.J., Nijhuis, F.J.N., Van Boxtel, M.P.J. and Knottnerus, J.A., 1999, Flexible work schedules and mental and physical health. A study of a working population with non-traditional working hours. *Journal of Organizational Behavior*, **20**, pp. 35–46.

Michelle, K., 2000, Asia's dirty secret. *Harvard International Review*, **22**, pp. 42–45.

Miller, G., 1978, *Odd jobs* (Englewood Cliffs, New Jersey: Prentice Hall Inc.).

Minichiello, V., Marino, R., Browne, J., Jamieson, M., Peterson, K., Reuter, B. and Robinson, K., 1999, A profile of the clients of male sex workers in three Australian cities. *Australian and New Zealand Journal of Public Health*, **23**, pp. 511–518.

Morris, J.A. and Feldman, D.C., 1996, The dimensions, antecedents and consequences of emotional labor. *Academy of Management Review*, **21**, pp. 986–1010.

Narayanan, L., Menon, S. and Spector, P.E., 1999, Stress in the workplace: A comparison of gender and occupations. *Journal of Organizational Behavior*, **20**, pp. 63–73.

Nieves, E., 2001, Anxious days in bordello country: Resort plan brings what Nevada brothels hate most: Attention. *New York Times* (August 19), p. 18.

Plumridge, L. and Abel, G., 2001, A 'segmented' sex industry in New Zealand: Sexual and personal safety of female sex workers. *Australian & New Zealand Journal of Public Health*, **25**, pp. 78–83.

Pratt, L.I. and Barling, J., 1988, Differentiating between daily events, acute and chronic stressors: A framework and its implications. In *Occupational stress: Issues and development in research*, edited by J.J. Hurel, L.R. Murphy, S.L. Sauter, and C.L. Cooper (London: Taylor Francis), pp. 41–51.

Prostitutes Collective of Victoria, n.d. Retrieved October 18, 2001, from the World Wide Web: http://www.arts.unimelb.edu.au/amu/ucr/student/1996/m.dwyer/pcvhome.html

Pyett, P.M. and Warr, D., 1999, Women at risk in sex work: Strategies for survival. *Journal of Sociology*, **35**, pp. 183–197.

Pyett, P.M., Haste, B.R. and Snow, J.D., 1996, Who works in the sex industry? A profile of female prostitutes in Victoria. *Australian and New Zealand Journal of Public Health*, **20**, pp. 431–433.

Rambo Ronai, C., 1992, Managing aging in young adulthood: The 'aging' table dancer. *Journal of Aging Studies*, **6**, pp. 307–317.

Rickard, W. and Growney, T., 2001, Occupational health and safety amongst sex workers: A pilot peer education resource. *Health Education Research*, **16**, pp. 321–333.

Romans, S.E., Potter, K., Martin, J. and Herbison, P., 2001, The mental and physical health of female sex workers: A comparative study. *Australian & New Zealand Journal of Psychiatry*, **35**, pp. 75–80.

San Francisco Coalition on Prostitution, n.d. Retrieved Oct. 18, 2001, from the World Wide Web: http://www.bayswan.org/COP.html

Scambler, G. and Scambler, A., 1999, Health and work in the sex industry. In *Health and work: Critical perspectives*, edited by N. Daykin and D. Lesley (New York, N.Y.: St Martin's Press, Inc.), pp. 71–85.

Schaubroeck, J. and Jones, J.R., 2000, Antecedents of workplace emotional labor dimensions and moderators of their effects on physical symptoms. *Journal of Organizational Behavior*, **21**, pp. 163–183.

Sells, B., 1994, What asbestos taught me about managing risk. *Harvard Business Review*, March–April, pp. 76–90.

Simmons, M., 1999, Theorizing prostitution: The question of agency. In *Sex work and sex workers*, edited by B.M. Dank and R. Refinetti (New Brunswick, NJ: Transaction), pp. 125–148.

The Sex Workers Alliance of Vancouver, n.d. Retrieved Oct. 18, 2001, from the World Wide Web: http://www.walnet.org/csis/groups/swav/swav.html

West, J., 2000, Prostitution: Collectives and the politics of regulation. *Gender, Work and Organization*, **7**, pp. 106–118.

Wilton, T., 1999, Selling sex, giving care: The construction of aids as a workplace hazard. In *Health and work: Critical perspectives*, edited by N. Daykin, and L. Doyal (New York, N.Y.: St Martin's Press, Inc.), pp. 180–197.

Wood, D., 1999, Missing. *Elm Street*, November, pp. 96–110.

Zohar, D., 2000, A group-level model of safety climate: Testing the effects of group climate on microaccidents in manufacturing jobs. *Journal of Applied Psychology*, **85**, pp. 587–596.

CHAPTER ELEVEN

Social Workers and Human Service Practitioners

Robert L. Lonne

11.1   INTRODUCTION

The concepts of 'burnout' and, more recently, 'vicarious trauma' have predominated in the research and literature concerning occupational stressors and strain for social workers and human service practitioners. Whilst there has been considerable research of the causes and effects of burnout and occupational stress, there is little evidence to suggest that these issues are being proactively or systemically addressed by most human service organisations. Rather, it is usually left to individual practitioners or managers to respond to its sometimes debilitating effects (Winefield *et al.*, 2000), despite the overwhelming empirical results indicating that organisational and job factors are the key influences on burnout (Barak *et al.*, 2001; Dollard *et al.*, 2001). Regrettably, the costs of inaction are evidenced in the high staff turnover that plagues the human services sector, and the decreased efficiency and effectiveness from otherwise committed and capable helping professionals (Barak *et al.*, 2001).

Nevertheless, it would be erroneous to paint the picture that a career in social work and the human services will inevitably be traumatic, short-lived and result in burnout that culminates in the physically, emotionally and psychologically destroyed practitioner leaving the helping professions forever. Whilst burnout and vicarious trauma are real and important issues to be tackled in the human services, they are sometimes overstated and portrayed as endemic in relation to their incidence. The reality is that the overwhelming majority of workers, although at times experiencing considerable occupational stress, generally cope quite well and productively use a range of coping strategies. They live life to the full, enjoy a challenging and stimulating career while experiencing the personal rewards that helping others can entail.

This chapter will review and summarise the research findings and literature with respect to occupational stress, burnout and vicarious trauma for social workers and human service practitioners, and will explore their sources, effects and outcomes. The similarities and differences found among the broad diversity of practice in the human services will be identified and critically examined. A range of micro and macro interventions and strategies will also be suggested and explored, along with the implications for stakeholders. The findings of a recent longitudinal study of factors affecting staff turnover and work stress for a group of Australian rural social workers will be outlined and discussed. Future research options for examining occupational stress in the human services will be suggested.

11.2   LITERATURE REVIEW

11.2.1   The Macro Context of Practice in the Human Services

There is considerable diversity among Western societies with respect to the educational qualifications of social workers and human service practitioners, and this can lead to significantly different status, remuneration, roles, ideologies and practice frameworks among staff. For example, the base qualification of social workers and human service practitioners can vary from those without formal qualifications, to a two-year diploma, a three- or four-year tertiary degree, or a Masters degree (McDonald, 1999; Pottage and Huxley, 1996; Zapf, 1993). In addition, different professional courses may focus their training on direct practice with individuals, community work, social policy development, or all of these and, hence, practitioners employ different approaches to interventions.

Notwithstanding this diversity and existing professional rivalries, much common ground exists among these social care practitioners, including many aspects of the social and community mandates for intervention, core knowledge and skill bases, professional values and ethics, and an overall 'helping and empowering' orientation toward the disadvantaged and marginalised. For this paper, social workers and human service practitioners are social care professionals who are employed to provide a diverse range of welfare and other helping services to communities, groups and individuals who experience distress, trauma and difficulties in meeting their social needs or in their social relationships. They undertake this work in order to enhance social functioning and facilitate social change.

The diversity found in human service practice does not merely reflect the need to respond to a range of social problems and needs. It has also developed as a result of a period of profound and rapid social, technological, economic and political change that occurred in the latter half of the 20th century. There have been many changes to the structures, processes, policies, organisations and approaches used to deliver social care programmes and services, frequently driven by conservative political ideologies that foster individualism rather than communalism (Harris and McDonald, 2000; Howe, 1994).

Across OECD countries there has generally been increased social care expenditure yet simultaneous fiscal tightness, usually accompanied by a decreased service delivery role for government that has been operationalised by a separation of 'purchaser' and 'provider' roles (Kalisch, 2000). There has also been increasing use of service delivery contracts to fund non-government and for-profit human service agencies (Healy, 1998; Howe, 1994). Furthermore, there has been major restructuring of service delivery mechanisms and processes which have led to a plethora of narrowly targeted programmes and services (Harris and McDonald, 2000; Williams, 1994). Whilst aiming to meet the diverse social care needs of communities and service users better, these structural changes have significantly contributed to an increasingly complex and fragmented social welfare systems (Healy, 1998; Howe, 1994, Parton, 1994, 1996).

The advent of 'managerialism' has also fundamentally altered the organisational contexts in which social workers and human service practitioners operate. Managerialism has seen a marked paradigm shift away from traditional professional values to embrace instead business principles such as efficiency and effectiveness, and practices like cost–benefit analysis and strategic planning (Jones and May, 1992). Moreover, there has been a simultaneous shift of power away from professionals to managers, who have marked their domain with a managerialist discourse that largely ignores notions of social justice and empowerment of service users (Fook *et al.*, 2000; Howe, 1994). These influences have significantly impacted on the ways in which services are delivered, as they have been accompanied by expanded policy directives, decreased professional autonomy, increased accountability, altered ethical values, creeping proceduralism within proliferating case management systems, tighter eligibility requirements for service users, and greater emphasis on social compliance and social control (Howe, 1994; Lonne *et al.*, in press; Parton, 1994, 1996; Pottage and Huxley, 1996; Zunz, 1998).

Overall, these changes have resulted in a convergence of job titles and descriptions, and increased flexibility in the types of qualifications deemed suitable for professional positions in the human services (Healy, 1998; McDonald, 1999; McDonald and Jones, 2000). Many in the sector have welcomed this, as the very broad array of practice roles and fields requires a diversity of practice orientations, skills and knowledge in order to appropriately respond to the complex social problems and issues stemming from the globalised and capitalist world economic and financial systems. However, as noted above, the current context in which human service workers practise is one characterised by rapid change, contested values and mandates, complex structural arrangements and relationships, and uncertainty about what exactly 'best practice' entails across the different community, organisational and practice field contexts.

11.2.2 Anxiety, Stress and Burnout in the Human Services

It is not be surprising that social welfare work has been found to be especially anxiety provoking due to the macro context of the human services, its focus on human beings, their complex problems and relationships, and emotionally-charged situations (Coholic and Blackford, 1999; Jones *et al.*, 1991; Strozier and Evans,

1998). However, the empirical research is not consistent in this regard as some comparative studies have found similar levels of anxiety in other occupations. Social workers' anxiety has been found to be positively associated with depression (Jones *et al.*, 1991) and burnout (Jayaratne *et al.*, 1986), and is moderated by supervisors' emotional and other supports (Koeske and Koeske, 1989). However, conceptualising the differences between anxiety and stress has been problematic, and most studies have focussed on the 'burnout' concept, often using it interchangeably with occupational stress, work stress, distress and vicarious trauma (Adams *et al.*, 2001; Dollard *et al.*, 2001; Strozier and Evans, 1998).

'Work stress' has been defined as a 'condition or intermediate arousal state between objective stressors and strain', with strain being 'reactions to the conditions of stress' (Dollard *et al.*, 2001, p. 15). On the other hand, the concept of 'burnout' describes the longer-term process where chronic stressors lead to occupational stress experienced by those in people-oriented professions that results in them being unable to cope with their work psychologically and emotionally (Barak *et al.*, 2001; Maslach, 1976). It has been conceptualised as involving the three components of:

- Emotional exhaustion;
- Depersonalisation of clients; and
- Decreased personal accomplishment (Maslach, 1998).

Emotional exhaustion is characterised by feeling emotionally overextended and depleted, while depersonalisation entails a practitioner becoming overly cynical and detached to others, particularly clients, and diminished personal accomplishment is evidenced when workers demonstrate reduced self-efficacy and productivity (Maslach, 1998). Emotional exhaustion is the burnout component that is closest to occupational stress (Koeske and Koeske, 1989, 1993). However, most research of burnout has involved human service workers, inviting the criticism that it is a flawed concept (Demerouti *et al.*, 2001; Jones *et al.*, 1991; Soderfeldt *et al.*, 1995) that merely re-labels emotional exhaustion (Wallace and Brinkerhoff, 1991) and reflects the predominance of 'feeling type' people in the human services (Garden, 1989).

Nevertheless, some authors (Maslach, 1998; Powell, 1994) argue that maintaining the three dimensions of burnout is critical because this ensures that burnout is an interpersonal framework rather than a narrow concept of stress reactions to work or other stressors. However, a re-examination of the dimensions of burnout found evidence linking work stress only with emotional exhaustion (Wallace and Brinkerhoff, 1991) and this is supported by other researchers (Koeske and Koeske, 1989, 1993; Koeske and Kirk, 1995a) who noted that *detachment*, a dimension of the depersonalisation aspect of burnout, is also an effective tool against it, as detached concern can assist a worker to prevent emotional overload.

Social work practitioners have generally been found to experience higher stress and lower well-being levels than other occupations (Bennett *et al.*, 1993;

Bradley and Sutherland, 1995; Collings and Murray, 1996; McLean and Andrew, 2000; Strozier and Evans, 1998; Um and Harrison, 1998), particularly statutory workers (Balloch *et al.*, 1998; Dollard *et al.*, 2001). In addition, there is evidence of social workers having higher levels of depression than others (Gibson *et al.*, 1989) although the results are inconsistent (Jones *et al.*, 1991).

On the other hand, some studies have found that social workers and human service practitioners have lower levels of stress than others (Demerouti *et al.*, 2001; Soderfeldt *et al.*, 1995). It has also been argued that the reported levels of stress, anxiety and burnout experienced by human service practitioners are relatively low, especially when viewed alongside the generally high levels of job satisfaction (Pottage and Huxley, 1996; Poulin, 1994; Powell, 1994). As Marriott *et al.* (1994, p. 204) note, 'One could speculate that social workers complain about or downgrade various aspects of their jobs but are nevertheless reasonably content.'

11.2.3 Vicarious Trauma

Over the past decade the concept of 'vicarious trauma' has been increasingly studied because of its perceived adverse consequences for professionals working with traumatised victims, although its prevalence remains unclear. Vicarious trauma has been defined as 'the enduring psychological consequences for therapists of exposure to traumatic experiences of victim clients' (Schauben and Frazier, 1995, p. 53), although it is arguable that it is an equally valid phenomenon for all human service workers who are exposed in a secondary fashion to client trauma.

Its symptoms may include a decreased emotional and physical energy, bystander guilt, anxiety, depression and PTSD, disconnection from loved ones, social withdrawal, and increased sensitivity to violence (Dane, 2000; Sexton, 1999). Physical signs of stress can also accompany these symptoms. Other signs include disruption to self-protective and safety beliefs, changed worldviews, helplessness, cynicism and despair (Cunningham, 1999; Sexton, 1999). There is said to be a generally faster recovery from these symptoms than from burnout (Sexton, 1999). Although it remains a nebulous concept and has been used interchangeably with terms such as countertransference, secondary traumatisation, burnout, and compassion fatigue (Dane, 2000; Illiffe and Steed, 2000; Sexton, 1999), there is some empirical evidence to suggest that it is conceptually different from burnout (Schauben and Frazier, 1995).

In either event, it is a normal reaction to empathic engagement with clients who have suffered significant and traumatic events and results from an accumulation of exposure experiences (Sexton, 1999). It goes, so to speak, with the territory of trauma counselling and possibly with many other human service worker roles and functions. Whilst there has not been a lot of research of vicarious trauma, it has been found among domestic violence workers (Coholic and Blackford, 1999; Illiffe and Steed, 2000), trauma therapists (Perlman and Mac Ian, 1995) and sexual assault counsellors (Schauben and Frazier, 1995). Higher levels have been found to be associated with burnout (Adams *et al.*, 2001) and higher caseloads of traumatised clients (Schauben and Frazier, 1995), but not with a prior history of trauma or assault for the worker (Adams *et al.*, 2001; Schauben and Frazier, 1995).

11.3   SOURCES OF STRESS IN THE HUMAN SERVICES

The overwhelming conclusion of the research is that structural work stressors, rather than personal characteristics, are the most influential factors affecting the incidence and intensity of burnout (Barak *et al.*, 2001; Bennett *et al.*, 1993; Dollard *et al.*, 2001; Zastrow, 1999), although there are contradictory findings about which factors are involved. Furthermore, it has been found that burnout can be moderated by organisational, professional and personal social support (Um and Harrison, 1998; Zunz, 1998). Interestingly, whilst client experiences are theoretically identified as contributing factors in the process of burnout, investigations have usually failed to establish a link (Barak *et al.*, 2001; Dollard *et al.*, 1999).

11.3.1   Work Factors

There are numerous work-related factors identified in the literature as contributing to work stress and burnout, and these can be placed into the following general groups: particular practice fields, work role issues, organisational structure and culture, high workloads, low levels of control and autonomy, supervision practices, lack of accomplishment and efficacy, violence and conflict, and racism within the organisation. However, there is increasing evidence that a combination of high job demands and low supports is the critical factor in burnout (Dollard *et al.*, 2001).

Studies of burnout and work stress have been conducted across various positions, occupations and practice fields in the human services such as child welfare and protection, mental health, hospitals and community health, disability, generic social services, and support services for the military. Very few studies have addressed the issue of burnout and work stress in non-government, not-for-profit and for-profit agencies. There have been a number of findings indicating that child welfare and child protection work are associated with higher levels of burnout and perceived stress (Gold, 1998; Jayaratne *et al.*, 1986; Jones *et al.*, 1991; Le Croy and Rank, 1987; Samantrai, 1992; Thorpe *et al.*, 2001). It is suggested that the contested and conflictual nature of this sort of statutory work leads to elevated stress for the workers involved who often tread a difficult path in being accountable, balancing the interests of all parties, and ensuring children are protected from abuse.

Social workers have also been found to have higher burnout and stress levels than other human service workers (Bennett *et al.*, 1993; Dollard *et al.*, 2001; McLean and Andrew, 2000) and this has been suggested to result from their statutory roles, somewhat limited autonomy and the complex decisions they are often required to make. Role ambiguity and role conflict for social workers and human service workers have consistently been found to be positively associated with levels of emotional exhaustion, burnout and work stress (Balloch *et al.*, 1998; Barak *et al.*, 2001; Dollard *et al.*, 2001; Siefert *et al.*, 1991). However, Um and Harrison (1998) recently used LISREL linear structural relation techniques to examine the process of stress–strain (burnout)-outcome (job satisfaction) in 166 Florida social workers, and concluded that role conflict, but not role ambiguity, intensified burnout and job dissatisfaction.

The organisational structure, culture and climate have also been found in some studies to contribute to higher levels of burnout and stress (Sundet and Cowger, 1990). For example, Bradley and Sutherland (1995) used a quantitative and qualitative research design to compare a group of British social workers and home help workers. The social workers reported stress resulting from the organisational structure and climate including factors such as inadequate resources, time pressures, paper work and lack of performance feedback. Dollard *et al.* (2001) also found organisational processes such as unfair job selections were related to perceptions of work stress. Poor employer–employee relations have also been identified as problematic (McLean and Andrew, 2000). Perhaps more important is the organisational culture with respect to work stress. Thompson *et al.* (1996) found in a comparative study of three British local authority social service departments that work groups could develop and embrace a 'culture of stress' that reflects and contributes to higher levels of stress. They concluded that management had a critical part to play in creating the vision and circumstances to prevent this.

The relationship between large caseloads and stress, although intuitively linked, has not been consistently born out by the empirical evidence, perhaps due to measurement difficulties (See Egan and Kadushin, 1993; Koeske and Koeske, 1989). For example, LeCroy and Rank (1987) found no significant relationship between social workers' job performance variables such as the amount of overtime, direct client contact, practice experience and burnout, but Barak *et al.* (2001) identified it as a significant factor in their review of 25 studies of the antecedents to turnover in human service practitioners. There is, however, stronger evidence regarding the negative impact of overall workloads on work stress (Bradley and Sutherland, 1995; Collings and Murray, 1996; Sundet and Cowger, 1990). Furthermore, when practitioners have limited control and autonomy over their work, burnout and stress levels tend to be higher and job satisfaction lower (Barak *et al.*, 2001; Balloch *et al.*, 1998; Guterman and Jayaratne, 1994; Koeske and Kirk 1995b; McLean and Andrew, 2000; Poulin 1994, 1995). However, high workloads appear to be particularly problematic when they translate to high job demands that are combined with low control and autonomy by the worker, and low resources and support (Demerouti *et al.*, 2001; Dollard *et al.*, 2001; Maslach, 1998; Soderfeldt *et al.*, 1995). This combination has been demonstrated to be associated with significant negative outcomes for the individuals and organisations concerned.

The importance of effective and supportive supervision has been consistently identified as of pivotal importance with respect to job satisfaction and moderating the effects of work stress while poor supervision has been found to contribute to higher work stress (Dollard *et al.*, 1999, 2000; Wilcoxon, 1989; Winefield *et al.*, 2000). Dissatisfaction with supervision has been found to be associated with higher levels of stress and may be a potent source of it, and trigger a decision to leave (Collings and Murray, 1996; Samantrai, 1992). Furthermore, the quality of supervision has been found to be an important element in both supporting workers and controlling the demands placed on them (Jones *et al.*, 1991). For example, Grasso (1994) undertook a two-year study of management styles and found that supportive and participatory supervision of staff was related to higher job satisfaction. However, collegial support may be even more important than supervisor support as a moderator of burnout (Dollard *et al.*, 2001; Himle *et al.*, 1989a, 1989b; Um and Harrison, 1998).

Some writers have claimed that the strongest factor causing burnout for clinical workers is lack of therapeutic success (Ratliff, 1988), but others have found it not to be a factor (Himle and Jayaratne, 1990; Wallace and Brinkerhoff, 1991). However, most studies have identified that practitioners' self-efficacy is inversely related to levels of burnout and job dissatisfaction, although its precise contribution as a preventative measure, moderating factor or consequence of burnout remains unclear (Barak *et al.*, 2001; Dollard *et al.*, 2001; Koeske and Koeske, 1989, 1993; Zunz, 1998).

There has been increasing attention given to the influence of violence and conflict in the workplace and its effects on work stress, especially when associated with a lack of recognition and support (Smith and Nursten, 1998). This issue has been particularly important in statutory services in Great Britain. For example, Balloch *et al.* (1998) conducted a longitudinal study using quantitative and qualitative methodologies and found a strong association between violence, threats of violence and verbal abuse, and stress levels. They concluded that rising stress levels may well be associated with dealing with increasing levels of workplace violence. The issue has also been recently identified as contributing to work place stress in a large-scale study of a statutory child welfare and juvenile justice agency in Australia (Dollard *et al.*, 2001).

The issue of racism and its effects on occupational stress in human service agencies has also been given increasing attention, although the evidence is mixed. A British study (Collings and Murray, 1996) found no relationship while a U.S. study found that perceptions of cultural sophistication within the organisation were associated with occupational stress levels for Latino social workers (Gant and Guitierrez, 1996).

11.3.2 Non-Work Factors

The research findings indicate that personal and demographic characteristics play a limited role in the work stress and burnout process, with these factors mostly acting as moderating influences rather than causal factors or sources of stress (Gutek *et al.*, 1988; Jayaratne *et al.*, 1983, see Section 11.3). There are equivocal findings with respect to the influence of age on work stress with Collings and Murray (1996) finding that older workers tended to have higher levels, whereas Barak *et al.* (2001) found the opposite in a review of burnout and turnover research. Most studies have not found gender to be an influential factor (Barak *et al.*, 2001; Dollard *et al.*, 2001; Collings and Murray, 1996).

However, there is evidence to suggest that a practitioner's pre-existing level of psychological well-being is negatively related to subsequent levels of emotional exhaustion and job dissatisfaction (Koeske and Kirk, 1995a). Zunz (1998) examined the influence of resiliency on burnout and found that those workers who were more resilient and also received social support tended to experience lower levels of burnout. With respect to other personal characteristics, Dollard *et al.* (2001) found that 'trait anxiety' and 'financial worries' had a significant positive relationship with psychological strain.

Practitioners with a high degree of personal commitment to their profession and work have been found to be more likely than others to have greater control of their work, experience less stress, and have higher job satisfaction (McLean and Andrew, 2000), and to also be more likely to stay with their employer (Barak *et al.*, 2001; Rycraft, 1994). Commitment may well be an important personal and professional value base in practitioners' perceptions of work stress.

Spousal support has also been found to be negatively associated with burnout (Barak *et al.*, 2001; Davis-Sachs *et al.*, 1985; Jayaratne *et al.*, 1986), and positively related to job satisfaction in human service workers (Dollard *et al.*, 2001). However, whilst home-work conflict, such as work intruding on family life, has been found to be associated with psychological strain for human service workers (Dollard *et al.*, 2001), Barak *et al.* (2001) found that it was not central to turnover considerations. The lack of influential non-work factors identified in the many studies of burnout and turnover in the human services has been consistent over the two decades of research and reinforces the primacy of organisational responses to address these issues.

11.4   CONSEQUENCES OF WORK STRESS IN THE HUMAN SERVICES

11.4.1   Organisational Outcomes

Despite the conceptual and methodological problems in stress research, there is strong empirical evidence associating work stress with adverse organisational outcomes that are costly in human resource and financial terms (Barak *et al.*, 2001). The direct costs of turnover such as expenses incurred during recruitment, training and separations, are compounded by the indirect costs such as the reduced efficiency of co-workers and new staff, as well as the potential loss of trust and confidence by agency clients unhappy with changed services and staff (Barak *et al.*, 2001). However, quantifications of the extent of the overall costs to individual organisations or the sector as a whole are difficult to obtain, frequently because public disclosure may be embarrassing and counterproductive to recruitment strategies or the public profile of agencies.

In many senses, the burnout-job satisfaction-turnover problems are circular in nature because organisational factors and high workloads increase work stress, which contributes to staff turnover that places greater burdens on the staff who remain, who in turn become more stressed and dissatisfied, and more likely to consider moving on etc. Alternatively, unhappy workers may stay on and thereby create a cumulative agency problem (Koeske and Kirk, 1995a). Unhappy and stressed workers who feel trapped in their agency may contribute to the mythology and overstatement of the burnout problems that occur, as well as the understatement of the career benefits of human service practice. Nevertheless, the occupational stress issues are substantial and lead to many organisational difficulties, and cause considerable personal distress to those affected.

Work stress and strain in the human services have been found to be associated with a range of adverse factors and outcomes including decreased job satisfaction and work performance (Barak *et al.*, 2001; Cooper *et al.*, 1988; Gutek *et al.*, 1988; McLean and Andrew, 2000) and increased absenteeism (Kirk *et al.*, 1993; Koeske and Koeske, 1989; Lyons *et al.*, 1995; Ramanathan, 1992). Higher staff turnover (Barak *et al.*, 2001; Lonne, 2001) and elevated rates of intention to quit have also been identified as adverse consequences (Barak *et al.*, 2001; Ramanathan, 1992; Koeske and Kirk, 1995b), although the relationship between intention to quit and actually leaving is not a linear one (Kirk *et al.*, 1993; Koeske and Kirk, 1995a).

However, there is a paucity of evidence with respect to demonstrating and quantifying the extent to which the work performance of those who experience burnout is adversely affected. We remain unsure about exactly how stressed practitioners respond to their situations and the ways in which this affects their work within the organisational and community contexts, and with client outcomes. These all remain critical issues that demand further research.

11.4.2   Personal Outcomes

The adverse outcomes for social workers and human service practitioners from work stress and burnout can be profound and debilitating, elevating it to a major occupational health and safety issue (Dollard *et al.*, 2001). Those affected are vulnerable to a number of adverse consequences for their physical and psychological well-being (Koeske and Kirk, 1995a). There is empirical evidence associating high work stress for practitioners with poorer health outcomes as measured on the GHQ (Balloch *et al.*, 1998; Dollard *et al.*, 2001; McLean and Andrew, 2000). Furthermore, a range of somatic symptoms and stress-related complaints has been linked to work stress including insomnia, exhaustion; stomach problems, dietary complaints, high blood pressure etc. (Adams *et al.*, 2001; Bradley and Sutherland, 1995; Gold, 1998). For some, this leads to increased alcohol and drug intake (Fewell *et al.*, 1993; Strozier and Evans, 1998).

Moreover, elevated rates of psychological distress such as anxiety and depression have also been associated with work stress, especially when the practitioner receives lower social support (Bennett *et al.*, 1993; Gibson *et al.*, 1989; Jones *et al.*, 1991). Increased levels of cynicism and negativity have also been identified as a consequence of work stress (Corcoran, 1987; Kirk *et al.*, 1993; Maslach, 1998), along with depersonalisation of clients (Koeske and Koeske, 1993) and feeling alienated (Powell, 1994). Apart from these serious complaints, high stress and burnout has been found to be a significant factor leading to job dissatisfaction and turnover (Barak *et al.*, 2001; Maslach, 1998; Um and Harrison, 1998).

A number of adverse personal outcomes are cited in the literature with respect to vicarious trauma. Besides burnout-like symptoms of exhaustion, irritability and somatic complaints, there is evidence that practitioners may experience anxiety and depression (Adams *et al.*, 2001; Schauben and Frazier, 1995; Sexton, 1999). Psychological effects can include altered religious or spiritual beliefs, changed world views, increased sensitivity to, and fear of, violence, intrusive thoughts, PTSD symptoms, despair and hopelessness, and poor concentration (Dane, 2000; Illiffe and Steed, 2000; Sexton, 1999). When the effects of burnout and vicarious trauma are considered, it is little wonder that practitioners caught in these situations consider leaving their jobs (Barak *et al.*, 2001), with some deciding to change their careers altogether.

11.5 ADDRESSING WORK STRESS

There is much that can be done about work stress, burnout, job dissatisfaction, turnover and their consequences. However, interventions remain the primary responsibility of human service employers rather than employees because the major influences on these factors are within their control. Nevertheless, despite these issues clearly being within the domain of occupational health and safety, responsibility for remedies mostly remains with the individuals directly affected, perhaps reflecting the individualistic orientation of Western society and the advice of some scholars (see Zastrow, 1999).

11.5.1 Organisational Strategies

Despite longstanding and repeated calls for job redesign to address work stress in the human services (Davis-Sachs *et al.*, 1985; Dollard *et al.*, 2001; Poulin, 1994, 1995; Thompson *et al.*, 1996; Winefield *et al.*, 2000), it has not been taken up to any large degree. Indeed, variety and autonomy for many workers have decreased as case management procedures have increased (McDonald, 1999). In view of the plethora of research findings indicating the significant moderating influence of increased worker control and autonomy on burnout, the reluctance of employers to embrace this strategy is surprising. It is fundamentally a management responsibility to establish and maintain an organisational climate and culture that deals openly with work stressors, vicarious trauma and burnout and does not individualise or label those staff who experience these phenomena.

Job redesign would be further enhanced through the proactive use of regular, supportive and good quality professional and administrative supervision (Grasso, 1994; Rauktis and Koeske, 1994; Rycraft, 1994). Management styles that embrace collaborative, participatory approaches that facilitate effective team work and collegial support have been shown to increase job satisfaction and moderate work stress (Smith and Nurston, 1998; Sexton, 1999; Thompson *et al.*, 1996; Zunz, 1998). Sadly though, many managers and supervisors remain unskilled to deal with the social and emotional requirements, or are simply not interested in making this a high organisational or personal priority (Winefield *et al.*, 2000). Similarly, access to debriefing, whilst demonstrated as an effective tool against stressful incidents and vicarious trauma, remains under utilised (Dollard *et al.*, 2001; Illiffe and Steed, 2000).

Organisations have significant roles to play in assisting their staff to manage appropriately and effectively the chaotic situations and stresses that can arise in some human service practice contexts. Having orderly processes and procedures in place can aid the management of work stress but, if overly rigid, can also contribute to it. Staff training in time management, workload planning and prioritising, and how to decrease interpersonal stress and conflict in work teams (e.g. improved communication) can be very productive. However, this should never be used as a manipulative ploy to shift responsibility for dealing with excessive workloads form the organisation and funding authority to individual practitioners.

Other effective measures to increase job satisfaction and decrease burnout include the provision of challenging and stimulating training (Dollard *et al.*, 1999, 2001), orientations for new workers (Barak *et al.*, 2001), appealing general working conditions including physical surroundings (Dollard *et al.*, 2001; Vinokur-Kaplan, 1991), ensuring there are adequate resources to meet the workload and closely monitoring this (Illiffe and Steed, 2000; Sexton, 1999), and addressing financial remuneration issues, which are frequently a key source of worker dissatisfaction (Marriott *et al.*, 1994; Rauktis and Koeske, 1994; Vinokur-Kaplan, 1991). Latting (1991) notes that while the human services attract idealistic practitioners to low paid positions, it is wrong to assume that are not motivated by extrinsic rewards such as salary.

11.5.2 Personal Strategies

There are a number of self-care strategies that social workers and human service practitioners can employ in order to prevent and remedy work stress and vicarious trauma, apart from leaving their positions or occupations. Many of these appear to be common sense, but in practice, with high workloads and the often-superhuman expectations of themselves that many human service workers hold, these measures can be under utilised. Perhaps the most commonly utilised strategy is accessing emotional and instrumental support from one's personal and professional networks, and supervision. As noted in Section 11.3.1, these are effective responses to dealing with work stress. As the saying goes: A problem shared is a problem halved. However, this assumes that worker are willing to disclose and share their issues with others, but some have been found to prefer to protect their loved ones and others from the distressing events and issues they confront (Illiffe and Steed, 2000; Jayaratne *et al.*, 1986), perhaps leading to less support being proffered.

Emotional distancing has also been cited as a useful, and perhaps necessary, measure to address occupational stress in the human services (Bennett *et al.*, 1993). In essence, it entails workers setting clear boundaries between themselves and their client and work situations in order to prevent themselves being emotionally overwhelmed by the distressing situations in which they often work. This strategy can be enhanced by reflective practice, which involves the worker in a continual process of self-exploration and improvement. Furthermore, some writers suggest that workers will benefit from developing and maintaining a sense of humour and focus on their own spirituality (Sexton, 1999). Practitioners can increase self-awareness and restore a sense of meaning, hope and connection with others through spiritually-oriented activities like meditation, being in nature, prayer and religious activities, existential explorations etc. This can help one to gain perspective about events bigger than oneself (Sexton, 1999). In a similar fashion, positive thinking can be a useful strategy for dealing with stressful events and issues that can eat away at one's self-esteem and confidence (Zastrow, 1999).

Other self-care strategies include emotional intimacy and sex, which can act as a physical and emotional release and connection with others, and help one to feel well grounded and refreshed. In a similar fashion, relaxation techniques, regular exercise and sporting activities can help workers to release the stress in a healthy and productive way (Dollard *et al.*, 2001; Illiffe and Steed, 2000; Sexton, 1999). Having relaxing times at the movies, concerts, playing or listening to music, sporting events and engaging in pleasurable hobbies can also be productive in handling stress.

Some workers address the workload pressures through planning, goal setting and time-management of their work tasks. This can also facilitate greater variety at work, and if possible and practicable, allow for particularly stressful work to be spaced by less demanding duties. Similarly, being proactive and positive about changing or adapting to distressing events through committed action (e.g. organisational lobbying, political activity) can prove to be an effective way of addressing the systemic factors and processes that affect the human condition and human service practice (Illiffe and Steed, 2000). These sorts of approaches can militate against feeling powerless in the face of macro events and abuses of power by those in positions of authority. Collective action is a powerful tool for political and social change.

11.6 THEORETICAL, METHODOLOGICAL AND POLITICAL ISSUES

Published reviews of the social work and human services work stress, burnout, job satisfaction and turnover literature have identified questionable quality, many varied definitions of stress and burnout, weak controls or failure to use controls, different operationalisations, employment of subjective measures and numerous independent variables, and different statistical procedures (Barak *et al.*, 2001; Ratliff, 1988; Soderfeldt *et al.*, 1995). Cross-sectional designs have been frequently utilised, thereby limiting causal inferences (Dollard *et al.*, 2001; Koeske and Kirk, 1995a). Shinn (1982) found that burnout research relies heavily on self-reports and suggested that different methodologies could assist with causality inferences.

In addition, the great diversity of practice contexts found in the human services makes research of occupational stress inherently difficult and many assumptions have been made with respect to the comparability of research results and their implications. Whilst the available studies have been conducted across a variety of organisational and practice contexts, with different staff qualifications, positions, roles and responsibilities, they have arguably not been representative of the broad spectrum of practice across the human services. Rather, they have tended to concentrate on the statutory government agencies dealing with child welfare, mental health, physical health and the social services. Practice fields with clientele such as the aged, young people and those with a disability, and community work and education, advocacy, domestic violence, non-government, for-profit and rural agencies have largely been neglected. Large sample studies in USA have tended to use the NASW members data-base and have often not reported any differences among practitioners from these practice fields or agency contexts. So the extent to which burnout and stress are problematic for all practitioners and practice contexts in the human services remains unclear.

A variety of stress theories and simplistic models of burnout have been used (Dollard *et al.*, 2001; Koeske and Koeske, 1989). Despite Maslach's (1998) claims to the contrary, burnout remains a nebulous concept (Shinn, 1982; Soderfeldt *et al.*, 1995) that has been used interchangeably with stress (Gibson *et al.*, 1989; Jayaratne and Chess, 1984), and may be confounded with the concept of vicarious trauma, which shares some symptoms. Conceptual differences and theoretical inconsistencies have no doubt contributed to the sometimes contradictory findings. These methodological and conceptual problems have also made comparison of results quite difficult.

Nevertheless, as noted above, some key themes and knowledge have been ascertained. However, whilst we can demonstrate that practitioners experience psychological and physical strain resulting from occupational stress, there has been little research about how this affects their organisational, interpersonal and client work outcomes. Are there, for example, adverse consequences for the clients of burnt out workers? Or are workers' able to mask their stress and to compartmentalise it so that it does not materially affect their clients and colleagues? Or are clients and colleagues on the receiving end of significantly poorer helping processes and working relationships? These issues require investigation.

Bearing this in mind, there has, in many respects, been a lack of perspective and an overstatement of the extent of burnout and work stress in the human services. Perhaps more correctly, there has been an understatement of the position for the majority of practitioners who cope exceedingly well despite the widespread issues of high workloads, role ambiguity and conflict, ad hoc supervision and support, high professional commitment yet inadequate resources, and social mandates and work tasks that can bring them into conflict with clients the general public. Studies have overwhelmingly identified only relatively small proportions of burnt out workers and these have received most of the attention.

The mythologizing of burnout may well play a role in occupational stress not getting the remedial interventions that it requires and deserves. The mythology contributes to a general perception that it is a universal issue, which therefore is too large to address with the available resources. In an increasingly litigious environment, employers may conclude that it is better to 'let sleeping dogs lie' and therefore largely ignore the issues for fear that focussing on them may lead to a rash of costly claims for compensation or demands for organisational and job redesign.

However, this view is based on incorrect assumptions, as practitioners appear to attach significant stigma to stress claims (Dollard *et al.*, 2001). Furthermore, despite major staff concerns about the strain they experience, they are willing to contribute to the development and implementation of remedial strategies (Dollard *et al.*, 2001). Burnout and strain are occupational health issues that demand attention from employers, not only because of the financial and human resource costs they entail, but because of the significant distress and consequences they have for those who are affected.

Work stress research occurs within a political context at the agency and sector levels. Many agencies are reluctant to engage in a detailed study of stress because of the possible ramifications stemming from findings that highlight the primary organisational responsibility for the extent and seriousness of the occupational health problems. There is often a great deal of sensitivity in human service organisations to well-founded, soundly-researched criticisms of management or government responsibility or inaction. Understandably, unions and staff associations are eager for information that supports a case for job redesign and other changes to reduce occupational stress, and this may pit them against a management orthodoxy that seeks to downplay resource inadequacies, workload issues and management failures.

The use of participatory processes and action research designs has much to offer (Dollard *et al.*, 2001), but also requires a willingness by stakeholders to uncover findings that may be 'bad news'. Processes are also needed to keep all stakeholders informed and involved, and to facilitate the resolution of conflicts that may arise between them. Furthermore, longitudinal designs are necessary to examine fully the relevant factors and enable causal inferences to be soundly based (Dollard *et al.*, 2001; Koeske and Kirk, 1995a). However, this requires organisations not only to find the necessary finances but also to engage in a lengthy and time-consuming process in an environment that often demands quick fixes. Moreover, there may be resistance to further change because staff are fatigued from the continual professional and organisational transformation processes outlined in Section 11.3.1. These factors mean that sector-wide studies are likely to be seen as less politically sensitive as the responsibility for stress is shared among many stakeholders. However, unless there is a preparedness to finance large-scale studies, some practice fields and contexts will continue to be under-represented, thereby providing an incomplete picture of the work stress in the sector. It is likely that well-conducted research studies will pay for themselves via the savings brought about through the efficiencies achieved by higher job satisfaction and decreased turnover.

11.7 WORK STRESS IN RURAL PRACTICE

Despite a paucity of empirical research, there is considerable anecdotal evidence to indicate that rural and remote human service practitioners experience high work stress, as evidenced by the high staff turnover found there (Cheers, 1998; Lonne, 1990; Zapf, 1993), notwithstanding evidence that the rural environment may be stress-reducing (Sundet and Cowger, 1990). The author conducted a two-year longitudinal study of the retention and adjustment of 194 rural social workers newly appointed to Australian rural and remote positions during 1994–95 (Lonne, 2002).

11.7.1   Design and Methodology

To practise in Australia, a social worker requires a four-year tertiary qualification or equivalent, and for this study, they were undertaking paid employment in a human services position with a role or duties in a practice field recognised in the social work profession. A rural community was defined as a settlement more than 100 km from the state capital city with a population of not more than 76,750 people.

Respondents were recruited through their employers' referrals, and self-referral following publicising of the study in professional journals and newsletters, and word of mouth. The study aimed to:

- Investigate and describe the characteristics of rural social workers, their practice, positions and communities; and
- To examine the factors affecting their length of stay, in order to increase retention.

The study design consisted of initially surveying respondents with a mailed questionnaire and subsequently when they either left their position or two years later, whichever came first. Data were collected on more than 80 independent variables concerning themselves, their practice, positions, communities, and events during their tenure. Response rates for both questionnaires were over 95%. The 123 respondents who relocated immediately prior to commencing their jobs also participated in a second panel design survey in which they completed an identical questionnaire every three months regarding the manner in which they adjusted to their new communities and jobs. In a test–retest reliability procedure with a sample of 29 rural social workers all instruments demonstrated acceptable alpha values (i.e. >0.60 with most >0.80).

11.7.2 Descriptive Results

The characteristics of these social workers and their practice are reported elsewhere (Lonne and Cheers, 1999, 2000), but in summary, most were females (75.6%) who were likely to be young (<30 years) and unmarried, whereas the males tended to be older, and married or separated. There was a mean age of 34.3 years and a mean prior practice experience of 3.5 years. However, 25.1% were young and inexperienced (<2 years); 11.5% were young, though experienced; 37.3% were older and experienced; and 26.2% were older, though inexperienced, indicating a fairly even distribution of age and experience.

These practitioners mostly (58.8%) lived in and worked in communities with less than 25,000 people, and, on average, rated themselves as moderately visible within their communities. Those in the smaller communities reported higher visibility within the community, which was positively related to higher self-perceptions of community acceptance and success on the job, and high levels of emotional exhaustion. These practitioners were usually very geographically isolated from larger centres ($M = 691$km), their line managers ($M = 101$km), and their primary social, emotional and collegial supports ($M = 1093$km).

Nevertheless, they were very positive about their rural lifestyles, with only 12.7% initially being dissatisfied, although this increased to 26.0% by the time of the second questionnaire. Perceptions of well-being were holistically assessed by respondents. It was positively related to factors such as their disposition to rural living and practice, numbers of local friends, level of community acceptance, usefulness of administrative supervision and degree of impact of social work services, but was negatively associated with high visibility, lack of privacy and increased after-hours work.

These practitioners tended to be solitary workers or were based in small work units, mostly in state government agencies (74.2%) rather than NGO workers (9.9%). Respondents were more likely to have generic casework roles (43.2%) or be in positions with a combination of roles (22.7%), rather than have community work (5.2%), specialist (11.9%), supervisor/managerial (12.9%) or 'other' (4.0%) positions. They also tended to undertake generalist practice across many practice fields ($M = 13$).

Employer support was patchy. Only 31.1% of workers received material or financial incentives to compensate for the rural location, 38.6% were provided with information and advice about the position or community prior to commencing duties, and most practitioners (52.8%) received less than two hours training per month, with isolated practitioners tending to receive less. Employers tended to either provide a range of supports or no supports, particularly NGOs.

Administrative and professional supervision was also patchy and least accessible by NGOs. The usefulness of the administrative supervision was variable and was negatively related to the perceived level of emotional exhaustion. That is, the more useful the supervision, the lower the level of emotional exhaustion.

Overall, for these respondents, rural social work was embedded in the fabric of the community. They frequently experienced high levels of visibility, long working hours, and involvement in community activities, all of which were positively associated with a sense of self-efficacy demonstrated in higher perceived work success and acceptance by the community. However, rural practice also involved regular after-hours disruptions to their privacy, which was related to higher levels of emotional exhaustion.

11.7.3   Emotional Exhaustion, Length of Stay and Job Satisfaction

Respondents indicated their degree of emotional exhaustion on a 10 cm visual analogue scale that was dichotomously anchored with 'I am not burned out at all' and 'I am totally burned out'. The mean rating was 4.1cm ($SD = 2.8$cm, median = 3.7), which was relatively low. One in six respondents (14.6%) indicated the 7.5cm point or higher, recoded as a *high* level of perceived emotional exhaustion, and 3.8% scored at the 9.0cm point or higher, which was recoded to be the *severe* level. Comparisons with previous studies are difficult, as most do not publish the cut-off levels or the proportions of social workers suffering burnout, but this level is within the range most studies cite (see Siefert *et al.*, 1991; Strozier and Evans, 1998).

Higher levels of emotional exhaustion were associated with the personal factors of being younger, having little social work and *rural* social work practice experience, and establishing fewer local friends. Practitioners who had increased numbers of stressful life events, later negative dispositions toward their rural lifestyles and lower levels of well-being also tended to have higher levels.

With respect to work factors, higher levels of emotional exhaustion were associated with full-time status, increased after-hours work, carrying mandatory authority and, in particular, for those working with young offenders, youth and women. It was also highly more likely for practitioners whose line managers were based closer and for practitioners who experienced work troubles. However, lower ratings were associated with receiving useful administrative, having greater work success and higher community acceptance.

For these rural social workers, emotional exhaustion was related to doing too much after-hours work in complex social work areas. Overall, greater life, practice and rural experience, along with good social support and fewer life and work stresses were likely to carry over to less likelihood of emotional exhaustion, and higher perceived well-being and disposition toward rural living. Provision of sufficient, high-quality administrative supervision alleviated emotional exhaustion and contributed to successful social work interventions and community acceptance.

Despite the highly positive attitudes toward rural living and practice, a significant staff turnover problem was identified. The mean actual length of stay was 16.1 months (SD = 8.1 months) compared to the 24 month mean initial anticipated length of stay. Around one-third of practitioners respectively stayed less than 12 months (34.4%), between 13 and 24 months (32.8%) and more than two years (32.8%). Furthermore, 39.7% of workers left both their position and employer during the study period and 29.3% of respondents left their initial positions but remained with their employers. Moreover, 49.8% of those who left their positions were classified as 'premature departures', which was defined as the departure of a practitioner from their position before at least 75% of their initial expected duration of employment.

Multiple and logistic regression analyses were conducted to identify the partialled-out factors affecting the dependent variables of 'length-of-stay', 'dissatisfied premature departure' and 'enhanced retention status'. The length-of-stay regression analysis accounted for 64% of the variance, with 'severe burnout' accounting for 3.3%. Employer/position variables accounted for 33.3% of the variance, including temporary tenure (13.1%), troubles at work (3.3%) and being with an NGO (2.0%), whilst personal factors totalled a further 17.6%, with rural variables playing only a minor role (3.9%).

The logistic regression equation for dissatisfied premature departure was able to correctly classify 89.3% of cases. Key influences that made this a more likely outcome were having high burnout (34.3 times), NGO employer (103.8 times), having negative well-being (12.4 times) and high levels of after-hours work (1.1 times), whereas those who received useful administrative supervision were one and a half times less likely than others.

However, personal factors were predominant in achieving enhanced retention (staying more than 25% longer than originally intended). The logistic regression equation correctly classified 86.5% of cases and influences that made this more likely included having an overall positive well-being (4.7 times), having a positive initial disposition toward rural practice (4.5 times), being aged young (<30 years) and having more than two years experience (8.6 times), and having a moderate amount of involvement in the community (6.7 times).

Those practitioners who relocated to take up their positions also completed the same questionnaire on their personal and professional adjustment every 3 months for 18 months. The SPSS General Linear Model (GLM) analytical procedure was used to analyse the time-series data, including job satisfaction and community satisfaction, which were measured on a 7-point Likert scale. This is a MANOVA procedure that can determine whether there is a statistically significant change in the dependent variable means over time, as well as whether the between-subject factors demonstrate statistically significant differences in means.

Job satisfaction decreased 20.0% over the first 9 months, then levelled out for the last 9 months. The overall 18-month decrease in job satisfaction was statistically significant, as was the initial drop, which supports previous Australian findings for rural social workers (Dollard *et al.*, 1999). However, there was a significant 28.6% drop in mean satisfaction for those workers who left their positions.

Workers who tended to have large decreases in job satisfaction included those that did not have high community acceptance, had lower community visibility, low numbers of local friends, high levels of emotional exhaustion and poor administrative supervision. On the other hand, higher satisfaction levels were more likely for those who were positively disposed toward rural practice and living, received two hours or more employer adjustment briefing and worked in locations where higher incentives were paid.

Job productivity was measured on a 10cm visual analogue scale, which was anchored at each end with 'I am not productive at all' and 'I am fully productive'. Job productivity initially increased in line with practitioners' familiarity with their jobs but tended to level off after 6 to 12 months, before decreasing over the last six months. The statistically significant inverted U-curve pattern consisted of a 15.6% increase in the first six months, followed by an 8.3% drop. Larger rises in mean productivity levels were found for practitioners who had more than 12 months practice experience, who were based, or residing in, communities with 10,000–25,000 population, who received relocation assistance, and perceived themselves to be better off after receiving initial employer incentives, or had received later employer incentives. Employer incentives clearly improved productivity levels. Importantly, productivity rose at the same time as job satisfaction decreased.

The level of satisfaction with the current rural community was measured on a similar 7-point Likert scale to job satisfaction. There was a statistically significant 13.0% decrease in mean satisfaction up to the six-month point, before increasing to the 12-month point whereupon it levelled off before decreasing again: An overall 18-month change that was not statistically significant. Practitioners who left their positions experienced a 21.5% drop in mean satisfaction. Practitioners who tended to have statistically significant decreases in satisfaction during the initial 6-month period did more after-hours work, received poor administrative supervision and experienced high levels of emotional exhaustion. Lower levels of satisfaction were experienced by those with no local friends, no community involvement, and who received one hour or less of employer training. Conversely, those who received two or more hours employer adjustment briefing, had a positive relocation attitude, and had an initial positive attitude toward rural living and practice tended to have higher community satisfaction levels. Satisfaction with the local community was clearly influenced by employer practices and personal attitudes rather than locale features.

Overall, these data further support the proposition that retention problems result from job, rather than rural community, factors, and that burnout plays a critical part in influencing decision making about length of stay. The study results showed that having a positive disposition toward the job and practice are important prerequisites but that work stress is a critical issue that has a cumulative negative effect over time on well-being, particularly when employers do not provide sufficient appropriate supports. The combination of high demands and low supports are critical with regard to job stress, job satisfaction and turnover. However, when appropriate supports are provided workers are much more likely to appropriately adjust to their positions and communities, become productive and stay longer than they otherwise might have. In this sense, employer supports provide the foundation for increased efficacy and enhanced retention.

11.8 DISCUSSION

Social work and the human services sector continue to undergo profound and rapid change to their structures, processes and practices as a result of ideological, fiscal and managerial pressures to become more efficient and effective. There has been considerable undermining of the broad social care mandate as a result of an increasing emphasis on economic and social values that promote individualist rather than communal responses to social problems. Furthermore, social care agencies have had to restructure themselves radically in order to survive in an increasingly competitive sector. The advent and expansion of case management has resulted in increasing proceduralism and decreased autonomy for practitioners.

Within this broad context, occupational stress and burnout in the human services has received considerable attention. In a rapidly changing world, practitioners have been required to alter their priorities and methods substantially in accordance with ideological values with which they may strongly disagree. Severely limited resources, increasing demands for help and a fragmented service delivery system have contributed to a work environment that is exceedingly pressured. Social work and human service practice are direct interventions into private situations that frequently involve a high degree of trauma, distress, conflict and unhappiness for service-delivery recipients. It is, by its nature, a difficult task with a significant amount of emotionally confronting and morally based work.

Given the degree of difficulty of the work, it is surprising that only a relative few workers experience burnout and vicarious trauma. This is testament to the resiliency and coping skills of practitioners who mostly continue to enjoy their chosen career despite the considerable stress that they can experience. Despite this, there is a tendency in the literature to over-focus on those who burn out and to offer solutions that are fundamentally the responsibility of the individual practitioner to address.

This flies in the face of the empirical evidence, which has consistently demonstrated that the occupational stress that practitioners experience primarily has its roots in the organisational and job-role aspects, rather than client or worker characteristics. It is fundamentally a systemic issue that involves serious conflicts and tensions, but which manifests itself in psychological and health strains for individual workers. Furthermore, it is very costly in terms of decreased efficacy and job satisfaction, and increased staff turnover.

So what can be done about it? There are a range of proactive responses that organisations and managers/supervisors can take, including job redesign to increase the work variety and worker autonomy. Giving staff greater control enhances their ability to deal with high work demands. Furthermore, a range of effective supports exists that not only facilitate professional development but also moderate the negative effects of work stress. These include time management and planning strategies, useful administrative and professional supervision, orientation programmes for new workers, training opportunities, supportive team structures and processes, debriefing for critical incidents, financial and material incentives and improved physical work conditions. These can be supplemented, but not replaced, by the various self-care strategies outlined earlier such as receiving social support, and relaxation strategies including intimacy, spiritual development, sport and other recreational activities.

The outcomes of the research of rural social workers' retention highlights the advantages of longitudinal study in being able to understand the processes leading to work stress and its adverse consequences. Moreover, it is apparent that practitioners assess their well-being after holistically taking into account a range of factors, including job satisfaction and their work environment. They assess how well the job is meeting their overall expectations. In this sense, the absence of employer supports is problematic. But the converse is also true. That is, useful employer supports provide a solid foundation on which job satisfaction and productivity, and adjustment to the work and community environments can build. These then influence subsequent decision making about leaving and staying.

11.9   FUTURE RESEARCH DIRECTIONS

Whilst there has been considerable research of work stress and burnout in the human services, collating the findings into a coherent knowledge base is difficult due to a range of design and methodological problems, variations and inconsistencies, and conceptual differences. For example, the similarities and differences between the concepts of burnout and vicarious trauma need to be better understood and empirically verified. Despite the vagaries of research inconsistencies, some consistent findings and key themes are evident. Whilst we know that structural and employer-controlled factors are primary influences, we do not know enough about the stress–strain-outcome process and what effects remedial steps may have. In order to make stronger causal inferences, longitudinal designs are necessary, along with statistical analyses that can partial out the effects of intervening variables.

There is also a need to use samples that are much more representative of the broad diversity of human service practice fields and methods. To date, studies have concentrated on only a few practice contexts. In particular, the not-for profit and for-profit sectors need to be targeted for comparative purposes, and NGOs should be used more for case studies. The differences in urban, rural and remote practice also demand further attention. Work roles other than direct-practice ones should also be examined. In addition, differences in qualifications, practice frameworks, and workers' coping strategies need to be thoroughly examined. Too little is currently known about the consequences of burnout on productivity, efficacy, organisational outcomes and clients' and service users' well-being.

The political sensitivity of these occupational health and safety issues should not be under-estimated. In a human services environment that is severely resource constrained, dealing holistically with occupational stress can be viewed as too expensive, and therefore to be left in the 'too hard basket'. This no doubt contributes to the common tendency to see burnout as an individual problem, explained away as the inability of those that are affected to cope. It probably also contributes to a general reluctance by human service agencies to research their staff's work stress and job satisfaction levels. No news may be perceived as good news and, for some managers, results that identify their own or their organisation's failures may not be seen in a positive light. Staff and financial pressures to address issues are often a prerequisite before research and concerted actions take place, but this may result in serious conflict and distrust between key stakeholders. Participatory action research has much to offer in managing the difficult process of involving all stakeholders (Dollard *et al.*, 2001).

In conclusion, social work and human service practice is a difficult occupation because it involves dealing with complex social issues and problems, and often requires face to face contact with trauma, unhappiness and conflict, whilst not always having the resources to respond adequately. Despite this, most practitioners enjoy their work and jobs. However, a range of organisational and role factors, more so than personal characteristics, contribute to work stress. This leads, for some, to burnout and vicarious trauma, and can result in job dissatisfaction and job turnover. When there is a combination of high job demands, low control and autonomy, and poor resources, the likelihood of significant work stress is significantly increased, along with the costs associated with decreased work performance and higher staff turnover.

But this situation does have some brighter aspects. For example, we are increasingly clear about the contributing factors as well as the systemic and personal strategies with which to respond. What is needed is the political and organisational willingness to recognise the nature of the problems and to prioritise the allocation of resources to address these costly issues. Efficiency and effectiveness dividends from reduced work strain for the human services workforce should cover the financing of further research, job redesign and practitioner supports. Continued failure to respond appropriately to this systemic occupational health and safety issue will see the existing high rates of work stress claims further increase (Dollard *et al.*, 2001). Ignoring these problems will only exacerbate the distress for those affected and unnecessarily divert resources to be spent on litigation defending inadequate and ineffective management of the primary occupational health and safety issue is the human services. It is time for a different approach.

11.10 REFERENCES

Adam, K.B., Matto, H.C. and Harrington, D., 2001, The traumatic stress institute belief scale as a measure of vicarious trauma in a national sample of clinical social workers. *Families in Society: The Journal of Contemporary Human Services*, **82**, pp. 363–371.

Balloch, S., Pahl, J. and McLean, J., 1998, Working in the social services: Job satisfaction, stress and violence. *British Journal of Social Work*, **28**, pp. 329–350.

Barak, M.E., Nissly, J.A. and Levin, A., 2001, Antecedents in retention and turnover among child welfare, social work, and other human service employees: What can we learn from past research? A review and metanalysis. *The Social Service Review*, **75**, pp. 625–645.

Bennett, P., Evans, R. and Tattersall, A., 1993, Stress and coping in social workers: A preliminary investigation. *British Journal of Social Work*, **23**, pp. 31–44.

Bradley, J. and Sutherland, V., 1995, Occupational stress in social services: A comparison of social workers and home help staff. *British Journal of Social Work*, **25**, pp. 313–331.

Cheers, B., 1998, *Welfare bushed: Social care in rural Australia* (Aldershot, England: Ashgate).

Coholic, D. and Blackford, K., 1999, Vicarious trauma in sexual assault workers: Challenges of a northern location. In *Promoting inclusion – redressing exclusion: The social work challenge. Conference proceedings of the Joint Conference of the AASW, IFSW, APASWE & AASWWE, Brisbane, Australia, 26–29 Sept. 1999*, edited by the Australian Association of Social Workers (Barton, Canberra: AASW), pp. 242–247.

Collings, J.A. and Murray, P.J., 1996, Predictors of stress among social workers: An empirical study. *British Journal of Social Work*, **26**, pp. 375–387.

Cooper, C.L., Cooper, R.D. and Eaker, L.H., 1988, *Living with stress* (Chichester, England: Penguin).

Corcoran, K., 1987, The association of burnout and social work practitioner's impressions of their clients: Empirical evidence. In *Burnout among social workers*, edited by D.F. Gillespie (New York: Haworth Press), pp. 57–66.

Cunningham, M., 1999, The impact of sexual abuse treatment on the social work clinician. *Child and Adolescent Social Work Journal*, **16**, pp. 277–290.

Dane, B., 2000, Child welfare workers: An innovative approach to interacting with secondary trauma. *Journal of Social Work Education*, **36**, pp. 27–38.

Davis-Sachs, M., Jayaratne, S. and Chess, W., 1985, A comparison of the effects of social support on the incidence of burnout. *Social Work*, **30**, pp. 240–244.

Demerouti, E., Nachreiner, F., Bakker, A.B. and Schaufeli, W.B., 2001, The job demands-resources model of burnout. *Journal of Applied Psychology*, **86**, pp. 499–512.

Dollard, M.F., Winefield, H.R. and Winefield, A.H., 1999, Burnout and job satisfaction in rural and metropolitan social workers. *Rural Social Work*, **4**, pp. 32–42.

Dollard, M.F., Winefield, H.R. and Winefield, A.H., 2001, *Occupational strain and efficacy in human service workers: When the rescuer becomes the victim* (London: Kluwer Academic Publishers).

Dollard, M.F., Winefield, H.R., Winefield, A.H. and de Jonge, J., 2000, Psychosocial job strain and productivity in human service workers: A test of the demand-control-support model. *Journal of Occupational and Organisational Psychology*, **73**, pp. 501–510.

Egan, M. and Kadushin, G., 1993, Burnout and job satisfaction in rural hospital social workers. *Human Services in the Rural Environment*, **16**, pp. 5–10.

Fewell, C.H., King, B.L. and Weinstein, D.L., 1993, Alcohol and other drug abuse among social work colleagues and their families: Impact on practice. *Social Work*, **38**, pp. 565–570.

Fook, J., Ryan, M. and Hawkins, L., 2000, *Professional expertise: Practice, theory and education for working in uncertainty* (London: Whiting and Burch).

Gant, L.M. and Gutierrez, L.M., 1996, Effects of culturally sophisticated agencies in Latino social workers. *Social Work*, **41**, pp. 624–644.

Garden, A., 1989, Burnout: The effect of psychological type on research findings. *Journal of Occupational Psychology*, **62**, pp. 223–234.

Gibson, F., McGrath, A. and Reid, N., 1989, Occupational stress in social work. *British Journal of Social Work*, **19**, pp. 1–16.

Gold, N., 1998, Using participatory research to help promote the physical and mental health of female social workers in child welfare. *Child Welfare*, **LXXVII**, pp. 701–724.

Grasso, A.J., 1994, Management style, job satisfaction, and service effectiveness. *Administration in Social Work*, **18**, pp. 89–105.

Gutek, B.A., Repetti, R.L. and Silver, D.L., 1988, Nonwork roles and stress at work. In *Causes, coping and consequences of stress at work*, edited by C.L. Cooper and R. Payne (London: John Wiley and Sons), pp. 141–173.

Guterman, N.B. and Jayaratne, S., 1994, 'Responsibility at-risk': Perceptions of stress, control and professional effectiveness in child welfare practitioners. *Journal of Social Service Research*, **20**, pp. 99–120.

Harris, J. and McDonald, C., 2000, Post-Fordism, the welfare state and personal social services: A comparison of Australia and Britain. *British Journal of Social Work*, **30**, pp. 51–70.

Healy, J., 1998, *Welfare options: Delivering social services* (St Leonards, NSW: Allen & Unwin).

Himle, D. and Jayaratne, S., 1990, Burnout and job satisfaction: Their relationship to perceived competence and work stress among undergraduate and graduate social workers. *Journal of Sociology and Social Welfare*, **17**, pp. 93–108.

Himle, D., Jayaratne, S. and Thyness, P., 1989a, The buffering effect of four types of supervisory support on work stress. *Administration in Social Work*, **13**, pp. 19–34.

Himle, D., Jayaratne, S. and Thyness, P., 1989b, The effects of emotional support on burnout, work stress and mental health among Norwegian and American Social Workers. *Journal of Social Service Research*, **13**, pp. 7–45.

Howe, D., 1994, Modernity, Postmodernity and Social Work. *British Journal of Social Work*, **24**, pp. 513–32.

Illiffe, G. and Steed, L.G., 2000, Exploring counsellor's experience of working with perpetrators and survivors of domestic violence. *Journal of Interpersonal Violence*, **15**, pp. 393–412.

Jayaratne, S. and Chess, W., 1984, Job satisfaction, burnout, and turnover: A national study. *Social Work*, **29**, pp. 448–453.

Jayaratne, S., Chess, W. and Kunkel, D., 1986, Burnout: Its impact on child welfare workers and their spouses. *Social Work*, **31**, pp. 53–59.

Jayaratne, S., Tripodi, T. and Chess, W., 1983, Perceptions of emotional support, stress, and strain by male and female social workers. *Social Work Research and Abstracts*, **19**, pp. 19–27.

Jones, A. and May, J., 1992, *Working in human service organisations: A critical introduction* (Melbourne: Longman Cheshire).

Jones, F., Fletcher, B. and Ibbetson, K., 1991, Stressors and strains amongst social workers: Demands, supports, constraints, and psychological health. *British Journal of Social Work*, **21**, pp. 443–469.

Kalisch, D.W., 2000, *Social policy directions across the OECD region: reflections on a decade of change* (Canberra, Australia: Department of Family and Community Services, Commonwealth of Australia).

Kirk, S.A., Koeske, G.F. and Koeske, R.D., 1993, Changes in health and job attitudes of case managers providing intensive services. *Hospital and Community Psychiatry*, **44**, pp. 168–173.

Koeske, G.F. and Kirk, S.A., 1995a, The effects of characteristics of human service workers on subsequent morale and turnover. *Administration in Social Work,* **19**, pp. 15–31.

Koeske, G.F. and Kirk, S.A., 1995b, Direct and buffering effects of internal locus of control among mental health professionals. *Journal of Social Service Professionals*, **20**, pp. 1–28.

Koeske, G.F. and Koeske, R.D., 1989, Workload and burnout: Can social support and perceived accomplishment help? *Social Work*, **34**, pp. 243–248.

Koeske, G.F. and Koeske, R.D., 1993, A preliminary test of a stress–strain-outcome model for reconceptualising the burnout phenomenon. *Journal of Social Service Research*, **17**, pp. 107–135.

Latting, J.K., 1991, Eight myths on motivating social services workers: Theory based perspectives. *Administration in Social Work*, **15**, pp. 49–66.

LeCroy, C. and Rank, M., 1987, Factors associated with burnout in the social services: An exploratory study. In *Burnout among social workers*, edited by D.F. Gillespie (New York: Haworth Press), pp. 23–39.

Lonne, B., 1990, Beginning country practice. *Australian Social Work*, **43**, pp. 31–39.

Lonne, B., 2002, *Retention and adjustment of social workers to rural positions in Australia: Implications for recruitment, support and professional education.* Unpublished doctoral thesis (University of South Australia: South Australia).

Lonne, B. and Cheers, B., 1999, Recruitment, relocation and retention of rural social workers. *Rural Social Work*, **5**, pp. 13–23.

Lonne, B. and Cheers, B., 2000, Rural social workers and their jobs: An empirical study. *Australian Social Work*, **53**, pp. 21–28.

Lonne, B., McDonald, C. and Fox, T. (in press), Emerging ethical issues in the human services: The times they are a changing? In *Managerialism, contractualism and professionalism in human* services edited by L. Briskman and R. Muetzelfeldt (Melbourne: Deakin University Press).

Lyons, K., La Valle, I. and Grimwood, C., 1995, Career patterns of qualified social workers: Discussion of a recent survey. *British Journal of Social Work*, **25**, pp. 173–190.

Marriott, A., Sexton, L. and Staley, D., 1994, Components of job satisfaction in psychiatric social workers. *Health and Social Work*, **19**, pp. 199–205.

Maslach, C., 1976, Burned-out. *Human Behaviour*, Sept., pp. 16–21.

Maslach, C., 1998, A multidimensional theory of burnout. In *Theories of organisational stress* edited by C.L. Cooper (Oxford: Oxford University Press), pp. 68–85.

McDonald, C., 1999, Human service professionals in the community services industry. *Australian Social Work*, **52**, pp. 17–25.

McDonald, C. and Jones, A., 2000, Reconstructing and reconceptualising social work in the contemporary milieu. *Australian Social Work*, **53**, pp. 3–11.

McLean, J. and Andrew, T., 2000, Commitment, satisfaction, stress and control among social services managers and social workers in the UK. *Administration in Social Work*, **23**, pp. 93–117.

Parton, N., 1994, Problematics of government: (Post) Modernity and social work. *British Journal of Social Work*, **24**, pp. 9–32.

Parton, N., 1996, *Social theory, social change and social work* (London: Routledge).

Perlman, L.A. and Mac Ian, P.S., 1995, Vicarious traumatisation: An empirical study of the effects of trauma work on trauma therapists. *Professional Psychology: Research and Practice*, **26**, pp. 558–565.

Pottage, D. and Huxley, P., 1996, Stress and mental health social work: A developmental perspective. *International Journal of Social Psychiatry*, **42**, pp. 124–131.

Poulin, J.E., 1994, Job task and organisational predictors of social worker job satisfaction change: A panel study. *Administration in Social Work*, **18**, pp. 21–38.

Poulin, J.E., 1995, Job satisfaction of social work supervisors and administrators. *Administration in Social Work*, **19**, pp. 35–49.

Powell, W.E., 1994, The relationship between feelings of alienation and burnout in social work. *Families in Society*, **75**, pp. 229–235.

Ramanathan, C.S., 1992, EAP's response to personal stress and productivity: implications for occupational social work. *Social Work*, **37**, pp. 234–239.

Ratliff, N., 1988, Stress and burnout in the helping professions. *Social Casework*, **69**, pp. 147–154.

Rauktis, M.E. and Koeske, G.F., 1994, Maintaining social worker morale: When supportive supervision is not enough. *Administration in Social Work*, **18**, pp. 39–60.

Rycraft, J.R., 1994, The party isn't over: The agency role in the retention of public child welfare case workers. *Social Work*, **39**, pp. 75–80.

Samantrai, K., 1992, Factors in the decision to leave: Retaining social workers with MSWs in public child welfare. *Social Work*, **37**, pp. 454–458.

Schauben, L.J. and Frazier, P.A., 1995, Vicarious trauma: The effects on female counsellors of working with sexual violence survivors. *Psychology of Women Quarterly*, **19**, pp. 49–64.

Sexton, L., 1999, Vicarious traumatisation of counsellors and effects on workplaces. *British Journal of Guidance and Counselling*, **27**, pp. 393–403.

Shinn, M., 1982, Methodological issues evaluating and using information. In *Job Stress and Burnout: Research, Theory and Intervention Perspectives* edited by W.S. Paine (Beverley Hills: Sage Publishers), pp. 61–79.

Siefert, K., Jayaratne, S. and Chess, W.A., 1991, Job satisfaction, burnout, and turnover in health care social workers. *Health and Social Work*, **16**, pp. 193–202.

Smith, M. and Nursten, J., 1998, Social workers' experience of distress – Moving towards change? *British Journal of Social Work*, **28**, pp. 351–368.

Soderfeldt, M., Soderfeldt, B. and Warg, L., 1995, Burnout in social work. *Social Work*, **40**, pp. 638–646.

Strozier, A.L. and Evans, D.S., 1998, Health and distress in social workers: Results of a national survey. *Smith College Studies in Social Work*, **69**, pp. 60–77.

Sundet, P.A. and Cowger, C.D., 1990, The rural community environment as a stress factor for rural child welfare workers. *Administration in Social Work*, **14**, pp. 97–110.

Thompson, N., Stradling, S., Murphy, M. and O'Neill, P., 1996, Stress and organisational culture. *British Journal of Social Work*, **26**, pp. 647–665.

Thorpe, G.L., Righthand, S. and Kubik, E.K., 2001, Brief report: Dimensions of burnout in professionals working with sex offenders. *Sexual Abuse: A Journal of Research and Treatment*, **13**, pp. 197–203.

Um, M. and Harrison, D.F., 1998, Role stressors, burnout, mediators, and job satisfaction: A stress–strain-outcome model and an empirical test. *Social Work Research*, **22**, pp. 100–115.

Vinokur-Kaplan, D., 1991, Job satisfaction among social workers in public and voluntary child welfare agencies. *Child Welfare*, **70**, pp. 81–91.

Wallace, J. and Brinkerhoff, M., 1991, The measurement of burnout revisited. *Journal of Social Service Research*, **14**, pp. 85–111.

Wilcoxon, S.A., 1989, Leadership behaviour and therapist burnout: A study of rural agency settings. *Journal of Rural Community Psychology*, **10**, pp. 3–14.

Williams, F., 1994, Social relations, welfare and the post-Fordism debate. In *Towards a post-fordist welfare state* edited by R. Burrows and B. Loader (London: Routledge), pp. 49–73.

Winefield, H., Dollard, M. and Winefield, A., 2000, The role of supervisors in managing occupational stress for service professionals. *Journal of Occupational Health and Safety*, **16**, pp. 343–349.

Zapf, M.K., 1993, Remote practice and culture shock: Social workers moving to isolated northern regions. *Social Work*, **38**, pp. 694–704.

Zastrow, C.H., 1999, *The practice of social work*, 6th edn (Pacific Grove, CA: Brooks/Cole Publishing Company).

Zunz, S.J., 1998, Resiliency and burnout: Protective factors for human service managers. *Administration in Social Work*, **22**, pp. 39–54.

CHAPTER TWELVE

Clergy in Crisis

Sarah J. Cotton, Maureen F. Dollard, Jan de Jonge
and Paul Whetham

12.1 INTRODUCTION

Clergy have not typically been perceived as human service professionals (Warner and Carter, 1984). Yet, they play a vital human service function in society. Numerous studies over the past 35 years (e.g. Chalfant *et al.*, 1990; Gurin *et al.*, 1960; Weaver, 1995) have demonstrated that approximately 4 out of 10 who have mental health needs, seek assistance from members of the clergy in times of personal distress (Weaver, 1997).

Our society is facing enormous change, and in turn this places extra pressures upon helping professionals in our community. Much research has focused on professionals such as social workers, yet although similar, clergy have uniquely different stressors that need to be examined individually. There is little or no systematic work examining the impact or nature of stress on clergy and therefore no clear direction for the development of interventions or organisational recommendations during this challenging time. Weaver *et al.* (1997) in a review of the published research literature reveals that in terms of empirical research in psychology, examination of clergy in relation to psychological and behavioural topics is extremely limited.

A recent review by Hall (1997) of the available empirical research, points out that in recent years, clergy have become an increasing focus of attention for two main reasons.

1. The increasing recognition of the extremely high demands of the pastorate.
2. The increasing awareness of the impact of clergy personal dysfunction on their ministries.

Research shows that clergy often confront high levels of occupational stress due to excessive demands (e.g. time demands, financial demands, family boundaries). These problems are often deepened by idealism and high expectations among clergy. High levels of occupational stress may result in burnout similar to that observed among human service professionals. Other consequence of ministry may include relational deficits, loneliness and sexual misconduct. This in term may lead some clergy to leave the ministry altogether (Krause *et al.*, 1998).

Given these disturbing consequences of ministry, it is of concern that there has been little research inspiring effective interventions for change. Following a review of current issues in clergy well-being this chapter will focus on a study commissioned by The Salvation Army (Southern Territory – Australia). Finally the implications of the research, politics of working within the clergy population and future directions for research will be discussed.

12.2   UNDERSTANDING THE CLERGY VOCATION

12.2.1   Definition

Finding a workable definition for 'clergy' is a challenge. Ordination is practiced and defined in different ways in various denominational traditions. For the purpose of this chapter we have used the definition of clergy developed by the National Council of Churches in the mid-1980s.

> Ordained clergy are defined as those clergy having 'full ministry', or holding that office 'having the most complete and unrestricted set of functions relating to the ministry of the Gospel, administering the Word and Sacrament or carrying out the office of pastor or priest in the church (Hartford Institute for Religion Research, 1994: 1).

Krause and Ellison (1998) make three distinct differences that separate clergy from either lay leaders (elders) or rank and file members. First, unlike other roles in the church, years of formal *training* are required to become a member of the clergy. Educational requirements for entry into the clergy vary greatly between different denominations and not all training will necessarily lead to qualification. Many denominations require that clergy complete a bachelor's degree and a program of theological study; others will admit anyone who has been called to the vocation. In the Australia (1991) census, 16 per cent of ministers had no recognised post-school qualification, 28 per cent held a degree and 12 per cent had postgraduate qualifications as their highest post-school qualification (Australia Bureau of Statistics, 1991).

Second, being a pastor is a *full time job* that occupies the majority of one's waking hours. In fact, Whitney (1998) argues that 'Running the church to meet the ever-growing demands is itself more than a full time job' (p. 3). Australian clergy for example, compared to all other occupations, work the longest hours (Australian Bureau of Statistics, 1991). A US survey of pastors by Fuller Institute of Church Growth (1991) found that 90 per cent of clergy work more than 46 hours a week (London and Wiseman, 1993). They are frequently called upon at short notice to visit the sick, comfort the dying and their families and provide counselling and social support to those in need. Involvement in community administrative and educational activities often requires clergy to work evenings, early mornings, weekends and during their holidays.

Finally, the third defining aspect is that many clergy have a sense of being chosen or *called by God* to assume the clerical role (Hood *et al.*, 1996). Ordained ministry is not just another job. Clergy seek ordination because they have a sense of call and a conviction that God has a plan for their lives. They accept their ministry as a gift from God, believing that God and the church have called them to serve.

> I often encounter objections when I refer to the ministry as a job.
> It is a calling, I'm told (Hulme, 1985: 39).

12.2.2   The Role of Clergy

Results of the Australian National Church Life Survey (NCLS) indicate that church leaders carry out a wide range of roles (Kaldor and Bullpit, 2001). In an attempt to understand the diverse and demanding nature of these roles, Australian researcher Blaikie (1979) identified a range of facets of a church leader's vocation. These include:

- *Educator* – training, instructing and leading study groups
- *Evangelist* – converting others to faith
- *Organiser* – organising and supervising the work of the parish and the congregation
- *Pastor* – visiting and counselling
- *Preacher* – delivering sermons, expounding the work of God
- *Priest* – conducting worship and administering the sacraments
- *Scholar* – reading, studying and writing
- *Social Reformer* – involved in attacking social injustices

Blizzard (1956) and subsequent researchers (e.g. Coates and Kistler, 1965) have found that many clergy enter the ministry with both personal and theological biases towards certain roles. Unfortunately, what many clergy find is that their preferences and priorities do not match the actual day-to-day demands of the ministry (Kaldor and Bullpitt, 2001; Kieren and Munro, 1988). The NCLS suggests that 43 per cent of senior clergy feel they waste time on tasks not central to their role (Kaldor and Bullpitt, 2001). The question of the clergy role is clearly a vexed one (Whetham and Whetham, 2000).

12.3   THE VALUE OF CLERGY TO SOCIETY

> It is the clergy who deal directly with the crises of meaning
> implicit in many of these crises of pain. The clergy are involved
> in many of the same crises in which physicians and nurses are

involved, and they continue to be involved after physicians and nurses have left the scene (Hulme, 1985: 3).

Historically, society has expected its religious leaders to be involved in a helping ministry to persons experiencing difficulty. Numerous studies (Bissonette, 1977; Clemens *et al.*, 1978; Rumberger, 1982) have identified clergy as the 'gatekeepers' of mental health services as most help seekers consider clergy as the first available and dependable line of defence in times of crisis and trouble. A survey conducted in America by Gurin *et al.* (1960) indicated that of those surveyed who experienced psychological distress severe enough to cause them to seek professional help, 42 per cent consulted clergy. In rural areas where mental health resources are generally scarce and under used, ministers serve a particularly vital function (Bradfield *et al.*, 1989).

Ministers also play an important role in offering emotional support to individuals suffering a personal tragedy. In fact Hulme (1985) believes that clergy are involved in the pains and crises of life and death, more than any other helping professionals. When people are confronted with losses, such as death or divorce, they frequently turn to the clergy for guidance and counselling. The recent American crisis is an example of how when faced with such loss people turn to the church. A recent article in an Australian paper highlighted this point. 'The Wilson family was never religious. That was until the terrorist attacks on the US...all of a sudden it seemed the world wasn't a safe place anymore...I thought it would be best for us to go to church and see if they can help us' (Kemp *et al.*, 2001, p. 66).

Although it might be expected that counselling carried out by pastors would decrease with the increased availability of professional mental health services, this does not appear to be happening. With the development of other mental health disciplines clergy are no longer the only persons involved in the 'care of souls'. Possibilities for referral are readily available. However for a variety of reasons many people are still inhibited to consult health professionals because of differences in value orientation, lack of awareness of services, the stigma involved and financial considerations. The frequent use of the clergy by the public should not be a surprise, given the clergy's availability, accessibility and the high trust that many members of society have in them (Weaver *et al.*, 1997). Clergy therefore are likely to continue to be involved in a significant amount of counselling in the years ahead (Fuller and Edwards, 2000; Virkler, 1979).

To have healthy and productive clergy who are able to offer the highest level of care and service to the community is vital. As in other traditional human service professions, the emotional maturity and stability of pastors should provide the foundation for their ministerial effectiveness (Hall, 1997).

12.4   THE MINISTRY: A VOCATION UNDER STRESS

Clergy are not immune to the occupational hazards endemic to all who serve as helping professionals (Virginia, 1998 and Richmond *et al.*, 1985). A recent significant review of the literature by Hall (1997) into the personal functioning of clergy reveals increasing recognition of the extremely high demands of the pastorate. In fact a recent study of clergy in Wales revealed that 38 per cent felt overwhelmed by the complexity of the pastoral demands they face each day (Francis and Kay, 2000). Although providing assistance to others can be a source of great personal reward, research indicates that excessive involvement in the help-provider role is a major source of occupational stress (Hall, 1997) and 'takes a toll' on the well-being of help-providers (Kessler *et al.*, 1985). It is therefore important that the unique stressors and strains of ministry are examined (Olsen and Grosch, 1991). Effective interventions depend upon it! (Richmond *et al.*, 1985).

12.4.1 Identifying Clergy Stressors

Time Demands

There is never a time when clergy are not on call to function in their pastoral role. Lack of time (time demands) is one of the most frequently cited work related stressors among pastors (Ellison and Mattila, 1983; Morris and Blanton, 1994b; Warner and Carter, 1984). The major factors influencing the degree of disruption created by time demands are the unpredictable nature and or uncertainty of being on call 24 hours a day, 7 days a week (Boss, 1980; Kierren and Munro, 1988; Norrell, 1989; Seamonds and Seamonds, 1981).

Financial Demands

Despite the enormous time committed to ministry, clergy are among the lowest paid professions in Australia (Australian Bureau of Statistics, 1991). Working long hours for comparatively low pay (financial constraints) is a frequently cited stressor for clergy and their families (Blanton, 1992; Herin, 1981; Houts, 1982; Hunt, 1978; Kaldor and Bullpitt, 2001; Lee, 1999; Lee and Balswick, 1989; Mace and Mace, 1982; Mattis, 1977; Oden, 1977; Presnell, 1977; Scherer, 1965). Lavender (1983) found that 95 per cent of all clergy were grossly underpaid. In 1991, 41 per cent of ministers had an annual income of $16,000 or less with only 6 per cent earning over $35,000 per annum. Considerable variation between the different religious groups was found and low incomes need to be interpreted in the context of some clergy receiving income in-kind such as food, car and housing (Australian Bureau of Statistics, 1991). In addition to the lack of adequate financial resources, there are also fears created by the low retirement benefits available and lack of equity in a home (Herin, 1981).

Coupled with these realities is the guilt that clergy may experience for being concerned with material matters (Chikes, 1968). Many clergy families are caught in the 'double bind' of attempting to address their personal and family needs while not seeing themselves as materialistic (Chikes, 1968). Because of this, many clergy find it difficult to assert their financial needs to both congregations and leadership

(Blanton, 1992). Unlike some other professionals, clergy do not have an organized channel (i.e. union) for collective bargaining, and or requesting greater remuneration for their services (Morris and Blanton, 1994b). Clergy are therefore often left to struggle between their guilt and an archaic system that requires of them to be far more that is humanly possible.

Identity and Image

Much attention is focused on the cleric's image or identity as the 'one set aside' (Birk, 2001). For this reason employment is often unreasonably dependent upon the clergyperson's image of being unerring and holier than most people. Because those in the ministry have a tendency to be extremely other-oriented, they find it difficult at times to engage in healthy levels of self-affirmation and are extremely dependent of the affirmations of others (Blanton, 1992). In fact, Walker (1978) pointed out that many clergy believe they must be affirmed as a pastor by *all* of their parishioners. They are unable to perceive the displeasure of a minority of parishioners in a reality based context. Such a belief can affect self-esteem since it sets an impossible goal. Often clergy have feelings of failure, when in reality they have been effective. Disillusionment and frustrations can therefore impact negatively on self-esteem (Blanton, 1992). In a survey of pastors by Fuller Institute of Church Growth (1991), 70 per cent said that they had lower self-esteem now compared to when they started in the ministry (London and Wiseman, 1993). In these and other circumstances, identity and image issues can be sources of stress for clergy (Malony, 1988; Richmond *et al.*, 1985).

High Expectations (Emotional Demands): 'The pedestal effect'

The ingrained stereotype (image) of clergy being God-like and capable of doing anything is commonly called the 'pedestal effect': the leader is treated as spiritually above others in the congregation (Whetham and Whetham, 2000). Although clergy are placed on a 'pedestal' by authorities and congregants, they often struggle to maintain an image of perfectionism (Birk, 2001). However, in a desire to serve the community well and faithfully, clergy often tolerate pressures, which lead to the development of these unrealistic expectations (McBurney, 1977). Deluca (1980) labelled the clergy profession a 'holy crossfire' as the cleric and his/her family attempt to juggle the expectations of self, family, congregation, denomination and God.

  Self: Research has consistently found that the majority of clergy have unrealistically high expectations of themselves (Birk, 2001; Episcopal Family Network, 1988; King, 1988; Rediger, 1982; Whittemore, 1991). Houts (1982) makes it clear that unhealthy self imposed expectations can contribute to a humourless and perfectionist lifestyle making it difficult to adjust to situational and/or normative stressors encountered in the ministry. Many clergy believe that they can indeed achieve perfection if only they try hard enough. Most are convinced that they must be good spouses and parents, competent administrators and skilled pastoral counsellors. They must also maintain a positive image with their peers, while handling all problems themselves and never seeking help (Devogel, 1986).

Family: Members of the pastor's family have been expected to make their home a showcase of Christian values and relationships e.g. children serving as role models for their peers (Lee, 1992; Lee and Balswick 1989; Mickey *et al.*, 1991). One of the most common expectations is the need to maintain the image of a perfect model of marriage (what is desirable, moral and strong in the institute of marriage) (Blanton, 1992; Mace and Mace, 1980; Noller, 1984; Presenell, 1977). As a result some clergy and/or their wives do not admit or recognise the warning signs of dysfunctional relationship patterns until severe deterioration has occurred e.g. viewing anger and conflict as unchristian (Niswander, 1982; Blanton, 1992). Indeed, their professional role may make it more difficult for couples to seek help with their marital problems (Morris and Blanton, 1994b). It seems that there is a tension between the covenants of marriage and the ministry (Noyce, 1980). Clergy must try to balance vows of ordination 'marrying' them to God and the church, as well as vows of marriage to their spouse (Hunt, 1978; Lee and Balswick, 1989).

Congregation: Church members (congregations) often also have unrealistic high expectations of the clergy's personal and professional competence (e.g. Blackbird and Wright 1985; Blanton, 1992; Devogel, 1986; Whitney, 1998). In fact Lee and Balswick (1989) found that clergy families ranked unrealistic congregational expectations as the second greatest stressor impacting on their families.

Denominations: Mace and Mace (1980) indicated that church organisations like secular employers, foster a highly competitive system with the emphasis upon success. As a result of these high expectations many clergy develop a workaholic pattern, seeking to mimic or duplicate previously recognised successes (Morris and Blanton, 1998).

God: Clergy frequently feel that not only must they please everyone in the congregation and denominational hierarchy but they must also satisfy the high expectations of God (Olsen and Grosch, 1991).

When clergy attempt to live up to these various expectations and as well, believe the myth of their own perfection they are at risk of confusing the role they play with who they really are (Melbane and Ridley, 1988; Sanford, 1982). Often they neglect their basic needs of emotional, physical and spiritual nurture, believing that they do not or should not have the same kind of needs, limits and weaknesses as other people (Rediger, 1982; Melbane and Ridley, 1988).

Family Boundaries

Coupled with these high expectations is a high level of intrusion of clergy work into family boundaries (Douglas, 1965; Lee and Balswick, 1989; Mace and Mace, 1980, 1982; Mattis, 1977; Morris and Blanton, 1994b). Clergy and their family operate in a unique 'built in community context', with ambiguous separations between their professional and private lives (Hulme, 1985; Lee and Balswick, 1989; Presnell 1977; Whetham and Whetham, 2000). The Australian NCLS found that the majority of leaders (65 per cent) found it hard to separate work and home life. Further findings revealed that more than 60 per cent of senior clergy see the lack of separation between work and personal life as a significant or highly significant pressure point (Kaldor and Bullpit, 2001). Unlike many professionals who are able to leave their work at the office and maintain clear limits regarding their private lives, clergy are overly visible and available to those whom they serve. Clergy often find themselves playing their professional role even during their free time: their neighbours may be church members, they frequently socialise with members of their congregation and their days off are often interrupted (day or night) by emergency concerns for which many clergy feel responsible (Jud *et al.*, 1970). These problems are often heightened if the parsonage (clergy home) is in close physical proximity to the church with church members literally on their doorstep much of the time (Blanton, 1992). In order to function effectively, the family must neither be disengaged from nor enmeshed in the church (Lee, 1988). Achieving such a delicate balance is not easy (Blanton, 1992).

> In no other profession are the philosophy and performance of the vocation so intimately entwined with the commitments, values and behaviour of one's private life, in the eyes of those who serve and those who are served (Presnell, 1977, p. 272).

Role Conflict and Ambiguity

When one's professional role invades so many areas of one's life as outlined above multiple, conflicting and ambiguous roles are inevitable (De Luca, 1980; Kay, 2000; Kunst, 1993; Malony and Hunt, 1991). The role conflict and ambiguity precipitated by competing professional and personal demands, and unrealistic expectations may be problematic for some clergy and, consequently, for their congregations. Some ministers may be able to manage conflicting personal and professional needs in an adaptive way, by maintaining healthy role boundaries, nurturing their personal lives and developing a realistic sense of their own potential and limitations (Jud *et al.*, 1970; McBurney, 1977; Rediger, 1982). However, it is becoming increasingly common for the blurring of roles to become problematic for ministers (Jud *et al.*, 1970; Whittemore, 1991; Kaldor and Bullpitt, 2001; Kay, 2000) particularly young clergy who are greeted with a multi-faceted role that they feel ill equipped to deal with (Whetham and Whetham, 2000).

First Few Years of Ministry

Literature worldwide has highlighted numerous problems associated with the first few years of full-time ministry (Brown, 1992; Pryor, 1982; Jud *et al.*, 1970; Mills and Koval, 1971). The majority of research suggests that church leaders who have spent less time in ministry are more prone to stress and burnout (e.g. Jud *et al.*, 1970; Pryor, 1982; Mills and Koval, 1971). One American study found that 50 per cent of the leavers sampled did so before two years of service (Wilson, 1971).

These findings are not surprising, given some of the stressors that clergy typically experience in their first parish. In contrast, the Australia NCLS found that burnout was highest among those who had been ordained between 6 and 20 years. Kaldor and Bullpit (2001) suggests that people new to the ministry may still be enthusiastic and committed enough to be able to survive what are potentially highly stressful placements. Despite the conflicting findings, Whetham and Whetham (2000) believe that the high level of stress experienced by new clergy in the transitional period is due to a variety of reasons e.g. difficulty of replicating college relationships, no regular supervision and relocation to new and sometimes remote areas.

Relocation

Relocation is not only a stressor faced by clergy new to the ministry but a stress producing experience (changes and losses) that all clergy and their families must deal with on a regular basis. Lavender (1983) estimated that nearly 200 clergy in America move every day of the year. In Australia church leaders have the second highest mobility rates of all professions next to the defence forces (Pryor, 1982). Frequent relocation undermines the family's sense of confidence in settling down, disrupts opportunities to establish social support networks, creates greater role demands and interruption of personal growth (Anderson and Stark, 1998). According to Alexander (1980) what seems to be unique to clergy families is the meaning they give to this stressor. It is difficult to accept and cope with the stress of moving because they feel the decision is not really made by the family. Often the choice to pull up roots and move to another location is a forced choice (Hausman and Reed, 1991) external to the family, and cannot be refused (Houts, 1982). Clergy families often feel a sense of powerlessness. In most instances they have limited involvement or are not consulted by church members and or denominational superiors concerning relocation decisions (Houts, 1982). In addition, the denominational hierarchy is often insensitive to the impact of the move upon the minister's family especially in regard to the impact of a move on the career involvement of clergy wives (Blanton, 1992) and the children's schooling (Cotton, 2000).

Lack of Social Support

Specific to the clergy population, lack of social support has been found to be one of the critical work related stressors (Morris and Blanton 1994a). Clergy's professional roles require them to continually engage intimately and intensely with others, but without a reciprocal contract in these relationships. Rediger (1982) described this stressful dynamic as 'half-intimacies' – unidirectional relationships in which church members relate to their pastor by disclosing deeply personal feelings and needs, without the pastor disclosing in the same way. Research amongst clergy has shown that deficient social support contributes to such consequences as marital maladjustment, loneliness, role overload (e.g. London, 1983; Richmond, 1991;

Warner and Carter, 1984), and inappropriate relationships with church members (e.g. Hall, 1997), burnout and depression (e.g. Virginia, 1998).

Gender 'The Shadow Ministry'

Because women have not been accepted equally within all Christian denominations (Australian Bureau of Statistics, 1991; Epstein, 1970; Rayburn, 1991a; Rayburn, 1991b; Rayburn, 1982), there are gender issues that can contribute to stress for women clergy (Birk, 2001; Piper, 1995; Rayburn, 1991a). These include few emphatic mentors to provide essential levels of social support (Rayburn, 1979), hostility and intolerance from others (Birk, 2001), lack of esteem for their person and their work (Birk, 2001), injustice in hiring and promotion situations (Rayburn, 1981) and the general attitude and suspicion that women are not quite good enough, competent enough or not holy or pure enough to be consecrated to do the work of 'real' clergy (Birk, 2001; Rayburn, 1981). Many women find that they are honoured in rhetoric but not in reality (Cotton, 2000).

Congregational Tensions

Congregations can also create their own set of tensions for the clergy. Conflicts with members and conflict between members flare up continuously (Blaikie, 1979; Croucher and Allgate, 1994; Dempsey, 1973; Dowdy and Lupton, 1978; Hulme, 1985). A survey of pastors by US Fuller Institute of Church Growth (1991) found that 40 per cent of clergy reported serious conflict with a congregational member at least once a month (London and Wiseman, 1993). In another study examining 243 ex-church leaders, the most significant reason for half the sample leaving was conflict with laity and other denominational leaders. A further half of the sample also felt that there was a lack of support from the congregation (Croucher and Allgate, 1994). Among church leaders, the NCLS Leader Survey found that a quarter of leaders found it hard to deal with difficult attenders. This factor was also strongly related to high levels of burnout (Kaldor and Bullpit, 2001).

Problems with the congregation are usually heightened by the fact that clergy frequently lack the necessary group work skills. Factors such as ingrained group dynamics, powerful members in the congregation, poor communication channels and unchallenged church expectations can plague clergy (Whetham and Whetham, 2000). Clergy striving to be good role models and not wanting to bite the hand that feeds them can further complicate relationships within congregations. Since clergy are ultimately dependent on the offering from the congregation they have to be careful how they present and confront others. All of these factors tend to compound the leaders' isolation and at times frustrate their relationships with those around them. In fact recent Australian literature suggests that many church attendees feel that clergy are 'out of touch'. For example, the NCLS found that 42 per cent of people were not aware of a clear vision within their church and

22 per cent agreed or strongly agreed that 'clergy are out of touch with the concerns of ordinary church attendees' (Kaldor and Bullpitt, 2001).

Changes in Society

Recent research suggests that the community at large does not have the same loyalties towards clergy that they used to (Neuchterlein, 1997). Before World War II people in communities looked up to clergy if for no other reason than that they were part of the small minority with professional credentials. In a time when relatively few people held college degrees they earned the respect accorded to all professionals. That educational gap between laity and clergy has largely disappeared. Today's laity typically have reaming credentials of their own while many clergy are by earlier standards shockingly uneducated (Neuchterlein, 1997).

Paralleling the declining respect for the clergy role is the declining numbers in church attendance. In the 1950s surveys of the time found that 44 per cent of the population were claiming to attend church at least once a month. Through the 60s church attendance declined. By 1970, it was 33 per cent and in 1990 it was 23 per cent (Bentley *et al.*, 1992). One reason given for this decline in church attendance is due to an increase in technology i.e. why go to church to hear an average sermon when you can watch a fantastic pastor on TV without leaving the comfort of your home? Meyrick (2000) found that a lot of clergy stress has to do with church decline. There is a desire to be successful and have a full church and when a church is struggling to survive that can produce stress.

Many factors behind the changing nature of ministry can be attributed to broader changes in society at large. For example, local communities no longer function as they used to. Women with families who a generation ago, stayed at home and were available as Church volunteers are now at work. Modern society is both more fluid in structure and less committed to moral and religious certainties (Meyrick, 2000). Kaldor and Bullpit (2001) found that the church leaders who saw good in society's changes were less likely to burn out.

Church Structure: 'A Unique Organisation'

To make matters even more complex clergy have to work within a church structure that often lacks clarity. The nature of the church organisation is different from that of most other organisations since attendance is voluntary and the people who participate expect no material reward (Malony, 1986). Whereas many organisations use people to develop products, the church uses people to develop people (Kunst, 1993). Unlike professional organisations, there are typically no clearly accountability mechanisms such as job descriptions, explicit time commitments, productivity expectations, distinct chains of command or grievance procedures (Whetham and Whetham, 2000).

Furthermore, while church members' contributions are different from those typically found in a business, so too is the form of leadership. Unlike any other organisation, churches adopt a seemingly paradoxical servant–leader model (Kunst,

1993; Whetham and Whetham, 2000). This type of leadership is based on Christ's teachings, that those who seek to be the greatest should be the least. In theory this means that leaders are to become servants and follow the example of Jesus. Although this is difficult to operationalise in a hierarchical organisation, it is nevertheless widely accepted at least as a principle of church leadership (Whetham and Whetham, 2000). Anecdotal studies of clergy have suggested that perceptions of denominational governance remain relatively unchanged as clergy still complain about the rigidness, authoritarianism and lack of connection between clergy and their denominational superiors (Gilbert, 1987: Harris, 1977).

Lauer (1973) in a article on the structure of the church, had this to say:

> Churches expect their ministers to do the impossible. His primary calling is spiritual, says the layman, but the minister is judged on organizational rather than spiritual criteria. The minister is a social being but tends not to have meaningful relationships with church members. The minister should not worry about money yet salary schedules may be inadequate. The ministry presents us with a case of structure punishment (p. 202).

Religious Upbringing

In addition to the unique church structure, research has shown that individuals who grow up in religious environments (compared to those from a non-Christian background) and later become church leaders are more prone to problems e.g. greater levels of loneliness (Whetham, 1997). While these individuals may feel safer and more in control of their familiar and immediate environment, it can come at a cost, suggesting that a life lived predominantly in a religious environment may not be a healthy one (Whetham and Whetham, 2000). Dunn (1965) found that emotional instability appears to increase with time in the religious life. The religious tend to be more perfectionistic, worrisome, introverted, socially inept and in more extreme cases, more isolated and withdrawn than the non religious. Dunn (1965) further observed that these patterns changed once church leaders leave the religious environment: 'once they have left this environment the personality test results of both religious men and women resemble more closely the results obtained from normals in the general population' (p. 134).

The literature supports the conclusion that due to the high number of stressors, fulfilling the role of clergy is difficult, and that some clergy and their families are not managing the stressors associated with the professional role very effectively (Noller, 1984).

12.5   THE CONSEQUENCES OF MINISTRY

Given the nature and extent of ministry demands, as well as theological orientation, ministers often tend to place professional demands over personal demands which although noble, often leads to serious consequences for both the minister and the church (Kunst, 1993; Rediger, 1982). Recent studies have found that three out of four ministers reported feeling stress severe enough to cause depression, anguish, anger, fear, and alienation (Daniels and Rogers, 1981). A study by Cumming of 90 Anglican clergymen found that 25 per cent of those interviewed had a stress profile of serious clinical concern, including significant higher levels of anxiety, depression, worry and anger-hostility (McGarry, 2001). It seems that despite the real spiritual aspects of the ministry, ministers are far from immune from developing significant psychological and emotional problems (Moy and Malony, 1987). This section specifically focuses on relational deficits (including marital dissatisfaction, effects on clergy family, clergy children, relationship with God), loneliness, sexual misconduct, burnout and retention and recruitment.

> "We've been on a perfection trip for the past 2000 years and it just doesn't work" "Now the clergy have to be accepting of their own humanity, and give themselves permission to say I'm struggling". Reverend Steven Ogden (McGarry, 2001, p. 5).

12.5.1 Identifying the Consequences of Ministry

Relational Deficits

As with other careers, the consequences of work-related stress may extend to other relational areas beyond the workplace (Lee and Balswick, 1989). Research confirms how important meaningful relationships are. Supportive and intimate relationships are a major factor in aiding adaptation to the unique and complex church environment and reducing stress levels (Benda and DiBlasio, 1992; Whetham and Whetham, 2000). Despite the benefits of having supportive relationships, it is clear from the research of clergy that all too often they have relational deficits at various levels.

Lack of Friendships: Whetham (1997) found extremely low levels of intimacy in church leaders' closest relationships (numerous denominations). This was found to be true regardless of denominational background, or whether participants were married or single. When asked about life at the moment, only 21 per cent of church leaders used the term friend and only 19 per cent expressed having close, deep, intimate or loving interactions with people. In other words four out of five church leaders did not mention meaningful relationships. This is consistent with international literature about clergy relationships.

Marital Dissatisfaction: Literature suggests that the spouse is the primary support person clergy look to for their intimacy needs to be met and to alleviate symptoms of loneliness (e.g. Kaldor and Bullpit, 2001). Ironically, however, for church leaders this important relationship is often impoverished. Church leaders appear to have more than their share of marital problems. Research is conclusive in its findings that marriages of church leaders are under severe strain (e.g. Bouma, 1979; Pryor, 1982; Whetham and Whetham, 2000). Clergy were ranked third highest of all professionals in America to seek divorces (Lavender, 1983). In fact divorce in clergy has been estimated to have quadrupled since 1960 (Stout, 1982).

Effects on Family: As the awareness of the professional and personal hazards of ministry increase, there has been a growing awareness of the needs of the minister's family (Lee, 1992; Morris and Blanton, 1994a; Moy and Malony, 1987). A fifth of all leaders agree or strongly agree that their family had been negatively affected by their role as a minister/pastor/priest (Kaldor and Bullpit, 2001). In a survey of pastors by Fuller Institute of Church Growth (1991), 33 per cent believed that ministry was a hazard to their family and 80 per cent believed pastoral ministry affected their families negatively (London and Wiseman, 1993). In Barna's (1993) study nearly half of the respondents agreed that pastoring had been difficult on their families. In fact, the impact of ministry on the pastor's home is an important reason given for leaving the ministry (Jud *et al.*, 1970). Literature involving clergy families has indicated that denominations may not fully consider the spill over effects upon all members of the clergy family (Morris and Blanton, 1994a).

Ministers' Children: While it is true that the entire clergy family shares the endemic stressors of ministry, children are also impacted by the endemic stressors of other systems e.g. school, part time work. The pile up of all these stressors can be an enormous weight on young lives, especially when coupled with developmental transitions (Morris and Blanton, 1994b). Ostrander and Henry (1989) found in a study of adolescents and adult children of clergy that demographic characteristics significantly influenced how they defined stress. Females reported higher stress ratings, as did those with lower incomes and those who had moved more frequently. In general however, interpersonal stressors affecting children in a clergy family have not been well documented.

Poor Relationship with God/Spirituality: Research has consistently shown that the same impoverishment in relationships with friends and family is often also experienced (paralleled) in clergy's relationship with God (Brokaw and Edwards, 1994; Hall and Brokaw, 1995; Hall *et al.*, 1996; Hatcher and Underwood, 1990; Kaldor and Bullpit, 2001; Keddy *et al.*, 1990; Whetham and Whetham, 2000). These findings are also consistent with the loneliness literature. One study found that religious university students, who were lonely, experienced less personal religious faith (Paloutzian and Ellison, 1982). Whetham and Whetham (2000) also found that leaders who expressed more loneliness were of particular concern. Not only were their relationships with people impoverished, but also they frequently had a poor relationship with God.

Loneliness

Given the high level of relational deficits, the clergy life is often very lonely (e.g. Brown, 1992; Warner and Carter, 1984; Whetham and Whetham, 2000). For many clergy loneliness is part of the job, not only because their role is a unique one but also because they often work alone and are isolated from their peers (Whetham and Whetham, 2000). Many clergy and their spouses report that they have few, if any genuinely close friends inside or outside the church (Lee and Balswick, 1989;

Whetham, 1997; Wright and Blackbird, 1986). In fact a US survey of pastors by Fuller Institute of Church Grown (1991) found that 70 per cent of clergy do not have someone they consider a close friend (London and Wiseman, 1993). A person who experiences loneliness feels a lack of quality in their relationships, regardless of whether they have people around them or not. The lonely person feels sad and distressed as a result of lack in relationships. For people who experience loneliness, there is a sense that there is no one with whom to share the deepest most vulnerable aspects of their lives (Whetham and Whetham, 2000). In fact literature worldwide states that loneliness and not being able to talk about feelings are key factors associated with burnout (Bach, 1979; Maslach and Jackson, 1982; Maslach, 1976, 1979; Pines and Aronson, 1981; Wubbolding and Kessler-Bolton, 1979). This has also been supported in research with clergy (e.g. Brown, 1992).

Burnout: 'Ministry Linked to Burnout'

Since its operationalisation in the early 1970s, burnout has been recognised as a serious threat, especially for human service professionals (Schaufeli *et al.*, 1993). Daniels and Rogers (1981) reviewed the research on burnout in human service professionals in general and applied these findings to clergy. They concluded that the symptoms of clergy who leave the ministry strongly resemble those of other helping professionals who experience burnout. In fact, clergy burnout is a concern to all religious denominations (Kaldor and Bullpit, 2001; Olsen and Grosch, 1991). Results from the NCLS (1996) conclude that 4 per cent of clergy – 1 in 25 – are suffering extreme burnout, with another 19 per cent finding burnout a major issue, 56 per cent borderline but coping and for just 21 per cent, burnout was not an issue (Kaldor and Bullpit, 2001). Furthermore, an Australian survey commissioned by the Anglican Church found that most of the 142 church leaders sampled were close to burnout. Five per cent were so affected that immediate remedial attention was thought necessary to restore their physical and mental health. A further 20 per cent said burnout was a factor in their lives, and 45 per cent said they were 'bordering on burnout' (Hay, 1995). Oswald (1991) describes burnout among church leaders as cyclic in nature, reinforcing itself. Starting with a high sense of call and commitment church leaders go out to serve in a sea of human need. Their commitment leads them to overdo their work, leading in turn to physical exhaustion and strain on family, marriage and other supportive relationships. A sense of being trapped starts to grow, alongside feelings of helplessness, hopelessness, resentment and guilt. In trying to get out of this trap church leaders commit themselves to working harder, drawing on their sense of call for energy to keep going. And so the cycle continues (Oswald, 1991).

Sexual Misconduct

The sexual misconduct literature reflects the burnout literature in that it identifies loneliness and a fear of intimacy as primary factors in sexual abuse by clergy

(Balswich and Thoburn, 1991; Laaser, 1991; Loftus, 1994; Muse *et al.*, 1993; Steinke, 1989; Thoburn and Balswick, 1993, 1994). The increasing incidence of sexually inappropriate acts committed by clergy has become a major concern to church communities and societies throughout the western worlds. While it is impossible to get exact figures there is disturbing evidence that a major problem exists (Hall, 1997). Various studies (e.g. Brock and Luckens, 1989; LaHaye, 1990) report the incidence of inappropriate sexual behaviour ranging from 14 to 25 per cent and of adultery ranging from 9 to 12 per cent (Hall, 1997). The toll of suffering caused by such behaviour is staggering (General Assembly of the Presbyterian Church, 1991). Sexual abuse has many damaging and long-lasting emotional and psychological consequences for the victims involved. There are many tragic stories of survivors of clergy sexual abuse in Australia (Ormerod and Ormerod, 1995). They highlight the inadequate mechanism for victims to deal with their trauma and bring clergy perpetrators to justice. Because clergy are highly visible symbols of the Christian gospel, such abuse can also do great spiritual harm both to the individuals involved and to the Christian community at large. Such traumatic experiences, as well as the possibility that clerical perpetrators may use God as a silencing strategy, have the potential to shatter survivor religious beliefs in a variety of ways, creating significant theological, spiritual and existential conflict (Farrell and Taylor, 2000). Research among secular psychotherapists indicates that 6 to 7 per cent admit sexual intercourse with clients (Pope and Vasquez, 1998). This figure is almost half that of clergy. It is such comparisons that led Hart (1993) (one-time dean of Fuller Seminary) to assert it would be safer to refer one of his daughters to a non-Christian psychologist than to a minister for counselling.

Retention and Recruitment

While little attention has been given to the clergy working conditions, or to the social and economic costs of clergy work stress, what research does tell us however is that clergy are leaving their 'calling' at alarming rates. Coupled with the high level of clergy leaving the ministry is the low number being recruited into the ministry. These issues are not only a concern for specific denominations but the decline in numbers of religious personnel is a phenomenon witnessed throughout the West in all religious communities (Henderson, 1997). Based on research carried out by John Mark Ministries, approximately one in two leave the ministry before retirement, constituting possibly one of the highest departure rates among all professions (Croucher, 1991a, 1991b). The NCLS Leader Survey adds further weight to these findings. Current leaders were asked how often they thought of leaving the ministry. Up to a quarter of senior clergy had sometimes or often thought about leaving the ministry (Kaldor and Bullpitt, 2001). Gallap and Poll (MacDonald, 1980) also revealed that one third of all pastors had thought about the implications of leaving what they thought would be a life long work. New research carried out in Wales has also found similar results with 53 per cent of clergy and other Church Leaders having thought about leaving their positions (Francis and

Kay, 2000). Kennedy and Sons (1997) reported that in the Roman Catholic Clergy (1997) only 521 men were ordained as clergy while the loss of clergy was 974, approximately a loss to gain ratio of 2:1 (Kennedy and Sons, 1997). A calculated projection of a continuation of this same statistical pattern of decline would mean that within little more than 50 years there would be no more Roman Catholic clergy (Virginia, 1998, p. 50).

In addition to the premature termination by choice is the premature termination by firing. For example, the Southern Baptist Convention (US) reported that during an 18-month period ending in 1989, some 2100 pastors were fired, reflecting a 31% increase since 1984 (Whittemore, 1991). In conjunction with these concerns are the enormous costs associated with losing qualified ministerial veterans who exit the ministry early in their careers. Thus it seems important for denominations to assist clergy in dealing effectively with the demands of parish ministry (Morris and Blanton, 1994a). Attrition can be lessened if clergy and their families are more adequately prepared for the demands they face (Blanton and Morris, 1999).

> That the question of resignation from the ministry could even
> arise should alert us to broader examination of the health of
> the whole organism (Johnson, 1963, p. 708).

12.6 THEORETICAL CRITIQUE OF THE WORK STRESS RESEARCH IN THE CLERGY PROFESSION

While empirical investigation into clergy has increased since the early to mid 1980s in some ways it is still in its infancy (Hall, 1997). Many of the studies are descriptive although some are correlational. However, overall there is very little methodologically sophisticated research in this area, and for the most part it is not theory driven. Empirical efforts that are available have been flawed by small sample sizes, denominational specificity, and other methodological problems (Houts, 1982). A systematic review of the literature on clergy in 8 major American Journals was conducted from 1991–1994 and only 4 out of 2468 (0.02%) quantitative studies considered clergy in their data (Weaver *et al.*, 1997). Virginia (1998) concludes that the absence of research into the clergy profession leaves open an entire subculture, which may well profit from gathering data and evidence pertaining to those factors which impact on the lives of clergy.

> Work with [in] religious organisations may represent one of
> the last great taboos of professional psychology (Pargament
> *et al.*, 1991, p. 403).

Given the limited theoretical work conducted within the clergy profession much can be learnt from contemporary work stress theories. Specifically, clergy health is a vocation of considerable interest for the Demands Control Support (DCS) Model (Karasek and Theorell, 1990), the Effort–Reward-Imbalance (ERI)

Model (Siegrist, 1996) and The Dual-level Social Exchange Model (Schaufeli *et al.*, 1996) and has not been examined in respect to any of them.

Demand-Control-Support Model: The DCS model is the dominant model in contemporary work stress research and attempts to explain and predict the relationship between psychological aspects of work environment and health outcomes (Muntaner and Schoenbach, 1994). The DCS model proposes that the psychosocial aspects of work environment namely, job control (over decision making and skill utilisation) and availability and quality of social support, can moderate the experience of strain arising from high job demands (Karasek, 1979; Theorell and Karasek, 1996). The DCS model has been tested in many different occupations and has received considerable empirical support (de Jonge and Kompier, 1997), but has not been examined in clergy populations as far as we know. The DCS model is particularly relevant when looking at workers in organisations where it is assumed that job design is an outcome of the inequitable power/control distribution in organisations. For example, within The Salvation Army many officers have high demands but a lack of control over organisational processes. Officers have limited control over when and where they will be next moved (lack of control). In fact, The Army has used frequent changes of appointment as a means of control. Unwanted influence and problems can be easily transferred far away. This encourages a tendency in the officers to 'Toe the Line' rather than challenge the 'status quo' (Bomberger 1974; Karlstrom, 1999). Further research is needed to explore the relationship between demands, control and social support in clergy (Lee, 1999).

Effort–Reward Imbalance Model: The ERI model has a more sociological focus and shifts from the concept of job control to the reward structure of work (Siegrist, 1996). It proposes that occupational strain arises when the effort invested in a job is high and the reward (money, esteem, security) is low. In ERI, high effort may arise from different kinds of job demands (i.e. time pressure, responsibility, and physical load). Furthermore, a specific pattern of coping with job demands and of eliciting rewards, termed 'overcommitment', is introduced (e.g. de Jonge *et al.*, 2000). Researchers have identified this distinct pattern of coping evolved from a critical analysis of the Type A global concept: individuals who score highly on overcommitment tend to misjudge both demanding and rewarding stimuli in their personal perception. For ministers, low rewards could result from a range of sources, including low income and forced mobility. Yancey (2001) raises the questions as to whether there is a profession that demands more and rewards less? Clergy spend many hours a week preparing a sermon and then hears at best on Sunday morning a polite 'Good Job' from a few parishioners at the door, that is as long as they have stayed within the 22 minutes allotted for preaching (Yancey, 2001). ERI would thus predict sustained emotional distress in ministers.

Dual-Level Social Exchange (DLSE): Similar in concept to the ERI Model, the DLSE model (Schaufeli *et al.*, 1996) proposes that reciprocity (the balance of investment and gains) plays a major role in the development of burnout. Lack of reciprocity can occur at the interpersonal level where emotion work is relevant and at the organisational level, where task-related and organisational stressors are relevant. Clergy for example cannot expect that drug addicts will return their feelings. Because of this lack of reciprocity, the emotional investments in clients have to be balanced by the organisation for which they work, perhaps in the form of recognition, avoiding unreasonable organisational problems or workload or by receiving social support. Zapf *et al.* (2001) outlines that if emotional work is high (the requirement to display organisationally desired emotions) it is likely that there is no reciprocity at the interpersonal level. If organisational and social stressors are high at the same time and particularly if the individuals believe that the organisation could do more to reduce such stressors, then the emotional demands at the interpersonal level may not be seen as balanced (at the organisational level) (Zapf *et al.*, 2001). It is this sense of having to wear a 'mask' and not being able to express how one really is (Cotton, 2000) that makes clergy a highly interesting population to study against the backdrop of the DLSE model.

12.7 INTERVENTIONS TRIED

Critics have suggested that denominational leaders have lost touch, are inaccessible and are failing to adequately acknowledge the enormous demands on clergy (Gilbert, 1987) and in return provide for them (Mace and Mace, 1980). However, anecdotal research in recent years seems to indicate that denominations are becoming more aware of a need to provide a wider range of financial, psychological, career and family support services (Morris and Blanton, 1994a). For instance some denominations have developed pastoral counselling programmes for their ministers (Kunst, 1993). Given the lack of systematic work examining the impact or nature of stress on clergy there is no clear direction for the development of interventions or organisational recommendations. A greater awareness of the stresses that impinge on clergy and their families could provide the basis for developing programmes and policies (Blanton, 1992). David Richards, a retired Episcopal bishop who runs a clergy counselling centre, commented, 'We need to install mechanisms for preventative care and guided growth, with procedures for intervention before stress leads to burnout and crisis' (Whittemore, 1991, p. 5).

> From the burnout studies researchers have made several recommendations for action on organisation, individual and training levels, which can be taken to prevent burnout. The ministry has none of these preventative factors built in on any of the three levels and the potential for burn out is strong (Daniels and Rogers, 1981: 244).

An example of a recent study that seeks to address these issues is one commissioned by The Salvation Army in Australia (Southern Territory) described below.

12.8   EXAMINING THE SALVATION ARMY OFFICER'S WELL-BEING: 'FLASH POINT IN THE THIRD SECTOR'

12.8.1   Background

In 2000 The Salvation Army commissioned a two-year study to examine the well-being of its officers (Cotton, Dollard and de Jonge). Despite the enormous benefits that The Salvation Army provides, the literature and anecdotal evidence from leadership, suggests high levels of burnout among officers. High stress levels seem to be adversely affecting officers and their families. Statistics reveal a trend of dramatic increases in the number of officers leaving the work (Golding, 2000). The age profile of officers is also increasing with the number joining too low to replace those retiring (Gledhill, 1998). The leadership of The Salvation Army is aware that there are many issues confronting clergy today, both nationally and internationally. Specifically Army leadership are concerned internally for their officers well-being and believe that this study will provide the insights, support and guidance. In addition, the research team were successful in gaining an Australian Research Council (ARC) Industry Linkage grant to focus further attention on the issue of relocation.

12.8.2   Aim

The primary aim of the project is to improve the organisation's understanding of how best to assist officers in providing the healthiest work environment for ministry. Not only will this study have significance for The Salvation Army, but it will also have significant implications for the church universally, in increasing the awareness and importance of clergy well-being. The project is innovative as it will extend and develop contemporary work stress theories and models (DCS, ERI and DLSE) in the area of clergy well-being. This will assist in the development of appropriate intervention programmes aimed at reducing baseline measures of minister's stress.

12.8.3   Research Methodology

The research methodology has been carefully designed from a combination of the latest research from the work stress literature and collaboration with the industry partner, through the utilisation of a steering committee (10 officers representative of the officer population e.g. age, gender) and experts in the field (e.g. key researchers in the field, psychologists). The guiding philosophy of the study is Participatory Action Research (PAR) as proposed by Landsbergis and colleagues (1993). Key stakeholders (Leadership, The Steering Committee and Experts in the Field) will be closely involved in every stage of the project. Triangulation will also be used as a method for strengthening the trustworthiness of the research findings (Breitmayer *et al.*, 1993).

Studies reporting occupational strain have been criticised for the fact that they rely only on self-report measures, utilise cross-sectional designs (Zapf *et al.*, 1996) and do not measure negative affectivity. This research plan addresses all three of these criticisms; introducing a longitudinal design, measures of negative affectivity and an objective measure of officer performance (Officer Development and Review Program). Furthermore, the longitudinal design will enable the researchers to monitor changes in well-being and productivity over time and possibly link changes to variations in the work environment (e.g. relocation) in specific locations as well as evaluate the effectiveness of any intervention/s. The one-year interval is chosen as several studies have indicated that the time appears to be long enough for possible changes in individual scores but not too long for too much attrition in the sample (de Jonge, 1995; Vermaat, 1994).

12.8.4   Conceptual Framework

As discussed previously (see Section 12.6), this study will draw on three contemporary theories of work stress (DCS, ERI and DLSE) in the organisational literature that are applicable to clergy. Kasl (1998) has recently argued that it is worth studying the relative contribution of each model to the explanation of health and well-being, in view of their differences and complementary aspects. Theorell (1998) also argued that a good exploration of the work environment should include components of both DCS and ERI models.

12.8.5   Structure of the Research Project

The project is being conducted in two stages as outlined below (Figure 12.1).

12.8.6   In-depth Interviews 'Informing the Research'

Background

In order to have a thorough understanding of the organisation it was first necessary to interview the members of the steering committee and the experts in the field. The information gained from the in-depth interviews assisted the research team in gaining a greater knowledge and understanding of the organisation climate/ culture. The in-depth interviews were also instrumental in designing the questionnaire for the pilot study. It was essential that the questions included in the questionnaire tapped the issues that confront officers of today.

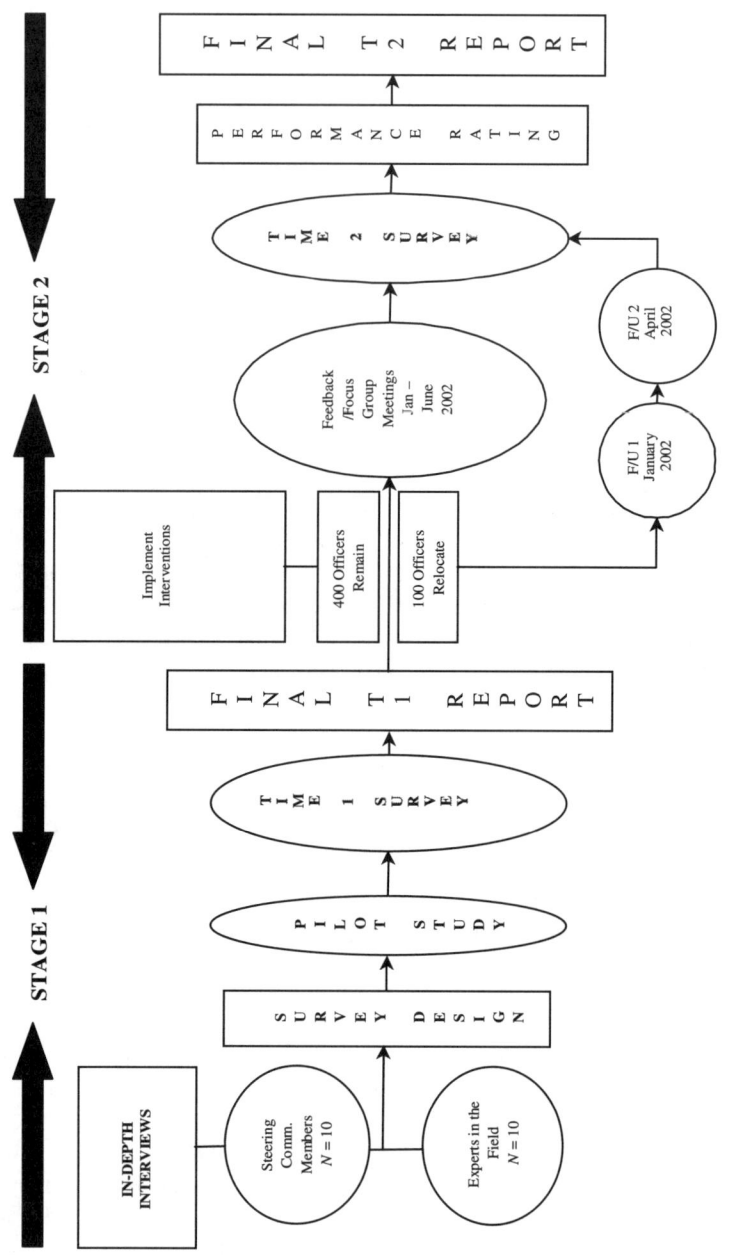

Figure 12.1 Overview of the Study (Stage one and Stage two).
Note: T1, Time 1; F/U, Follow-up; T2, Time 2.

Method

Participants

The research team interviewed 10 officers (members of the steering committee) and 10 experts in the field. The 10 members of the steering committee were selected in consultation with leadership to represent the diversity of the officer population i.e. age, gender, marital status, state. The 10 experts in the field consisted of a mixture of key players in the area of clergy well-being i.e. researchers, psychologists.

Procedure

A literature review and discussions with key players in the organisation assisted the research team in constructing an interview schedule. The interview schedule was then piloted and revised before being used to interview the 20 subjects. Participants were sent an outline of the questions enabling them to have the opportunity to reflect and think about their responses. Participants were then contacted and assigned a time that was suitable for them to be interviewed. The chief investigator interviewed participants (semi-structured) face to face or on the telephone depending on their location and availability. With the consent of each participant, the interviews were audio taped and then transcribed. Once the interviews were transcribed they were analysed using content analysis by the chief investigator for relevant themes.

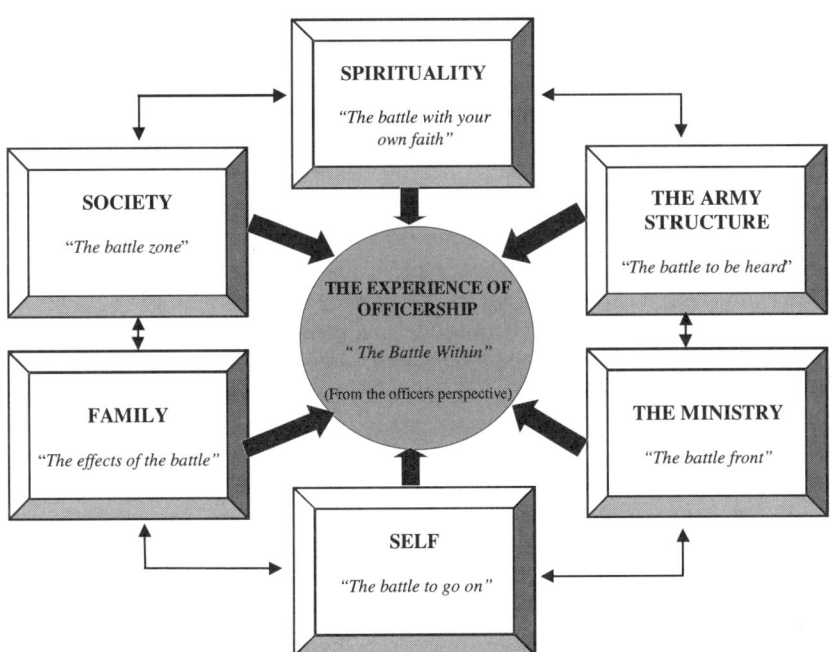

Figure 12.2 The Experience of Officership.

Results

Discussion
The results of the in-depth interviews were very clear in highlighting the areas of importance when considering officer's well-being. In fact, there was an overall census from The Steering Committee and Experts that the results obtained were highly reflective of the officer experience. From the results of the in-depth interviews, the pilot questionnaire was constructed around the six main dimensions of officership as shown in Figure 12.2.

12.8.7 Time 1 Study

Background

The 'Time One' survey was viewed as an instrument that would indicate to the officers by way of content that the research team understood many of the issues confronting them (face validity). If a successful survey could be conducted and interventions recommended then it could be an effective means with which to install confidence between leadership, the officers and the research team.

Method

Participants
Participants at Time 1 were 362 out of 521 (response rate 69%) Salvation Army officers from the Southern Territory of Australia. The sample consisted of 48% men and 52% women, between the ages of 26–75 (M = 47 years, SD = 9.83). Eighty nine percent of officers were married with 76% of officers indicating that they had children (average number of children being two per officer). Officers were stationed at various appointments, the majority (49%) were currently serving in a corps (church) appointment, 15% were in a social appointment, 18% were in an administration appointment, 12% were in a combination appointment i.e. corps and social, and 6% were in an 'other' type of appointment. The number of year's officers had served in an officer role ranged between 0 and 44 years (M = 15 years, SD = 10.75) and the total number of appointments officers had been placed in ranged between 1 and 31 appointments (M = 7 appointments, SD = 4.60). The sample was representative of the population with respect to gender and state.

Procedure
'Blind coded' questionnaires and information regarding the study were sent in personally addressed envelopes to the 521 officers and delivered through internal departmental mechanisms. Each officer received a stamped addressed return envelope. To ensure a high response rate, the research team used a modified approach to Dillman's (1978) method of survey follow-up: an e-mail reminder (2 weeks); a message in the Divisional newsletters (4 weeks) and a second copy of the questionnaire including another letter from the research team (5 weeks).

Measures
The final content of the questionnaire was influenced from a range of sources: In-depth interviews, discussions with leadership, the pilot study, discussions with the steering committee and from similar studies reported in the literature. Figure 12.3 shows the key variables measured in the survey.

Statistical Treatment
Pearson correlations were conducted to assess zero-order relationships between demographic (person and job), personal disposition, work environment (general and specific), support and coping and levels of strain, efficacy and performance variables. By comparing the findings with other studies (averages) we were able to assess the various levels of officers' well-being e.g. psychological distress and burnout. Frequency and descriptive data was also used to obtain averages (mean), the amount of variation in the results from the mean (*SD*) and percentages. Open-ended questions were analysed (e.g. Neuendorf, 2002) by grouping similar responses into themes.

Preliminary Results

Demanding Aspects of Ministry Identified
Salvation Army officers were asked to rank (in terms of the most demanding) the three most demanding aspects of ministry. The results obtained were consistent with the literature across all denominations. It was clear from the results (Table 12.1) that the three most demanding aspects of ministry specific for Salvation Army officers were unrealistic expectations followed closely by high levels of workload and administration demands.

Table 12.1 Demanding Aspects of Ministry

The 10 Most Demanding Aspects of Ministry	Freq
Unrealistic expectations (Emotional demands)	235
Workload	221
Administration 'The continuing paper war'	201
Constant availability 'always on duty' (24 hours/7 days a week)	180
People (contact/ministry)	179
Keeping a balance 'Juggling roles between work and home'	77
Poor Leadership	60
Finances (Financial strain)	57
Dealing with conflict	52
Working within the structure/system	51

Levels of Strain
Although clergy have been clearly identified as a stressful occupation in the literature it was important that levels of strain in this unique profession were compared with other human service professionals. One problem in stress research

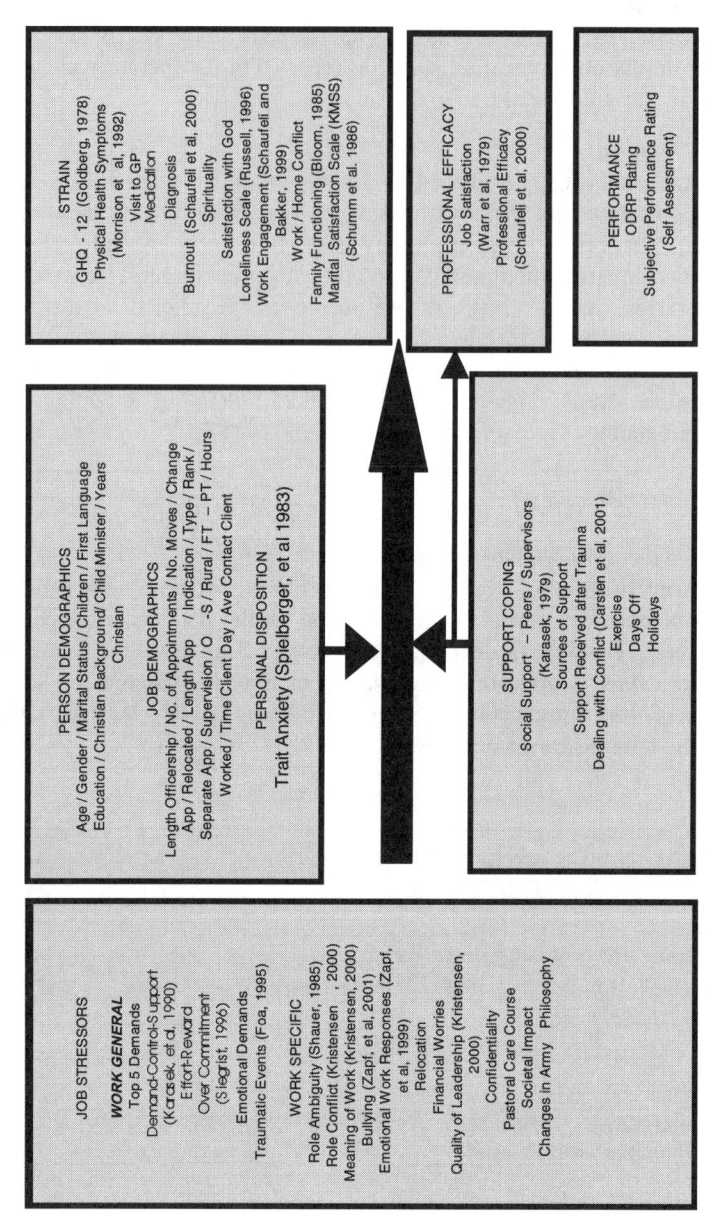

Figure 12.3 Key Variables Measured in the Survey.

is that measures vary from one study to the next. The following tables report means, standard deviations and internal consistencies of the standardised measures used, and compare measures of GHQ, burnout, and job satisfaction with other occupational groups. Caution must be exercised when comparing data against other human service groups, as they are not normative.

While it is evident that the average level of psychological distress experienced by Salvation Army officers was not significantly higher than the majority of other occupations (Table 12.2), it is important to acknowledge that officers were significantly higher than a sample of young South Australians (Winefield *et al.*, 1993), Psychiatric Nurses (Janman *et al.*, 1988) and British teachers (Parkes, 1990). In addition 22% of officers fall in the high to severe range of psychological distress (when binary scored). In comparison a recent Australian sample ($N = 10,600$) found 10.4 per cent of the population fall into this category (Andrews *et al.*, 1999).

Table 12.2 GHQ-12 Scores (Likert Scoring): Means, Standard Deviation and *t*-tests: Salvation Army Officers with Other Samples

	M	*SD*	*N*	α	*t*-test
NZ Prison Officers (Voges *et al.*, 1982)	12.50	5.50	332	–	−2.76**
SA Public Sector Workers (Macklin and Dollard, 2000)	12.25	5.77	84	.89	−1.36
University of Adelaide Staff (Jarrett and Winefield, 1995)	12.20	5.90	1961	–	−2.71**
Correctional Officers (SA) (Dollard, 1996)	12.18	7.22	414	.93	−1.78
SA Private Sector Workers (Macklin and Dollard, 2000)	11.76	6.29	143	.91	−.63
Human Service Workers (Dollard *et al.*, 2001)	11.53	5.83	798	.90	−.41
SA Nurses (South Australia) (Dollard, 1996)	11.49	5.71	106	.91	−.17
Salvation Army Officers	**11.39**	**5.07**	**359**	**.88**	
British University Staff (Daniels and Guppy, 1992)	11.30	4.70	221	–	.21
STA Clerical Administration (SA) (O'Mara, 1991)	11.25	5.20	176	–	.30
SA Youth Workers (Dollard *et al.*, 2001)	11.23	–	140	–	.32
SA Manager (Dollard *et al.*, 2001)	10.84	–	43	–	.67
Young South Australians (Winefield *et al.*, 1993)	10.70	5.20	8998	–	2.47**
Psychiatric Nurses (Janman *et al.*, 1988)	10.30	5.10	349	–	2.85**
British Teachers (Parkes, 1990)	9.13	5.21	157	–	4.62**

Note: **$p < .01$; SA, South Australia; STA, State Transport Authority; NZ, New Zealand.
Table adapted from Dollard *et al.*, 2001.

Burnout

The means and standard deviations in Table 12.3 indicate that the level of exhaustion for Salvation Army officers is significantly higher than that reported by the other occupations. Similarly, levels of cynicism were also significantly higher compared to that of the other occupations. Levels of professional efficacy on the other hand were higher than the National Sample (The Netherlands) (Schaufeli and

Dierendonck, 2000) and significantly lower than the Netherlands and White Collar workers (Schutte *et al.*, 2000).

Table 12.3 Mean and standard deviations of the MBI – General subscales

	M	SD	N	α	t-test
Exhaustion					
Salvation Army Officers	**2.83**	**1.45**	**356**	**.89**	
National Sample (The Netherlands)	1.78	1.21	1111	–	12.35**
(Schaufeli and Dierendonck, 2000)					
White Collar (Schutte *et al.*, 2000)	1.53	1.36	3378	.86	17.04**
Finland (Schutte *et al.*, 2000)	1.51	1.40	8529	.86	17.40**
Blue Collar (Schutte *et al.*, 2000)	1.46	1.43	5677	.86	17.52**
Sweden (Schutte *et al.*, 2000)	1.45	1.55	267	.76	11.41**
Netherlands (Schutte *et al.*, 2000)	.86	1.29	259	.76	17.74**
Professional Efficacy					
Netherlands (Schutte *et al.*, 2000)	4.87	1.61	259	.68	−2.66**
White Collar (Schutte *et al.*, 2000)	4.82	1.46	3378	.83	−4.30**
Sweden (Schutte *et al.*, 2000)	4.78	1.74	267	.70	−1.77
Finland (Schutte *et al.*, 2000)	4.65	1.68	8529	.83	−1.44
Salvation Army Officers	**4.57**	**.97**	**343**	**.83**	
Blue Collar (Schutte *et al.*, 2000)	4.55	1.80	5677	.82	.35
National Sample (The Netherlands)	4.29	.97	1111	–	4.67**
(Schaufeli and Dierendonck, 2000)					
Cynicism					
Salvation Army Officers	**1.85**	**1.40**	**353**	**.84**	
Blue Collar (Schutte *et al.*, 2000)	1.58	1.75	5677	.74	3.46**
Finland (Schutte *et al.*, 2000)	1.49	1.68	8529	.76	4.69**
Sweden (Schutte *et al.*, 2000)	1.42	1.86	267	.65	3.16**
National Sample (The Netherlands)	1.34	1.13	1111	–	6.23**
(Schaufeli and Dierendonck, 2000)					
White Collar (Schutte *et al.*, 2000)	1.31	1.56	3378	.78	6.82**
Netherlands (Schutte *et al.*, 2000)	1.11	1.92	259	.62	5.26**

Note: **$p < .01$; The comparison occupations from Schutte *et al.* (2000) are employees within a multi-national forestry organisation; National Sample (The Netherlands) with all kinds of workers.

When the burnout dimensions (UBOS) of officers as outlined in Table 12.3 are compared with the norms from the national sample (The Netherlands) (Schaufeli and Van Dierendonek, 2000) it is clear that levels of exhaustion are high (falling between the range of 2.20–4.19). While levels of cynicism fall into the average category (.50–1.99) they are close (only .15 out) to falling within the high range (2.00–3.49) (only .14 out). Consistent with Table 12.3, levels of professional efficacy fall into the average category (3.67–4.99).

Overall, Salvation Army officers report high levels of exhaustion and cynicism. These finding are consistent with recent literature in the area of clergy well-being indicating high levels of clergy burnout (e.g. Kaldor and Bullpit, 2001). Levels of professional efficacy however are significantly higher than a large

National Sample (The Netherlands) (all kinds of workers). This is consistent with the high levels of satisfaction reported in the clergy profession.

Levels of 'global' job satisfaction
Eighty three percent of officers reported vocational satisfaction in the range moderately satisfied to extremely satisfied. Thirteen per cent reported they were moderately dissatisfied to extremely dissatisfied with their vocation.

The results were as follows: *Extremely Satisfied 12.9%, Very Satisfied 42.1%, Moderately Satisfied 28.4%, Not sure 3.7%, Moderately dissatisfied 5.6%, Very dissatisfied 4.8% and Extremely Dissatisfied 2.5%*. On average, officers were moderately satisfied with their vocation.

In comparison with other samples (Table 12.4), job satisfaction of Salvation Army officers was significantly higher than family and community service workers (Dollard *et al.*, 2001) and correctional officers (Dollard, 1996).

Table 12.4 Global Job Satisfaction: Means, Standard Deviations and comparison with other samples

	M	*SD*	*N*	*t*-test
Salvation Army Officers	**5.29**	**1.40**	**356**	
British Blue Collar Male Workers (Warr *et al.*, 1979)	5.24	1.46	200	.40
Nurses (South Australian) (Dollard, 1996)	5.18	1.10	102	.84
Human Service Workers (Dollard *et al.*, 2001)	4.84	1.37	806	5.13**
Correctional Officers (SA) (Dollard, 1996)	4.20	1.60	416	10.10**

**p < .01; Table adapted from Dollard *et al.*, 2001.

To develop an understanding of clergy satisfaction, officers were asked to rank (in terms of the most satisfying) the three most rewarding aspects of ministry (Table 12.5). The top three factors were: the privilege of ministry; seeing people's lives change/grow – spiritually and personally; and, seeing people saved 'soul saving'. This reinforces the importance of intrinsic satisfiers in clergy work.

Factors associated with strain and efficacy
Specifically we used psychological distress (GHQ) and burnout (emotional exhaustion – EE) as indicators of strain, and professional efficacy (PA) and satisfaction (JS) to indicate efficacy (see also Figure 12.3). There was a large correlation between GHQ and EE ($r = .50$, $p < .01$) and a medium sized correlation between PA and JS ($r = .35$, $p < .01$). Correlations were larger within the strain and efficacy measures than between them, although strain and efficacy were [negatively] correlated [GHQ with PA ($r = -.29$, $p < .01$); GHQ with JS ($r = -.32$, $p < .01$); EE with PA ($r = -.20$, $p < .01$); EE with JS ($r = -.25$, $p < .01$)].

Table 12.5 The Ten Most Satisfying Aspects of Officership

Response	f
Privilege of ministry	331
Seeing people's lives change/grow – *spiritually and personally*	262
Seeing people saved 'Soul Saving'	211
Fulfilment of 'Calling'	164
Platform ministry i.e. *preaching*	140
People contact	130
Serving/being used by God/working for God	114
Making a difference – 'impacting lives'	73
Variety of work – availability of ministry opportunities	46
Personal/self development/fulfilment i.e. *Attend courses, learning*	41
Ministering to the saved 'Nurturing' – 'Growing the saints'	41

Although the data are explored more fully in the study, due to space we will present here only factors associated with strain and efficacy. Many correlations were observed so we chose Pearson $r > .20$ (>small) as the cut off point. [Correlation coefficients are defined as small, when $r = .20$, medium when $r = .30$, and large when $r = .50$ (Cohen, 1992)].

Strain
As shown in Table 12.6 there were 11 factors *commonly* associated with higher levels of strain. These were higher levels of: a tendency to over commit; imbalance between efforts and rewards; negative affectivity; emotional work responses; role conflict; role ambiguity; emotional demands, relocation demands, mobbing; and financial worries, and lower levels of rewards. While control showed a small to medium sized correlation with GHQ ($r = -.29$, $p < .01$), it showed only a small correlation with EE ($r = .11$, $p < .05$).

Efficacy
As shown in Table 12.7, 5 factors were commonly associated with higher levels of efficacy were higher: meaning associated with the work; control; rewards; and peer support, and lower levels of role ambiguity.

Discussion

While it is evident that the levels of psychological distress experienced by Salvation Army officers is average in comparison with other occupations, it is important to acknowledge that 22% of officers fall in the high to severe range of psychological distress. Results also indicate that officers experience higher levels of burnout compared with other occupations (exhaustion and cynicism). Despite these risky

Table 12.6 Strain: Risk factors identified

Psychological Distress	r	Exhaustion	r
A tendency to over commit	.33**	A tendency to over commit	.51**
Imbalance between efforts and rewards	.33**	Imbalance between efforts and rewards	.45**
Negative affectivity	.54**	Negative affectivity	.45**
Emotional work responses	.31**	Emotional work responses	.43**
Role conflict	.26**	Role conflict	.38**
Role ambiguity	.32**	Role ambiguity	.36**
Rewards	−.29**	Rewards	−.30**
Emotional demands	.22**	Emotional demands	.35**
Relocation demands	.22**	Relocation demands	.29**
Mobbing (bullying)	.21**	Mobbing (bullying)	.26**
Financial worries	.23**	Financial worries	.25**
Job control	−.29**	Demands common to ministry	.39**
Meaning	−.22**	Demands/effort	.37**
		Officers in a shared appointment	−.21**
		Quality of leadership	−.20**
		Younger officers	−.20**
		Officers in a Corps (Church) appointment	.20**

Note. Only significant correlations ≥.20 included in table. Factors above the solid line common to both psychological distress and emotional exhaustion.

Table 12.7 Efficacy: Risk factors identified

Job Satisfaction	r	Professional Efficacy	r
Meaning	.53**	Meaning	.41**
Job control	.39**	Job control	.35**
Role ambiguity	−.35**	Role ambiguity	−.26**
Rewards	.42**	Rewards	.24**
Social support from peers	.24**	Social support from peers	.22**
Relocation demands	−.31**	Problem solving (conflicts)	.26**
Imbalance between efforts and rewards	−.30**	Compromising (coping)	.26**
Quality of leadership	.30**	Working part time	.22**
Negative affectivity	−.26**		
Emotional work responses	−.25**		
Role conflict	−.21**		
Financial worries	−.20**		

Note. Only significant correlations ≥.20 included in table. Factors above the solid line common to both job satisfaction and personal accomplishment (efficacy).

consequences of ministry, officers report high levels of satisfaction and average levels of professional efficacy within their vocation in comparison with other occupations.

The results generally support the notion that the development and strain and efficacy in workers are separate and important processes, as suggested by Karasek (1979) (see also Dollard *et al.*, 2001). It makes good sense as researchers to be interested in both aspects because we not only want to see reduced strain among workers, but we also want to ensure that workers experience satisfaction and feel that they are making an effective meaningful contribution. The essence of the dynamism, the separate yet related aspects of the two processes is captured in a comment by an officer

> Saying yes to God has been a great blessing even though there have been sacrifices and hard times. Couldn't have found such satisfaction in any other vocation/calling.

Preliminary results further suggest the relevance of some of the critical components of the theories outlined earlier in explaining levels of strain and efficacy in clergy.

Strain
It is clear when looking at the correlations (Table 12.6) that there is strong support for Siegrist's Effort–Reward Imbalance Model. Rewards rather than control were most important in predicting higher levels of strain overall. Those with a high imbalance between demands and rewards plus a tendency to over-commit reported higher levels of strain (Siegrist, 1996). In appears that the high demands of ministry as reported in the literature (e.g. Hall, 1997, Morris and Blanton, 1994b) are not matched by the rewards officers receive i.e. inadequate salaries and insufficient long service leave. This probably also underpins the link between financial worries and strain. We also found other important demands in addition to those articulated by Siegrist. The demands of emotional work and those of relocation, central to the ways of Salvation Army work showed important correlations with strain. Role conflict and role ambiguity on the other hand are frequently found in general organisational research to be linked to higher levels of strain. Finally being in the clergy does not lead to escape from bullying within the organisation, higher levels of the experience being associated with higher levels of strain.

The results lend small support to Karasek's (1979) original Demands Control Model, as demands (albeit different from those articulated in the model), and control were associated with strain (GHQ small-medium, EE small). The importance of social support (e.g. Theorell and Karasek, 1996) was not supported. It is important to note however that an exploration of possible interaction effects is yet to be conducted. Following Van der Doef and Maes, 1999 this study tested the strain hypothesis of the DCS model: that workers in jobs combining high demands, low control and low support would report higher levels of strain, shown by direct effects in a regression model (as opposed to the buffer hypothesis).

Furthermore, initial findings show support for the Dual-Level Social Exchange model by Schaufeli *et al.*, 1996. High levels of emotional work responses are evidently linked with increased strain. This lack of reciprocity that Schaufeli talks about is very evident in the literature and in interviews we conducted with experts in the field. Measures of loneliness (not reported here) also showed officers had higher levels of loneliness in comparison to other samples, indicating their lack of opportunity to develop reciprocal relationships generally. This lack of reciprocity at a personal level is possibly exacerbated when clergy also experience a lack of reciprocity at an organisational level (high imbalance between efforts and rewards) shown here to be clearly linked to higher levels of strain.

Efficacy

Theorising within the Demand–Control model suggests that high levels of demands, combined with congruent levels of control will lead to situations of motivation, satisfaction, efficacy and productivity and there is some evidence to support this (see Dollard *et al.*, 2001, p. 138). Preliminary results here show the importance of control and support, but not high demands as expected. This may have something to do with the nuances of clergy work, that is, clergy may expect high demands as part of their professional role and therefore high demands may not be as important as other factors in its association with efficacy.

Strain and efficacy

The results support the idea of separating out the factors associated with strain on the one hand, and efficacy on the other. The constructs are associated with different aspects of the work environment, but they are also associated (albeit negatively) giving some support to the dynamic associations between job strain and feelings of mastery outlined by Karasek and Theorell, 1990, and supported in other research of 800 human service employees ($r = -.42$, $p < .01$) (Dollard *et al.*, 2001).

12.8.8 Next steps in the journey

The second stage of the study is outlined in Figure 12.1. At the completion of the Time 1 study five short-term – relatively easy to implement – interventions were put forward to Salvation Army leaders for their consideration. Out of these five interventions, four were implemented: (1) feedback/focus group mornings held throughout the territory; (2) copy of the 'Time 1' executive summary distributed to all officers; (3) pilot relocation resource package provided to officers who relocated; and (4) a phone call from the pastoral care department within the first three months of officers relocating. The Time 2 survey (June 2002) will provide an opportunity to evaluate the effectiveness of these interventions as well as provide a longitudinal snap shot of the officer vocation. It is also hoped that an objective measure of performance will be obtained from officers by using the Officer Development and Review Program average rating. This will provide an opportunity to further assess the important link between well-being and performance. A final report will then be given to Salvation Army leaders at the conclusion of the project outlining recommendations and further areas of research.

As an adjunct to the officers' well-being study, the Australian Government has provided further financial support through an Australian Research Council (ARC) grant to focus further attention on the issue of relocation of Salvation Army officers. Approximately 150 officers who relocated in January 2002 to new appointments will be followed up over 3 10-week intervals. Examination of relocation will in effect be a natural experiment and its impact will be evaluated for its effect on officers and their families who relocate, compared with those officers who remain in their appointment.

12.9   POLITICS OF WORKING WITH CLERGY

Despite a unique and demanding role it is interesting to note that many clergy do not seek support from others (Whetham, 1997; Whetham and Whetham, 2000). There seems to be a general unwritten rule that clergy should be strong and God-like, being all things to all people. Consequently, for clergy to publicly acknowledge dissatisfaction or question the establishment may be interpreted as being unfaithful or unfit to fulfill one's calling. Social desirability plays an important and complex role here.

For these reasons gaining access to clergy populations or working toward organisational change can prove to be difficult. Feedback and communication are desperately lacking in these monologue institutions as is being revealed in this study (lack of reciprocation). This is unfortunate since open communication channels are an essential tool in challenging ingrained clergy stereotypes and lessening unrealistic expectations of clergy among the broader church community.

Although research in the past has helped provide critical feedback about clergy health symptomatology, it appears to have had little impact at an organisational level. As discussed, there are many reasons for this. In addition, it may be that traditional researchers have not directly involved church hierarchy in the research design or recommendations. The participatory action research method used in our study set out to address these concerns. By engaging different stakeholders of the system throughout the research process greater ownership of the issues raised was encouraged. It is hoped that by including these different stakeholders that greater organisational change will result.

12.10   FUTURE RESEARCH DIRECTIONS

While the current Salvation Army Study seeks to address many of the limitations of previous clergy research, the need for more research in this area is readily apparent. Such areas include the need to identify coping strategies that relate to the effective adaptation of clergy and their families (Ostrander *et al.*, 1994), the process of parenting in the context of the clergy family, as well as the types of stressors and coping strategies that characterise the experience for clergy children (Blanton,

1992; Stevenson, 1982). In fact, literature involving clergy families has indicated that denominations may not fully consider the spill over effects of clergy work upon all members of the clergy family (Morris and Blanton, 1994a). For this reason Morris and Blanton (1994b) urge researchers to further the work in the understanding of clergy/family well-being.

One of the most notable findings from Hall's (1997) review of the literature is that there are virtually no studies on pastors' spirituality.

> It is critical that pastor's spiritual maturity be specifically examined since it is intricately related to their psychopathology and psychological maturity as well as to their vocational function as spiritual leader (Hall, 1997, p. 250).

More recent research (Whetham, 1997; Whetham and Whetham, 2000; Kaldor and Bullpitt, 2001) has examined clergy's relationship with God. However, it would be worthwhile to examine the link between clergy's relationship with God and their well-being.

Another area of research that needs further research is the area of relocation amongst the clergy population. Although the ministry is an occupation that demands frequent relocation it has rarely been studied in this regard compared to other groups e.g. business executives (Anderson and Stark, 1988) and military populations (Merchant and Medway, 1987). Very little research has been directed at attempts to identify factors associated with successful adaptation to relocation and few studies have compared the well-being of relocators with that of similar non-relocators (Frame and Shehan, 1994). Stevenson (1982) also recommended greater attention to be given to the needs of children in the moving process. In addition, there have been few recent empirical studies of psychopathology among clergy (Hall, 1997; Keddy *et al.*, 1990).

In general, more sophisticated theoretical bases for understanding pastor's personal functioning are needed and should be tested in more methodologically rigorous ways i.e. longitudinal studies. Methodologically multivariate analyses such as structural equation modelling could also be employed to provide a more sophisticated understanding of the relationships between variables (Hall, 1997).

> The disturbing degree of silence in the church community regarding the plight of clergy, combined with a lack of preventative strategies at individual and structural levels, suggesting that the problems of clergy within the organised church continue and need to be further investigated (Whetham, 1997: 4).

12.11 CONCLUSION

The research approach although guided by theory to some extent is also grounded in the realities of the everyday experiences of clergy work. Initial ground work to the development of the survey instrument involved in-depth interviews with experts in the field, a pilot study, and the influence of a representative group of clergy (the Steering Committee). Preliminary analysis shows the importance of each of the DCS, ERI and Dual-Level Social Exchange models but each has its own limitations in capturing the variables intrinsic to Salvation Army clergy. A better understanding of the levels of strain can be given by exploring local variables not just the key variables specified in the dominant models. For example issues such as relocation, poor quality of leadership (related to aspects of strain and efficacy) and meaning of work are not predicted as being a source of strain in the three models. These findings therefore, underscore the importance of using a grounded approach to uncover the unique dimensions of this profession.

We have provided here a preliminary snap shot of the levels of strain and efficacy clergy are experiencing, and have attempted to tease out some key factors associated with strain and efficacy. Further statistical analyse will seek to examine the relationships in more detail. For example, structural equation modelling will enable the research team to develop and test a model of clergy stress and well-being for Salvation Army officers and clergy more generally. In addition the study is a live event as outlined above, collecting data over a 2 year period.

12.12   ACKNOWLEDGEMENTS

The authors would like to thank Mr John Petkov for calculating the t-test scores.

12.13   REFERENCES

Alexander, C., 1980, The interface of marriage and ministry: A survey of conference directors of pastoral care. Paper presented to The Consultation on Clergy Marriage, Lake Junaluska, NC.

Anderson, C. and Stark, C., 1988, Psychosocial problems of job relocation: Preventive roles in industry. *Social Work*, **33**, pp. 38–41.

Andrews, G., Hall, W., Teeson, M. and Henderson, S., 1999, Mental Health of Australians. Canberra: Mental Health Branch, Commonwealth Department of Health and Aged Care.

Australian Bureau of Statistics, 1991, *In Australian social trends (1994)* (Canberra: Australian Bureau of Statistics.).

Bach, G., 1979, The George Bach self-recognition inventory for burned-out therapists. *Voices*, **15**, pp. 73–76.

Barna, G., 1993, Today's *pastors*. Ventura (CA: Regal Books).

Benda, B.B. and DiBlasio, F.A., 1992, Clergy marriages: A multivariate model of marital adjustment. *Journal of Psychology and Theology*, **20**, pp. 367–375.

Bentley, P., Blombery, T. and Hughes, P., 1992, Faith *without the Church? Nominalism in Australian Christianity* (Kew: Christian Research Association).

Birk, J.M., 2001, Religious Occupations and Stress Questionnaire (ROS): Instrument Development. *Counselling and Values*, **45**, pp. 136–145.

Bissonette, R., 1977, The mental health gatekeeper role: A paradigm for conceptual pretest. *International Journal of Social Psychology*, **23**, pp. 31–34.

Blackbird, T. and Wright, P.H., 1985, Pastors' Friendships, Part 1: Project overview and an exploration of the pedestal effect. *Journal of Psychology and Theology*, **13**, pp. 274–283.

Blaikie, N.W.H., 1979, *The plight of the Australian clergy: To covert, care or challenge?* (St Lucia, Queensland: University of Queensland Press).

Blanton, P., 1992, Stress in clergy families: Managing work and family demands. *Family Process*, **26**, pp. 315–330.

Blanton, P. and Morris, M.L., 1999, Work-related predictors of physical symptomatology and emotional well-being among clergy and spouse. *Review of Religious Research*, **40**, pp. 331–348.

Blizzard, S.W., 1956, The minister's dilemma. *The Christian Century* (April 25), pp. 508–510.

Bloom, B.L., 1985, A factor analysis of self-reporting measures of family functioning. *Family Process*, **24**, pp. 225–239.

Bomberger, H., 1974, The parsonage: A way of Christian family living. *Lutheran Quarterly*, **26**, pp. 58–63.

Boss, P., 1980, *Family stress management* (Newbury Park, CA: Sage).

Bouma, M., 1979, *Divorce in the parsonage* (Minneapolis, MN: Bethany Fellowship).

Bradfield, C.B., Wylie, M.L. and Echterling, L.G., 1989, After the Flood: The response of Ministers to a natural disaster. *Sociological Analysis*, **49**, pp. 397–407.

Bradshaw, S.L., 1977, Ministers in trouble: A study of 140 cases evaluated at The Menninger Foundation. *The Journal of Pastoral Care*, **31**, pp. 230–242.

Brief, A.P., Burke, M.J., George, J.M., Robinson, B.S. and Webster, J., 1988, Should negative affectivity remain an unmeasured variable in the study of job stress? *Journal of Applied Psychology*, **73**, pp. 207–214.

Breitmayer, B, Ayres, L. and Knafl, K., 1993, Triangulation in qualitative research: Evaluation of completeness and confirmation purposes. *Image*, **35**, pp. 237–343.

Brock, R.T. and Lukens, H.C., 1989, Affair prevention in the ministry. *Journal of Psychology and Christianity*, **8**, pp. 44–55.

Brokaw, B.F. and Edwards, K.J., 1994, The relationship of God image to level of object relations development. *Journal of Psychology and Theology*, **22**, pp. 352–371.

Brown, M., 1992, *Loneliness and repressed anger as correlates of burnout amongst clergy* (Unpublished honours thesis: University of Southern Queensland).

Chalfant, H.P., Heller, P.L., Roberts, A., Briones, D., Aguirre-Hochbaum, S. and Farr, W., 1990, The clergy as a resource for those encountering psychological distress. *Review of Religious Research*, **31**, pp. 305–313.

Chikes, T., 1968, Partners in the basic parish. *Pastoral Psychology*, **19**, pp. 8–14.

Clemens, N., Corradi, R. and Wasman, M., 1978, The parish clergy as mental health resource. *Journal of Religion and Mental Health*, **17**, pp. 227–232.

Coates, C. and Kistler, R., 1965, Role dilemmas of protestant clergymen in a metropolitan community. *The Review of Religious Research*, **6**, pp. 147–152.

Cohen, J., 1992, A power primer. *Psychological Bulletin*, **112**, pp. 155–159.

Collins, G., 1977, Burnout: The hazard of professional helpers. *Christianity Today*, **21**, pp. 12–14.

Cotton, S.J., 2000, Examining *The Salvation Army Officer's Vocations: The initial findings of the in-depth interviews* (The University of SA: Unpublished Results).

Craddock, A.E., 1996, Relational resources as buffers against the impact of stress: A longitudinal study of seminary students and their partners. *Journal of Psychology and Theology*, **24**, pp. 38–46.

Croucher, R., 1991a, Ex-Pastors. *The Australian Christian*, December, pp. 20–21.

Croucher, R., 1991b, *Recent trends among evangelicals: Biblical agendas, justice and spirituality* (2nd edn), (Victoria: John Mark Ministries).

Croucher, R. and Allgate, S., 1994, Why Australian pastors quit parish ministry. Reprinted from *Pointers. Bulletin of the Christian Research Association*, **4**, p. 1.

Daniels, K. and Guppy, A., 1992, Control, information-seeking preferences, occupational stressors and psychological well-being. *Work and Stress*, **6**, pp. 347–353.

Daniels, S. and Rogers, M.L., 1981, Burn-out and the pastorate: A critical review with implications for pastors. *Journal of Psychology and Theology*, **9**, pp. 232–249.

De Dreu, C.K.W., Evers, A., Beersma, B., Kluwer, E., Nauta, A., 2001, A *theory-based measure of conflict management strategies in the work place*. Unpublished Journal Article (The Netherlands: University of Amsterdam).

De Jonge, J., 1995, *Job Autonomy, Well-Being, and Health: A Study among Dutch health care workers* (PhD Thesis, Datawyse, Maastricht).

De Jonge, J., Bosma, H., Peter, R. and Siegrist, J., 2000, Job strain, effort–reward imbalance and employee well-being: A large-scale cross-sectional study. *Social Science and Medicine*, **50**, pp. 1317–1327.

De Jonge, J. and Kompier, M.A.J., 1997, A critical examination of the Demand-Control-Support model from a work psychological perspective. *International Journal of Stress Management*, **4**, pp. 235–258.

De Luca, J.R., 1980, The holy crossfire: A job diagnosis of a pastor's position. pastoral psychology, **28**, pp. 233–242.

Dempsey, K.C., 1973, Professionalism and conflict in minister-lay relationships. *Journal of Christian Education*, **16** (Oct.), pp. 121–138.

Devogel, S.H., 1986, Clergy Morale: The ups and downs. *The Christian Century*, **103**, pp. 1149–1152.

Dollard, M.F., 1996, *Work Stress: Conceptualising and implications for research methodology and workplace intervention* (Work and Stress Research Group: Whyalla).

Dollard, M.F., Winefield, H.R. and Winefield, A.H., 2001, *Occupational strain and efficacy in human service workers* (Kluwer Academic Publishers).

Douglas, W., 1965, *Ministers' wives* (New York: Harper and Row).

Dowdy, E.H.R. and Lupton, G.M., 1978, The survey of declining recruitment to the ministry: Final report prepared for the Australian Council of Churches. (mimeo), 15pp. In R.J. Pryor. *High calling high stress: The vocational needs of ministers: An overview and bibliography* (South Australia: Australian Association for the Study of Religions).

Dunn, R., 1965, Personality patterns among religious personnel: A review. *The Catholic Psychological Record*, **3**, pp. 125–137.

Ellison, C.W. and Mattila, W.S., 1983, The needs of evangelical Christian leaders in the United States. *Journal of Psychology and Theology*, **11**, pp. 28–35.

Episcopal Family Network, 1988, *Episcopal Clergy Families in the 80s*. Hartford (CN: Episcopal Family Network).

Epstein, C.F., 1970, Encountering the male establishment: Sex-status limits on women's careers in the professions. *American Journal of Sociology*, **75**, pp. 965–82.

Farrell, D.P. and Taylor, M., 2000, Silenced by God – and examination of unique characteristics within sexual abuse by clergy. *Counselling Psychology Review*, **15**, pp. 22–31.

Faulkner, B.R., 1981, *Burnout in ministry* (Nashville: Broadman).

Foa, E.B., 1995, *PTSD Scale Manual*. National Computer Systems: USA.

Frame, M.W. and Shehan, C.L., 1994, Work and well-being in the two-person career: Relocation stress and coping among clergy husbands and wives. *Family Relations*, **43**, pp. 196–205.

Francis, L.J. and Kay, W.K., 2000, *Pastoral Care Today: Practice, Problems and Priorities in Churches Today* (The Centre for Ministry Studies (CMS): Wales).

Freudenberger, H.J. and Richelson, G., 1980, *Burnout: The high cost of success – and how to cope with it* (London: Arrow).

Fuller, J. and Edwards, J., 2000, *Monograph: Responding to mental health problems; the rural and remote context* (SACRRH: South Australia).

General Assembly of the Presbyterian Church, 1991, Policies and Procedures on Sexual Misconduct, USA.

Gilbert, B., 1987, Who *ministers to ministers?* (New York: Alban Institute).

Gledhill, R., 1998, Salvation Army prepares to march to a different tune. *The Times*. 15th June/http://www.churchnet.org.uk/news/files2/news439.htm

Goldberg, D.P., 1978, Manual of the GHQ (Windsor: Nfer-Nelson).

Golding, B., 2000, *A five year review of the retention and recruitment trends* (The Salvation Army Southern Territory Internal Report).

Gurin, G., Veroff, J. and Feld, S., 1960, *Americans view their mental health: A nationwide interview survey* (New York: Wiley).

Hall, T.W., 1997, The personal functioning of Pastors: A review of empirical research with implications for the care of pastors. *Journal of Psychology and Theology*, **25**, pp. 240–253.

Hall, T.W. and Brokaw, B.F., 1995, The relationship between spiritual maturity and level of object relations development. *Pastoral Psychology*, **43**, pp. 373–391.

Hall, T.W., Brokaw, B.F., Edwards, K.J. and Pike, P.L., 1996, The relationship of spiritual maturity to level of object relations development and God image (Unpublished manuscript, Biola University, La Mirada, CA).

Harris, J., 1977, *Stress, power and ministry* (New York: Alban Institute).

Hart, A., 1993, Paper presented to the African Enterprise Pastors Conference. Sydney, Australia, August.

Hartford Institute for Religion Research, 1994, Clergy women: An uphill calling. retrieved October 8, 2001, from the World Wide Web: http://hirr. hartsem.edu/bookshelf/clergywomen_abstract.html

Hatcher, S.W. and Underwood, J.R., 1990, Self-concept and stress: A study of a group of Southern Baptist ministers. *Counselling and Values*, **34**, pp. 187–196.

Hausman, M.S. and Reed, J.R., 1991, Psychological issues in relocation: Response to change. *Journal of Career Development*, **17**, pp. 247–258.

Hay, I., 1995, Clergy 'burnt out' says study. *Southern Cross: The Newspaper of the Anglican Diocese of Sydney*, **1**, p. 8.

Henderson, A., 1997, New Orders. *Sydney Morning Herald*, 15th February: 4; Hovda, R.W. (1980). *Strong, loving and wise: Presiding in liturgy* (Washington, D.C.: The Liturgical Conference).

Henry, D., Chertok, F., Keys, C. and Jegerski, J., 1991, Organizational and family systems factors in stress among ministers. *American Journal of Community Psychology*, **18**, pp. 931–952.

Herin, M., 1981, *The interface of marriage and the ministry: Finding of the consultation on clergy marriages* (International Growth Centre: Lake Junaluska, NC).

Hood, R.W., Spilka, B., Hunsberger, B. and Gorsuch, R., 1996, *The psychology of religion* (New York: Guilford).

Houts, D.C., 1982, Marriage counselling with clergy couples. *Pastoral Psychology*, **3**, pp. 141–150.

Hulme, W.E., 1985, *Managing stress in ministry* (Harper and Row: USA).

Hunt, R.A., 1978, Clergy families under stress. Paper presented at the Seminar on Clergy marriage and divorce, United Methodist Church, Oklahoma City, OK.

Hutchison, K.R., Nichols, W.C., Jr. and Hutchison, I.W., 1980, Therapy for divorcing clergy: Implications from research. *Journal of Divorce*, **4**, pp. 83–94.

Janman, K., Jones, J.C., Payne, R.L. and Rick, J.T., 1988, Clustering individuals as a way of dealing with multiple predictors in occupational stress research. *Behavioral Medicine*, **14**, pp. 17–29.

Jarrett, R. and Winefield, A.H., 1995, Internal climate survey (University of Adelaide).

Jerden, L, 1980, The mission of marble retreat. *Home Missions*, **51**, pp. 21–28.

Johnson, R.N., 1963, Ministerial health and unhealth. *The Christian Century*, **80**, pp. 706–708.

Jud, G.J., Mills, E.W. and Burch, G.W., 1970, *Ex-pastors: Why men leave the parish ministry* (Philadelphia: Pilgrim Press).

Kaldor, P. and Bullpitt, R., 2001, Burnout in church leaders (Adelaide: Openbook Publishers).

Karasek, R.A., 1979, Job demands, job decision latitude, and mental strain: Implications for job redesign. *Administrative Science Quarterly*, **24**, 285–308.

Karasek, R.A. and Theorell, T., 1990, *Healthy work: Stress, productivity and the reconstruction of working life* (New York: Basic Books).

Karlstrom, M., 1999, Time for democracy? *The Officer*, October: 15–17.

Kasl, S.V., 1998, Measuring job stressors and studying the heath impact of the work environment. *Journal of Occupation Health Psychology*, **3**, 390–401.

Kay, W.K., 2000, Role conflict and British pentecostal ministers. *Journal of Psychology and Theology*, Summer **28**, pp. 119–125.

Keddy, P.J., Erdberg, P. and Sammon, S.D., 1990, The psychological assessment of Catholic clergy and religious referred for residential treatment. *Pastoral Psychology*, **38**, pp. 147–159.

Kemp, M., Devlin, R. and King, M., 2001, Falling Icons. The Advertiser, 22 Sept., pp. 65–66.

Kennedy, J.P. and Sons (eds), 1997, *The official Catholic directory* (New Providence, N.J.: Reed Publishing Company, Inc).

Kessler, R.C., McLeod, J.D. and Wethington, E., 1985, The cost of caring: A perspective on the relationship between sex and psychological distress. In I.G. Sarason and B.R. Sarason (ed.), *Social support: Theory, research, and applications,* (The Hague, The Netherlands: Martinus Nijhoff Publishers), pp. 491–506.

Kieren, D. and Munro, B., 1988, Handling greedy clergy roles: A dual clergy example. *Pastoral Psychology*, **36**, pp. 239–248.

King, W., 1988, The clergy family project of the diocese of Alabama (Birmingham, AL: The Episcopal Diocese of Alabama).

Krause, N., Ellison, C.G. and Wulff, K.M., 1998, Church-based emotional support, negative interaction, and psychological well-being: Findings from a national sample of Presbyterians. *Journal for the Scientific Study of Religion*, **37**, pp. 725–742.

Kristensen, T.S., 2000, The Copenhagen psychosocial questionnaire. *National Institute of Occupational Health, Copenhagen, Denmark.*

Kunst, J.L., 1993, A system malfunction: Role conflict and the Minister. *Journal of Psychology and Christianity*, **12**, pp. 205–213.

Laaser, M.R., 1991, Sexual addiction and clergy. *Pastoral Psychology*, **39**, pp. 213–35.

LaHaye, T., 1990, If ministers fail, can they be restored? (Zondervan Publishing House).

Landsbergis, P.A., Schurman, S.J., Israel, B.A., Schnall, P.L., Hugentobler, M.K., Cahill, J. and Baker, D., 1993, Job stress and heart disease: Evidence and strategies for prevention. *New Solutions*, Summer, pp. 42–58.

Lauer, R.H., 1973, Organizational punishment: Punitive relations in a voluntary association – a minister in a protestant church. *Human Relations*, **26**, pp. 189–202.

Lavender, L., 1983, *They cry too* (New York: Hawthorne Books).

Lee, C., 1988, Toward a social ecology of the ministers family. *Pastoral Psychology*, **36**, pp. 249–259.

Lee, C., 1992, *PK: Helping pastor's kids through their identity crisis.* Grand Rapids (MI: Zondervan Publishing House).

Lee, C., 1999, Specifying intrusive demands and their outcomes in congregational ministry: A Report on the ministry demands inventory. *Journal for the Scientific Study of Religion*, **38**, pp. 477–489.

Lee, C. and Balswick, J., 1989, Life in a glass house: The ministers family in its unique social context (Grand Rapids, MI: Zondervan).

Loftus, J., 1994, *Understanding sexual misconduct by clergy: A handbook for ministers,* (Pastoral Press).

London, H., 1983, *Clergy families and career paths in the United Methodist Ministry* (Nashville, TN: General Board of the United Methodist Church).

London, H.B. and Wiseman, N. B., 1993, *Pastors at risk* (USA: Victor Press).

MacDonald, G., 1980, Dear church, I quit. *Christianity Today*, **27**, pp. 17–21.

Mace, D. and Mace, V., 1980, What's happening to clergy marriages? (Nashville, TN: Abingdon Press).

Mace, D. and Mace, V., 1982, Marriage enrichment for clergy couples. *Pastoral Psychology*, **30**, pp. 151–159.

Macklin, D.S. and Dollard, M.F., 2000, Work stress in the public and private sector: A test of the Demands Control Support model. Unpublished honours thesis (Department of Psychology, University of South Australia).

Malony, H.N., 1986, *Church organization development* (Pasadena, CA: Integration Press).

Malony, H.N., 1988, Men and women in the clergy: Stresses, strains, and resources. *Pastoral Psychology*, **36**, pp. 164–168.

Malony, H.N. and Hunt, R., 1991, *The psychology of clergy* (Ridgefield, CT: Morehouse-Barlow).

Maslach, C., 1976, Burn-out. *Human Behaviour*, **5**, pp. 16–22.

Maslach, C. and Jackson, S., 1982, Burnout in health professionals: A social psychological analysis. In J. Sanders and J. Suls (eds), *Social psychology of health and illness* (Hillsdale, NJ: Erlbaum).

Mattis, M., 1977, Pastors' wives study. Report from the Research Division of the Support Agency (New York: United Presbyterian Church).

McBurney, L., 1977, *Every pastor needs a pastor* (Waco, TX: Word Books).

McGarry, A., 2001, Clergy feel unholy stress. *The Australian.* 12 March., p. 5.

McGregor, D., 1960, *The Human Side of Enterprise* (New York: McGraw-Hill).

Mebane, D.L. and Ridley, C.R., 1988, The role-sending of perfectionism: Overcoming counterfeit spirituality. *Journal of Psychology and Theology*, **16**, pp. 332–339.

Merchant, K.H. and Medway, F.J., 1987, Adjustment and achievement associated with mobility in military families. *Psychology in the Schools*, **24**, pp. 289–294.

Mickey, P.A., Wilson, R.L. and Ashmore, G.W., 1991, Denominational variations on the role of the clergy family. *Pastoral Psychology*, **39**, pp. 287–294.

Mills, E., 1968, Types of role conflict among clergy-men. *Ministry Studies*, **2**, pp. 13–15.

Mills, E., and Koval, J., 1971, Stress in the ministry (Washington: Ministry Studies Board).

Morris, M.L. and Blanton, P.W., 1994a, Denominational perceptions of stress and the provision of support services for clergy families. *Pastoral Psychology*, **42**, pp. 345–364.

Morris, M.L. and Blanton, P.W., 1994b, The influence of work-related stressors on clergy husbands and their wives. *Family Relations*, **43**, pp. 189–195.

Morris, M.L. and Blanton, P., 1998, Predictors of family functioning among clergy and spouses: Influences of social context and perceptions of work-related stressors. *Journal of Child and Family Studies*, **7**, pp. 27–41.

Morrison, D.L., Dunne, M.P., Fitzgerald, R. and Cloghan, D., 1992, Job design and levels of physical and mental strain among Aust prison officers. *Work and Stress*, **6**, pp. 13–31.

Meyrick, S., 2000, Clergy: low status, high stress. Retrieved March 7, 2002 from the World Wide Web: http://www.media.anglican.com.au/tma/2000/2000_02/stress. Html

Moy, S. and Malony, H.N., 1987, An empirical study of ministers' children and families. *Journal of Psychology and Christianity*, **7**, pp. 52–64.

Muntaner and Schoenbach., 1994, Psychosocial work environment and health in the U.S. metropolitan areas: a test of the demand-control and demand-control-support models. *International Journal of Health Services*, **24**, pp. 337–353.

Muse, J.S., 1992, Faith, hope, and the 'urge to merge' in pastoral ministry: Some countertransference-related distortions of relationship between male pastors and their female parishioners. *The Journal of Pastoral Care*, **46**, pp. 299–308.

Muse, S., Chase, E. and The Pastoral Institute, 1993, Healing the wounded healers: 'Soul' food for clergy. *Journal of Psychology and Christianity*, **12**, pp. 141–50.

Neuendorf, K.A., 2002, *The Content Analysis Guidebook* (Thousand Oaks, CA: Sage).

Niswander, B.J., 1982, Clergy wives of the new generation. *Pastoral Psychology*, **3**, pp. 160–169.

Noller, P., 1984, Clergy marriages: A study of a Uniting church sample. *Australian Journal of Sex, Marriage and Family*, **5**, pp. 187–197.

Norrell, J.E., 1989, Clergy family satisfaction. *Family Science Review*, **2**, pp. 337–346.

Noyce, G., 1980, The tensions of our calling. *Christian Ministry*, **11**, pp. 18–21.

Oden, M., 1977, Updating the minister's wife. *The Christian Ministry*, **8**, p. 1977.

Olsen, D.C. and Grosch, W.N., 1991, Clergy Burnout: a self psychology of systems perspective. *The Journal of Pastoral Care*, **45**, pp. 297–304.

O'Mara, N., 1991, A study of work stress among government employees (Perth, WA: Murdoch University).

Ormerod, N. and Ormerod, T., 1995, When ministers sin: Sexual abuse in the churches, (NSW: Millenium Books).

Ostrander, D.L. and Henry, C.S., 1989, An investigation of how clergy adolescents and adult children define sources of stress in their lives. Paper presented at the National Council on Family Relations, New Orleans, LA.

Ostrander, D.L., Henry, C.S. and Fournier, D.G., 1994, Stress, family resources, coping, and adaptation in ministers' families. *Journal of Psychology and Christianity*, **13**, pp. 50–67.

Oswald, R.M., 1991, *Clergy Self-Care: Finding a balance for effective ministry* (New York: The Alban Institute, Inc).

Paloutzian, R.F. and Ellison, C.W., 1982, Loneliness, spiritual well-being and the quality of life. In L.A. Replau and D. Perlman (eds). *Loneliness: A sourcebook of current theory research and therapy* (New York: Wiley).

Pargament, K.I., Falgout, K., Ensing, D.S., Reilly, B., Silberman, M., Van Haitsma, K., Olsen, H. and Warren, R., 1991, The congregation development program: Data-based consultation with churches and synagogues. *Professional Psychology: Research and Practice*, **22**, pp. 393–404.

Parkes, K.R., 1990, Coping, negative affectivity, and the work environment: Additive and interactive predictors of mental health. *Journal of Applied Psychology*, **75**, pp. 399–409.

Pines, A. and Aronson, E., 1981, *Burnout: From tedium to personal growth* (New York: Free Press).

Piper, J.L., 1995, Work stress among Lutheran clergy women in the United States of America and Norway. Dissertation Abstracts International, A56 (10).

Pope, K.S. and Vasquez, M.J.T., 1998, *Ethics in Psychotherapy and Counselling (2nd edn)* (San Fransisco: Josey-Bass).

Presnell, W., 1977, The minister's own marriage. *Pastoral Psychology*, **4**, pp. 272–281.

Pryor, R.J., 1982, *High calling high stress: The vocational needs of ministers: An overview and bibliography* (South Australia: Australian Association for the Study of Religions).

Rayburn, C.A., 1981, Some reflections of a female seminarian: Woman, whither goest thou? *Journal of Pastoral Counseling*, **16**, pp. 61–65.

Rayburn, C.A., 1982, Impact of non-sexist language and guidelines for women in religion. *Journal of Pastoral Counseling*, **17**, pp. 5–8.

Rayburn, C.A., 1991a, Some reflections of a female seminarian: Woman, whither goest thous? *Journal of Pastoral Counselling*, **16**, pp. 61–65.

Rayburn, C.A., 1991b, Further study of women seminarians' needs for counselling and consulting. A paper presented at the American Psychological Association Annual Convention, August 26.

Rayburn, C.A., Richmond, L.J. and Rogers, L., 1983, Stress among religious leaders: Though: *Fordham University Quarterly*, **58**, pp. 329–344.

Rediger, G.L., 1982, *Coping with clergy burnout*. Valley Forge (PA: Judson Press).

Richmond, L., 1991, Stress and single clergy women. *Psychotherapy in Private Practice*, **8**, pp. 119–125.

Richmond, L.J., Rayburn, C. and Rogers, L., 1985, Clergymen, clergy women, and their spouses: Stress in professional religious families. *Journal of Career Development*, **12**, pp. 81–86.

Rogers, A., 1991, Stress in married clergy. *Psychotherapy in Private Practice*, **8**, pp. 107–115.

Rumberger, D., 1982, Pastoral openness to interaction with a private Christian counseling service. *Journal of Psychology and Theology*, **10**, pp. 337–345.

Russell, D., 1996, UCLA Loneliness Scale V.3: Reliability, validity, and factor structure, *Journal of Per Ass*, **66**, pp. 20–40.

Sanford, J.A., 1982, *Ministry burnout* (New York: Paulist Press).

Seamonds, D. and Seamonds, H., 1981, The story of raising a pastoral family. *Leadership*, **2**, pp. 16–28.

Schaufeli, W. and Bakker, A., 1999, The Utrecht Work Engagement Scale (UWES).

Schaufeli, W.B., Maslach, C. and Marek, T. (eds), 1993, Professional burnout: Recent developments in theory and research (Washington, DC: Taylor and Francis).

Schaufeli, W.B., Van Dierendonck, D. and Van Gorp, K., 1996, Burnout and reciprocity: Toward a dual-level social exchange model. *Work and Stress*, **10**, pp. 225–237;

Schaufeli, W.B. and Van Dierendonck, D., 2000, UBOS: Utrecht Burnout Scale (Manual) (in Dutch), (Swets and Zeitlinger, Amsterdam).

Scherer, R., 1965, New light on ministerial compensation. *Pastoral Psychology*, **16**, pp. 45–52.

Schumm, W.R., Paff-Bergen, L.A., Hatch, R.C., Obiorah, F.C., Copeland, J.M., Meens, L.D. and Bugaighis, M.A., 1986, Concurrent and discriminant validity if the Kansas Marital Satisfaction Scale. *Journal of Marriage and the Family*, **48**, pp. 381–387.

Schutte, N., Toppinen, S., Kalimo, R. and Schaufeli, W., 2000, The factorial validity of the Maslach Burnout Inventory-General Survey (MBI-GS) across occupational groups and nations. *Journal of Occupational and Organizational Psychology*, **73**, pp. 53–66.

Shauer, J.C., 1985, Role stress resulting from role conflict and ambiguity related to organizational change. PhD Thesis. *UMI. Dissertation Information Service.*

Siegrist, J., 1996, Adverse health effects of high-effort/low reward conditions. *Journal of Occupational Health Psychology*, 1, pp. 27–41

Slack, S., 1979, Clergy divorce. *The Christian Ministry*, **10**, pp. 22–26.

Spielberger, C.D., 1983, *Manual for the state-trait anxiety inventory*. Palo Alto (CA: Consult. Psych Press).

Steinke, P.L., 1989, Clergy affairs. *Journal of Psychology and Christianity*, **8**, pp. 56–62.

Stevenson, R.M., 1982, Children of the parsonage. *Pastoral Psychology*, **3**, pp. 179–186.

Stout, R.J., 1982, Clergy divorce spills into the aisle. *Christianity Today*, **26**, pp. 20–23.

Swenson, D., 1998, Religious differences between married and celibate clergy: does celibacy make a difference? *Sociology of Religion*, **59**, pp. 37–44.

Taylor, C. and Goldsworthy, G., 1981, *Battle guide for Christian leaders* (Cudgen, NSW: Wellcare Publications).

Theorell, T., 1998, Job characteristics in a theoretical and practical health context. In C.L. Cooper (eds). *Theories of Organizational Stress* (Oxford; Oxford Uni Press).

Theorell, T. and Karasek, R.A., 1996, Current issues relating to psychological job strain and cardiovascular disease research. *Journal of Occupational Health Psychology*, **1**, pp. 9–26.

Thomas, D. and Cornwall, M., 1990, Religion and family in the 1980s: Discovery and development. *Journal of Marriage and the Family*, **52**, pp. 983–992.

Van der Doef, M. and Maes, S., 1999, The job demand-control (-support) model and psychological well-being: A review of 20 years of empirical research, *Work & Stress*, **13**, pp. 87–114.

Vermaat, K., 1994, *Flexible workload control in nursing* (Limburg University, Maastricht).

Virginia, S.G., 1998, Burnout and depression among Roman Catholic secular, Religious, and monastic clergy. *Pastoral Psychology*, **47**, pp. 49– 67.

Virkler, H.A., 1979, Counselling demands, procedures, and preparation of parish ministers: A descriptive study. *Journal of Psychology and Theology*, **7**, pp. 271–280.

Voges, K.E., Long, N.R., Roach, M.S. and Shouksmith, G.A., 1982, The perception of stress by prison officers and their wives: An occupational study. Unpublished report, (Department of Psychology, Massey University, New Zealand).

Walker, J.T., 1978, What's behind the stress in clergy families? Paper presented at the Summer Missouri Area Pastor's School, Columbia, MO.

Warner, J. and Carter, J., 1984, Loneliness, marital adjustment, and burnout in pastoral and lay persons. *Journal of Psychology and Theology*, **12**, pp. 125–131.

Warr, P.B., Cook, J.D. and Wall, T.D., 1979, Scales for the measurement of some work attitudes and aspects of psychological well-being. *Journal of Occupational Psychology*, **52**, pp. 129–148.

Weaver, A.J., 1995, Has there been a failure to prepare and support parish-based clergy in their role as front-line community mental health workers? A Review. *The Journal of Pastoral Care*, **49**, pp. 129–149.

Weaver, A.J., Samford, J.A., Larson, D.B. and Koenig, H.G., 1997, What do psychologists know about working with the clergy? An analysis of eight APA journals: 1991–1994. *Professional Psychology: Research and Practice*, **28**, pp. 471–474.

Whetham, P., 1997, Understanding the relationships of clergymen using personal construct psychology. Doctor of Philosophy (Clinical Psychology) Thesis, (University of Wollongong).

Whetham, P. and Whetham, L., 2000, Hard to be Holy (Adelaide: Openbook Publishers).

Whitney, D., 1998, The importance of spiritual formation in the training of ministers. *The Founders Journal*, **32** Spring.

Whittemore, H., 1991, Ministers under stress. Parade Magazine, April 14, pp. 4–6.

Wilson, R.L., 1971, Drop-outs and potential drop-outs from the parish ministry. Department of Research and Survey, National Division of the Board of Missions, United Methodist Church, New York. In R.J. Pryor (ed.), *High calling high stress: The vocational needs of ministers: An overview and bibliography* (South Australia: Australian Association for the Study of Religion).

Winefield, A.H., Tiggemann, M., Winefield, H.R. and Goldney, R.D., 1993, Growing up with unemployment: A longitudinal study of its psychological impact (London: Routledge).

Witcomb, N., 1979, The thoughts of Nanushka. Volume I–VI (Gillingham Printers: Underdale S.A.).

Wright, P.H. and Blackbird, T., 1986, Pastors' friendships, Part 2: The impact of congregational norms. *Journal of Psychology and Theology*, **14**, pp. 29–41.

Wubbolding, R.E. and Kessler-Bolton, E., 1979, Reality therapy as an antidote to burnout. *American Mental Health Counsellors Association Journal*, **1**, pp. 39–43.

Yancey, P., 2001, Replenishing the inner pastor. *Christianity Today*, **45**, p. 104.

Zapf, D., Seifert, C., Schmutte, B., Mertini, H. and Holz, M., 2001, Emotion work and job stressors and their effects on burnout. *Psychology and Health*, **16**, pp. 527–545.

Zapf, D., Dormann, C. and Frese, M., 1996, Longitudinal studies of organisational stress research: A review of the literature with reference to methodological issues. *Journal of Occupational Health Psychology*, **1**, pp. 145–169.

Zapf, D., Vogt, C., Seifert, C., Mertini, H. and Isic, A., 1999, Emotion work as a source of stress. The concept and development of an instrument. *European Journal of Work and Organizational Psychology*, **8**, pp. 371–400.

CHAPTER THIRTEEN

Stress in Psychological Work

Susan P. Griffiths

13.1 INTRODUCTION

The focus of this chapter is on work stress affecting psychologists who are engaged in providing clinically oriented services to individuals in one-to-one counselling and psychotherapy, and/or to families and groups. However, because of the limited quantity of literature in this domain, some material related to counsellors and psychotherapists engaged in similar types of work will be included where appropriate as will literature that refers to psychologists whose professional role is unclear.

Many commentaries on the demands of psychological work do not use occupational stress constructs, although the notion of work related stress is implicit in much of what is written. The literature on the demands and challenges of rural and remote psychological service provision is a good example of this (see Cohen, 1992; Griffiths and Andrews, 1995; Schank and Skovolt, 1997; Sterling, 1992). The broader domain of occupational stress however, has a comprehensive and well-established literature consisting of numerous, largely complementary, theoretical models and related research (Dunnette and Hough, 1992). As it is beyond the scope of this chapter to summarise the broader theoretical and empirical literature an integrated stress–strain model of work stress will be used consistent with the orientation of Kahn and Byosiere's (1992) integrative literature review. Specifically, a four-phase, transactional model as described by (Matteson and Ivancevich 1987; Ivancevich and Matteson, 1980) will be used to guide this review. This model consists of the following four phases: (1) 'the objective sources of stress indicative responses' (Kahn and Byosiere, 1992, p. 574); (2) the appraisal process; (3) immediate physiological, psychological and behavioural effects, and (4) longer-term strain related effects. To some extent a linear process through phases 1–4 is implied in this model, however, mediating and moderating effects, such as coping, personality and other individual differences are hypothesised as are transactional relationships between elements of the stress–strain process (Lazarus and Folkman, 1984).

Of the four phases outlined, three aspects will guide the structure of this discussion of work stress in psychological practice. First, the literature that identifies the 'objective sources of stress' (Kahn and Byosiere, 1992, p. 574) or 'stressors' as they are referred to by Lazarus and Folkman (1984) will be discussed. This section of the chapter will also review occupational stress scales designed to assess stress in psychologists as well as prevalence studies and correlates of stress. The second component will focus on the appraisal process. This aspect is related to psychologists' beliefs and aspects of professional identity that lead practitioners to appraise some phenomena as stressful yet other phenomena as benign. The third aspect pertains to the longer-term consequences of stress, alternatively known as strain. Strain incorporates a number of constructs, although burnout is probably the most widely investigated in human-service work and will be focused on in this chapter. However, other strain outcomes such as ill health, attrition and psychological difficulties have been investigated and will be noted where appropriate. Not included in this review is literature focusing on personal variables that mediate the 'stressor–stress reaction' process (e.g. Hellman *et al.*, 1987) and coping as a mediator of the 'stress–strain' process. Similarly, literature on the immediate physiological, psychological, and behavioural effects of stress will not be reviewed as research in this area is rarely profession specific.

13.2 SOURCES AND CORRELATES OF OCCUPATIONAL STRESS IN PSYCHOLOGICAL AND PSYCHOTHERAPEUTIC WORK

The research literature on stress in psychological and psychotherapeutic work has been expanding steadily since the early 1980s. The disciplinary areas that have been the subject of research interest include school psychologists (e.g. Huebner and Mills, 1997), clinical psychologists (e.g. Cushway and Tyler, 1994, 1996), educational psychologists (e.g. Freeman, 1987), counsellors working in counselling centres (e.g. Rodolfa *et al.*, 1988) and psychologists in private practice (e.g. Nash *et al.*, 1984). The focus of the research has been on identifying sources of stress (stressors), coping, and prevalence of stress affected practitioners. The literature on sources of stress, prevalence of stress affected practitioners and the effects of variables such as age, gender and level of experience will be reviewed in the following two segments. However, before commenting on this literature, a review of the measures used in stress research with psychologists and psychotherapists has conceptual value. Measures provide strong indications of the relevant aspects of the construct of interest. Therefore, stress scales designed to explore stress in psychological work offer an overview of the main sources of stress hypothesised to affect psychologists.

13.2.1 Psychologist and Mental Health Worker Occupational Stress Scales

A review of the professional stress literature revealed six measures used in research with psychologists. The first measure was designed to assess stress in school psychologists. Known as the School Psychologists and Stress Inventory (SPSI), it was originally developed by Wise (1985) and is comprised of 35 items. The items identify potentially stressful events in the work of school psychologists. The second measure developed by Deutsch (1984) was designed to assess stress in psychotherapists and is made up of two parts, one part measuring stress and the other therapist beliefs. The 'therapist stress scale' consists of 36 items, 21 of which were derived from an earlier measure developed by Farber (1979, in Deutsch, 1984). These items pertain to client behaviours and the therapist's role. The 'therapist beliefs' component consists of 13 irrational beliefs that might also contribute to therapist stress. Although beliefs properly belong to the appraisal process this part of the scale will be considered here as well as later, in the context of appraisal. A third scale by Rodolfa *et al.* (1988) is essentially a modified version of the Farber scale used by Deutsch with additional items relating to supervision particularly supervisee encounters with supervisors. The fourth scale, developed by Hellman *et al.* (1986) was designed to assess psychotherapist stress. Based on earlier work by Farber and Heifitz (1981, in Hellman *et al.*, 1986) the Therapeutic Stresses and Stressful Patient Behaviors (TSSPB) rating scale is comprised of two sets of items, the first focuses on therapeutic work and the second on patient/client behaviours. The fifth measure is the Mental Health Professionals Stress Scale (MHPSS) (Cushway *et al.*, 1996) was developed using samples of clinical psychologists and mental health nurses. It is comprised of 42 items organised into 7 sub-scales. The items mainly identify situations related to work such as 'too much to do' and 'inadequate staffing', although one sub-scale identifies negative affect and self-doubt. The sixth measure developed by Nash *et al.* (1984) to assess stress in American Psychological Association members who are in independent practice contains 15 items. These items are embedded in a broader survey designed to measure a range of work related variables related to work stress and job satisfaction.

To indicate how the domain of psychologist and mental health worker stress has been conceptualised by stress scale developers, items from the various scales have been organised into 10 thematic categories. The first column of Table 13.1 shows the thematic categories. Example items are shown in the second column and the number of different items per theme across all scales in column three. The one exception is Deutsch's (1984) 13 irrational beliefs, which because of their unclear status as stressors were treated as one item. The fourth column shows the scales from which the themes were derived.

The content of the measures reviewed above is consistent with the four-phase model described above with the stress scale items reflecting potential sources of stress (Phase 1) for psychologists and mental health personnel.

13.2.2 Sources of Stress

In addition to exploring the item content of scales other sources of stress in psychological and mental health services can be derived from literature specifically investigating or identifying sources of stress for psychologists, psychotherapists and related mental health providers.

Table 13.1 Sources of stress in psychological and psychotherapeutic work stress scales

Major themes	Example items	N of different items identified	Sources*
Workload	Too much to do	6	1; 2; 3
	Case-load uncertainty		4; 5; 6
	Boring work		
Client	Termination	21	1; 2; 3
Management	Limited change		4; 5; 6
	Suicidal clients		
Organisational	Unsupportive management	17	1; 3; 4
Issues	Poor communication		5; 6
	Limited consultation		
Professional	Multidisciplinary work	7	1; 2; 3
Relationships	Criticism from supervisor		4; 5; 6
	Limited emotional support		
Resources	Lack of finances	8	1; 5
	Inadequate staffing		
	Poor working environment		
Professional	Uncertain about own capabilities	14	1; 2; 3
Issues	Threat of legal action		4; 5; 6
	Doubts about effectiveness of work		
Beliefs and	Irrational beliefs about role	2	2; 3; 4
Expectations	Unrealistic sense of responsibility		
Support	Lack of recognition from colleagues	3	1; 2; 3
	Lack of recognition from clients		6
Public role	Case presentations	3	1; 3
	Conference presentations		
Spill-over	Inadequate time for recreation	13	2; 3; 4
	Limited time with family		5; 6
	Economic uncertainty		

*Key: 1 – Wise (1985); 2 – Deutsch (1984); 3 – Rodolfa *et al.* (1988); 4 – Hellman *et al.* (1986);
5 – Cushway *et al.* (1996); 6 – Nash *et al.* (1984).

Two research reports will be reviewed in this segment (Cushway and Tyler, 1994; Freeman, 1987) plus a review article (Cushway and Tyler, 1996) and a discussion paper (Basson, 1988). Two papers focus on practitioners delivering clinical services in public settings (Cushway and Tyler, 1994; 1996). One study focuses on educational psychologists (Freeman, 1987) and another pertains to school psychology (Basson, 1988).

In their research on sources of stress using a sample of 101 British clinical psychologists employed in the public sector, Cushway and Tyler (1994) listed the 11 most frequently reported stressors. These are presented in rank order, with the percentages of respondents reporting each shown in brackets:

(1) too much work (56%); (2) poor quality of management (50%); (3) too many different things to do (41%); (4) lack of resources (39%); (5) conflicting role and relationships with other staff (35%); (6) lack of money (30%); (7=) uncertainty about the future in the National Health Service (25%); (7=) feeling inadequately skilled (25%); (7=) lack of recognition and support (25%); (10=) paperwork/bureaucracy (15%) and (10=) difficult clients and distressing nature of the work (15%) (p. 38).

In a later publication Cushway and Tyler (1996) take a slightly different approach to identifying sources of stress in psychologists. In this instance they report a factor analysis of the MHPSS using data from clinical psychologists. The seven factors and the percentage of variance accounted for by each are as follows: '(1) Professional self-doubt (17.6%); (2) Home-work conflict (10.6%); (3) Organisational structure and processes (9.2%); (4) Relationships and conflicts with other professionals (5.4%); (5) Workload (5.0%); (6) Lack of resources (4.0%); and (7) Client/patient related difficulty (3.4%)' (p. 147). The factor structure of the scale is helpful in conceptualising the broad domains from which individual stressors derive.

Freeman (1987) reports a qualitative study in which she surveyed 20 educational psychologists (EPs) and interviewed another 7 regarding sources of stress in their work. Too much work, ineffective management, inadequate clerical support, external controls on work and role problems are listed as particularly stressful aspects of work. Freeman's (1987) findings are consistent with the sources of stress identified in the reported stress measures.

In a discussion paper, Basson (1988) identifies potential sources of stress for EPs working for the various education departments in the Republic of South Africa. Four domains of stressors are identified. First are those stressors associated with factors intrinsic to the job. Included in this domain are: (1) the itinerant nature of the service; (2) lack of educational facilities and resources; (3) inadequate assessment techniques for use in cross-cultural assessment and diagnosis; (4) clients with limited personal and financial resources; (5) reluctant and uncooperative teachers, and (6) working in dangerous and stressed environments. The second domain includes a variety of role related stressors including role conflict, role overload and role ambiguity. Third, Basson (1988) identifies 'professional and career-related issues' (p. 208). Most of these issues derive from South African EPs having an unclear professional identity as a newly emerging specialisation. In addition, industrial issues regarding the registration status of EPs was considered to be potentially deleterious in terms of stress. The fourth domain includes organisational aspects of the work. Basson (1988) identifies (1) working in a bureaucratic system; (2) not being consulted in professional decision-making; (3) working in an auxiliary service; (4) role constraints, and (5) having an unsupportive line-manager.

Basson's (1988) paper is interesting because the sources of stress are considered in relation to the broader social-political context. Although, Cushway and Tyler (1994) note the impact of the National Health System (NHS) in Britain as a factor in engendering stress for NHS employed clinical psychologists, Basson (1988) extends the discussion on contextual factors including the effects on practitioners of apartheid, rural work, and inequitable distribution of resources. These factors would have been particularly relevant in South Africa in the late 1980s.

13.2.3   Most Frequently Identified Stressors

Many studies on stress in psychological work seek to identify the most frequently reported sources of stress. As noted above Cushway and Tyler (1994) rated 'too much work, ... poor quality of management ... too many different things to do ... lack of resources ... and conflicting role and relationships with other staff' (p. 38) as the top five stressors affecting their sample of 101 clinical psychologists with 35% – 56% of the sample noting at least one of the five.

Role overload, in addition to a number of service related issues, is a concern for the school psychologists in Huebner and Mills' (1997) research using the SPSI. The five highest ranking stressors in order are: (1) insufficient time for adequate performance of job; (2) conflict between children's needs and organisational/job constraints; (3) report writing; (4) 'keeping your district "legal"; and (5) inadequate services for children.

In a study conducted by Burden (1988) also using the SPSI, school psychologists in USA, UK and Australia all ranked 'being notified of unsatisfactory job performance' in the highest position. However, similar to Huebner and Mills' (1997) 'insufficient time to perform job adequately' is also ranked by all groups in the top five stressors as is 'being caught between the child's needs and the constraints of the job'. Threat of legal action is also ranked in the top five by all three groups. Potential suicide was ranked in the top five by US and UK school psychologists but not by the Australian sample. Neither did Australian school psychologists rank working with uncooperative school administrators in the top five stressors, although this item was ranked highly by the other two groups.

Similar to all three studies reported above, workload features as a major stressor in Nash *et al.*'s (1984) study of psychologists in independent practice. However for these psychologists excessive workload was in fifth position behind 'time pressures', 'economic uncertainty', 'caseload uncertainty' and 'business aspects'. Slightly different concerns are to be expected for psychologists in the private sector.

Deriving a clear picture of from studies such as these is complicated. First, a variety of measures are used that may not be comparable in their design or content. Second, there is disparity in the level of analysis of stressors between studies. Role overload for example as noted in Huebner and Mills (1997) is a complex phenomenon consisting of a range of contributing experiences. Cushway and Tyler (1994), by comparison are more focused on specific stressors some of which may together add up to role overload. However, even when item content is refined and similar scales are used, variation in item content and scale utilisation across studies can influence the ranking of stressors. Three studies that utilised the scale developed by Farber (1979 in Deutsch, 1984) demonstrate this trend. The Farber scale focuses more on client behaviours and the therapist's experiences in the context of client work. As a result the five highest ranking items in Deutsch's study of stress in psychotherapeutic work are 'suicidal statements made by client, . . . expression of anger towards the therapist, . . . severely depressed client, . . . apparent apathy or lack of motivation, . . . [and, equal fifth] client's premature termination' (p. 841). These results are very similar to those of Farber (1979 in Deutsch, 1984). In a survey of 60 psychotherapists Farber found the following five items to be ranked most highly 'suicidal ideation, . . . aggression and hostility, . . . premature termination, . . . agitated anxiety [in the client] . . . [and] apathy and depression [in the client]' (p. 841). With scores totalled across three levels of counselling practitioners, professionals, interns and practicum students Rodolfa *et al.*'s (1988) results are difficult to compare with either Farber (1979 in Deutsch, 1984) or Deutsch (1984). Although Rodolfa *et al.* (1988) used a modified version of the Farber (1979 in Deutsch, 1984) scale they report client behaviours separately from therapist experiences and also include items on supervision. The five most highly rated items for 'client behaviours' are 'physical attack on therapist, . . . suicide attempt, . . . suicidal statements, . . . reporting a current crime . . . [and] expressing anger towards the therapist' (p. 45). 'Inability to help the client . . . receiving criticism from a supervisor, . . . professional conflict . . . seeing more clients than usual . . . [and] lack of client progress' (p. 46) constitute the top five rankings in the 25-item scale of therapist experiences. Suicidal threats and aggressive behaviour are both high-ranking stressors in the two previously noted studies using versions of the Farber (1979 in Deutsch, 1984) scale. The 'therapist experiences' ranking is more difficult to compare although it is interesting to note that an item related to work load ranked highly.

Only tentative conclusions can be drawn from studies such as these. Not only are results difficult to compare when different measures are used or when scales are modified but caution must be exercised in determining *what* the scales measure. As Burden (1988), noted about school psychologists,

> rating items as stressful does not necessarily mean that a person is subjected to a great deal of stress from any particular event (e.g. potential suicide cases are clearly stressful but may only occur very occasionally). We have no way of knowing . . . how this would balance in the day-to-day running of a school psychologist's job against weekly contact with, say, an uncooperative school principal (p. 58).

It is dilemmas such as these that have led some stress researchers to explore the distinctions between frequency, duration, and demand of identified stressors (Dewe, 1991). These distinctions are not explored in any of the studies reported here.

Another confounding aspect in interpreting this kind of research is that the meaning attributed to highly ranked stressors is unclear. For example, the highly ranked client behaviours identified in studies using the Farber (in Deutsch, 1984) scale may indicate that psychotherapists and psychologists find these client characteristics intrinsically unpleasant. However, the same items may be identified as stress provoking because they challenge the professional's competence. It is impossible to determine meaning from this kind of data. A finding reported by Cushway and Tyler (1996) sheds some light on the meanings of item rankings. Although their research (Cushway and Tyler, 1994) indicates 56% of their sample of clinical psychologists report 'too much work' as the highest stressor, the factor that accounted for the greatest proportion of the variance on the MHPSS is the 'professional self-doubt' factor (Cushway and Tyler, 1996). 'Too much work' in a prevailing climate of 'professional self-doubt' might suggest that workload impedes a sense of accomplishment. This is a different meaning from 'too much work' in the context of 'home-work conflict' where workload might be interfering with personal social support and recreation. The issue of the meaning of identified stressors is an important factor when considering support and stress management.

13.2.4   Prevalence

Despite the reservations expressed by Burden (1988) regarding the interpretation of stress scale data, attempts have been made to identify proportions of samples that fall into a high stress category according to the measure being used. These findings give some indication of the prevalence of distressed personnel. For example, in order to provide some indication of prevalence of distress in his sample of 30 Australian school psychologists based on their SPSI scores Burden (1988) states that 13% of this sample obtained a score greater than 7. This score, despite being a somewhat arbitrary 'cut-off', is considered 'indicative of relatively severe stress' (p. 49). Therefore 13% of the sample is relatively severely stressed.

Another approach to determining prevalence is to use a measure of strain as a dependent variable. Measures of burnout have been used in this way (e.g. Arvay and Uhlemann, 1996) as has the General Health Questionnaire (GHQ). This latter approach was reported by Cushway and Tyler (1996) to determine the prevalence of dysfunctional stress in British clinical psychologists. The GHQ provides an indication of 'caseness' described as a 'just significant clinical disturbance' (p. 142). An earlier study by Cushway (1992, in Cushway and Tyler, 1996) of clinical psychology trainees revealed a 49% level of GHQ caseness. However, a later study of qualified clinical psychologists (Cushway and Tyler 1992, in Cushway and Tyler, 1996) put 29.4% above the 'cut-off' for caseness. Another study by the same authors identifies 40% as qualifying for caseness although the sample in this study contains both clinical psychologists and mental health workers (Cushway *et al.*, 1996).

Another approach is to determine the stress/distress levels of psychologists either by measurement or self-report. This approach often fails to link the reported levels of stress/distress with specific work-related factors or indicate whether psychologists as a group are more or less distressed than other professional groups. For example, Cushway and Tyler (1996) report that both trainee and qualified clinical psychologists identified similar levels of stress with 24% reporting themselves as very stressed and half as moderately stressed.

Thorenson *et al.* (1989) provide a more comprehensive picture of the areas in which psychologist experience distress, but again it is unclear to what extent these findings are work and/or profession related. Using the Psychologist Health Questionnaire (PHQ) to assess distress in a sample of 379 American psychologists, five categories of distress were identified: depression, marital dissatisfaction, feelings of loneliness, recurrent physical illness and recognition of a drinking problem. Approximately 10% of the sample reported distress in these categories. Further analysis showed that 19% indicated distress in one category whereas 9% reported distress in 2–4 categories. Thorenson *et al.*, suggest that these figures 'constitute liberal and conservative estimates of distress' (p. 144).

Although there is considerable disparity across these estimates there is a clear indication that a proportion of psychological practitioners experience distress. Adopting the liberal and conservative estimates strategy, it appears, depending on the approach taken and the measures and samples used, that for qualified psychologists, 9% might be a reasonable conservative estimate of distress with a proportion around 30% (possibly higher) at the more liberal end.

13.2.5   Correlates of Reported Stress

Some of the research has attempted to identify trends in work stress of psychologists and other mental health workers. For example, stress levels have been explored in terms of both age and gender. Deutsch (1984), Huebner and Mills (1997) and Cushway and Tyler (1996) each report that stress is negatively correlated with age. Cushway and Tyler (1996) report this trend for both trainees and qualified psychologists. As might be expected experience appears to be correlated similarly (Huebner and Mills, 1997). However, Cushway and Tyler (1996) note that although both less and more experienced personnel report high levels of stress, the reported stress appeared more likely to translate into distress (as measured by the GHQ) for the less experienced group. Reports on a relationship between gender and stress are less consistent. Huebner and Mills report no gender effects for their sample of school psychologists. However, Cushway and Tyler (1996) found that 'in all [their] studies of British clinical psychologists women have consistently reported more psychological distress than men' (p. 143). Deutsch (1984) also found women to report more stress on four of seven factors of the scale used in her study of psychotherapists. Gender related differences in the reporting of feelings, and distress, as well as availability of social support and multiple role strain have all been suggested as possible factors in these findings. Finally, two other variables have been considered in relation to reported stress. Huebner and Mills noted trends in stress reporting change according to the time of the school year and Cushway and Tyler (1996) explored partnership status and reported stress. They found that psychologists who qualified for 'caseness' on the GHQ were less likely to be married or live with a partner. Some of these findings are replicated in the burnout research and will be considered again later in this chapter.

13.2.6   Sources and Correlates of Stress: Concluding Comments

There is a remarkable convergence of stressors across the reviewed studies. This is particularly evident in the correspondence between the items on Cushway *et al.*'s (1996) mental health workers stress scale and the individual stressors identified in the other research. In part this convergence may reflect conceptual biases in work stress research, however, the stressors identified in Table 13.1, with the possible exception of 'professional expectations and beliefs' provide a useful frame of reference for identifying sources of stress in psychological work.

Despite the appealing convergence of issues, the range of stressors identified in this review does not encompass the diversity of issues that might pertain if scope of the domain is extended. For example, stressors associated with high visibility, information management, service utilisation and relocation reported as demanding and challenging by rural and remote psychologists in countries such as Australia, Canada and the United States are generally not included in most measures. Burden (1988) highlights the importance of the cultural context in identifying stressors in the work of school psychologists, noting that the salience of individual stressors varies from context to context.

Although this review is not focused on 'person variables' in the stress–strain process, the demographic and person factors that appear to be related to increased reporting of stress and measured distress are of general interest. An understanding of the role of these variables in the stress–strain process might be helpful in developing recruitment and educational policy.

Another aspect that is of interest pertains to the theoretical underpinnings of stress research. The studies reviewed in this chapter are devoid of any clear theoretical framework or fail to make a theoretical framework explicit. Cushway and Tyler (1994) refer to a transactional model of stress that emphasises an appraisal process, although this orientation is introduced almost as an afterthought. Despite the substantive focus of the reviewed research, none is inconsistent with the model proposed at the beginning of this chapter. The focus on sources of stress or stressors is consistent with Phase 1 of Matteson and Ivancevich's, (1987; Ivancevich and Matteson, 1980) model. The one exception is Deutsch's (1984) study in which irrational thoughts are identified as a source of stress. Thoughts, irrational or otherwise, would appear to be related to appraisal, not to sources of stress. The exception may be intrusive thoughts and obsessive ruminating, which as a function of their intrusiveness (as opposed to content) might constitute a source of stress. Beliefs about the profession and/or professional work on the other hand serve to determine which stressors are experienced as stress. Therefore, an important focus in stress research needs to be on the meanings that underpin the identified stressors.

13.3 THE APPRAISAL PROCESS

According to Lazarus and Folkman (1984) appraisal is a 'complex, meaning-related cognitive activity' (p. 26) that reflects 'the unique and changing relationship between a person with certain distinctive characteristics (values, commitments, styles of perceiving and thinking) and an environment whose characteristics must be predicted and interpreted' (p. 24). The meanings, values, commitments and profession related perceptions and beliefs of psychologists are important as they are intimately related to that which is appraised as stressful.

Very little research has focused on the appraisal process. Although Deutsch (1984) proposed that irrational beliefs are related to the stress process, irrational thoughts, as already noted, were construed in her work as a source of stress.

13.4 STRAIN

Strain occurs as a consequence of prolonged and/or unmitigated and/or extreme stress at work. According to Matteson and Ivancevich (1987; Ivancevich and Matteson, 1980) strain may be manifested in workers as ill-health, absenteeism, job quitting, burnout and a range of other conditions that significantly compromise or impede the worker's performance. As noted above, Cushway and Tyler (1996) used the GHQ as a dependent measure for their work stress scale using the cut-off for caseness as a psycho-physical indicator of dysfunctional stress. Other researchers exploring the stress effects of trauma clients on counsellors and psychologists have developed theoretical constructs such as 'vicarious trauma' (McCann and Pearlman, 1990) and 'compassion fatigue' (Figley, 1995) which if present result in impaired performance of the practitioner. Some researchers have explored the links between job conditions and attrition in Australian rural and remote counsellor and/or psychological practice (see Ozanne, 1999; Wolfenden *et al.*, 1994). However, these orientations to strain are either discrete areas of research (e.g. Figley) or take the form of generalised dependent measures where the focus is not on the strain construct itself (e.g. Cushway and Tyler, 1996). In the work stress literature as it applies to psychologists the most frequently explored strain construct is burnout.

13.4.1 A Brief Review of Burnout Research

In this segment the main findings from the research on burnout pertaining to psychological work will be summarised. However, because burnout is a construct so firmly linked to generic helping work it is not always easy to tease out literature that pertains to specific professional groups within the helping domain. Therefore, both broad trends and literature specific to psychology will be noted.

Burnout research tends to fall into two main categories. First, there are empirical studies directed towards substantive and/or theoretical questions. Second, there is a substantial body of psychometric literature, much of which has focused on the Maslach Burnout Inventory (MBI) (Maslach and Jackson, 1986). The substantive literature is the focus of this segment. However, as so much of the burnout research uses the MBI, a brief description of the measure will be given.

The MBI is a self-report measure consisting of 22 items. It is based on a model of burnout proposed by Maslach and Jackson (1986) that hypothesised that over extended human service professionals initially became emotionally exhausted as consequence of their work stress. If the exhaustion is not addressed then workers become increasingly depersonalised from their clients as a form of coping. Finally, increased depersonalisation leads to a sense of inadequacy in the work role and a reduced sense of personal accomplishment.

The three sub-scales of the MBI, Emotional Exhaustion (EE), Depersonalisation (DP) and Personal Accomplishment (PA), reflect Maslach and Jackson's (1986) tripartite conceptualisation of burnout. Summing the items for each sub-scale generates three separate scores. These scores are treated separately and are not combined into a single indicator of burnout. High scores on EE (9 items) and DP (5 items) are consistent with burnout as are low scores on PA (8 items). There is no specified cut-off for the presence of burnout on any dimension. The scores are interpreted in terms of indicating 'high' 'medium' or 'low' levels of burnout based on norms derived from a variety of human service professionals.

The psychometric properties of the MBI are well documented (Byrne, 1993). Reliability is thought to be moderately strong (Byrne, 1993; Maslach and Jackson, 1986). Internal consistency is sound, as is test–retest reliability (Maslach and Jackson, 1986). There is also substantial evidence for convergent validity (Maslach and Jackson, 1986) although there is debate regarding discriminant validity (see Jackson *et al.,* 1986; Kahill, 1988).

13.4.2   Correlates of Burnout

Table 13.2 shows the main burnout correlates reported in the literature. The correlates have been categorised and the direction of their effects noted. The relationship between the variables identified in Table 13.2 and burnout is complex.

Many of these studies are cross-sectional designs and although significant correlations have been identified between variables of interest and burnout, causality cannot be inferred. In addition, many of the variables do not act independently of one another. Some have direct effects and buffering or moderating effects. Even when longitudinal designs are employed, and causal directions can be determined, the very nature of the stress process does not preclude burnout effects as being stressors in their own right. Given the complexity

Table 13.2 Correlates of burnout in health and human services*

Category	Variable	General trend
Demographic	Age	Negative correlation †
	Gender	Differential results on MBI
	Family status	Higher b/o for singles ††
Professional	Experience	Negative correlation †
	Education	Positive correlation
	Professional preparation	Negative correlation †
Job related attitudes	Unrealistic expectations	Positive correlation †
	Involvement with work but not job	Positive correlation
	Commitment	Positive correlation
Job characteristics	Perceived workload	Positive correlation †
	Time pressure	Positive correlation
	Role conflict	Positive correlation †
	Role ambiguity	Positive correlation †
	Feedback	Negative correlation
	Participative decision making	Negative correlation
	Autonomy	Negative correlation
	Supervisor support	Negative correlation †
	Boundary spanning	Positive correlation
	Job level	Differential for gender
	Resources	Negative correlation †
Clients	Severity of needs	Positive correlation †
	Immediacy of contact	Positive correlation
	Compliance	Negative correlation †
Extra-organisational factors	Personal social support	Negative correlation †
Personal factors	External locus of control	Positive correlation
	Self-esteem	Negative correlation
	Neuroticism	Positive correlation
	Trait anxiety	Positive correlation
	State anxiety	Positive correlation
	Type A	Positive correlation
	Extraversion	Differential results on MBI
	Empathic personality	Positive correlation
	Feeling 'type'	Positive correlation
	Helping 'type'	Positive correlation
	Control coping	Negative correlation

*Derived from reviews by Cordes and Dougherty, (1993); Heubner, (1993); Perlman and Hartman, (1982); Schaufeli, (1999); Schaufeli and Buunk, (1996).
†Denotes trend has been demonstrated for psychologists; ††Denotes contrary trend has been demonstrated for psychologists; b/o = burnout.

of the phenomena the division of this discussion into 'correlates' and 'consequences' may seem somewhat foolhardy. However, it serves as a useful organisational device in the face of a burgeoning literature.

The correlational literature is not always easy to interpret. Methodological problems, sample size, and limited identification and control of co-variates has, at times, led to contradictory and ambiguous results. However, in more than 20 years of research trends are apparent.

Reviews of the literature (e.g. Schaufeli, 1999; Schaufeli and Buunk, 1996; Cordes and Dougherty, 1993) tend to use similar frameworks for organising the literature in this area. A distinction is usually made between correlational research where the precise relationship between the variable of interest and burnout is unclear (usually cross-sectional research) and outcome studies where consequences of burnout have been identified via longitudinal research designs. Consequences of burnout are addressed in the next segment.

13.4.3   Consequences of Burnout

For reasons similar to those associated with identifying predictors or 'causes' of burnout, identifying the consequences of burnout is equally problematic. In the early, more clinically oriented, literature it was not uncommon for lists of burnout symptoms to be generated. In a sense symptoms can be understood as concurrent consequences. Medically speaking symptoms are co-morbid with disease entities and occur as indicators and as consequences of the disease process. In the burnout literature differentiating between indicators and consequences is not always possible and the exercise has tautological elements. For example, an emotional consequence of burnout might be indifference to recipients, which is also an indicator of depersonalisation on the MBI.

This exercise is also confounded by the lack of longitudinal research. The majority of burnout studies are cross-sectional designs employing correlational analyses, which, as noted in Section 13.4.2, do not indicate causality. Therefore caution should be exercised when speaking in terms of causes and consequences. Also, as noted previously, the transactional nature of stress and the likelihood of unknown factors driving both hypothesised and measured relationships adds to the complexity of the situation. However, there is theoretical and practical merit in at least identifying what might be the consequences of burnout.

Schaufeli and Buunk (1996) review six categories of consequences. They identify, mental; physical; behavioural; social; attitudinal and organisational manifestations of burnout. This structure is not unlike that proposed by Kahill (1988) who grouped the consequences into the following categories: physical; emotional; interpersonal; attitudinal and behavioural. Six categories and their associated consequences as indicated by the research are displayed in Table 13.3. However, it should be noted, that unlike with the 'causes' of burnout literature the application to psychologists as a distinct professional group of burnout consequences is less clear. Therefore, these will not be noted in Table 13.3.

Table 13.3 Consequences of burnout*

Thematic category	Specific consequences
Mental Consequences	Lowered self-esteem
	Depression
	Irritability
	Helplessness
	Anxiety
	Diminished frustration tolerance
	Cognitive impairment
Physical Consequences	Fatigue
	Insomnia
	Headaches
	Gastrointestinal disturbances
	Sexual problems
	Increased incidence of illness
	Increased risk of cardio-vascular disease
Behavioural Consequences	Personal neglect – poor diet and reduced physical exercise
	Aggression
	Increased use of caffeine and other substances
Attitudinal Consequences	Negative attitudes towards clients, job,
	Organisation, and self
Social Consequences	Withdrawal from social contact
	Decreased involvement with recipients
	Negative spillover from work to home
Organisational Consequences	Intention to quit
	Absenteeism
	Lowered productivity
	Decreased efficiency
	Lack of preparation

*Derived from reviews by: Cordes and Dougherty (1993); Kahill (1988); Perlman and Hartman (1982); Schaufeli (1999); Schaufeli and Buunk (1996).

Although there are methodological problems inherent in much of the research investigating consequences of burnout, a brief review of the general trends is helpful to this discussion. One of the main assumptions in exploring occupational stress and burnout in psychological work is that compromised practitioners are less able to fulfil their work roles effectively. In addition, as Mills and Huebner (1998) demonstrated in a prospective study of school psychologists 'not only may stressful occupational experiences predispose individuals to experience burnout, but also high levels of burnout may predispose individuals to experience additional occupational stress' (p. 103). This transactional feature of the stress–strain process should not be underestimated.

The brief review in the preceding section highlights the ways in which burnout can lead to both negative professional and negative personal outcomes.

13.5 INTERVENTIONS IN WORK STRESS AND BURNOUT

Interventions designed to reduce damaging levels of stress in psychologists are rarely reported. Forman (1981) reported a study evaluating a cognitive behavioural stress management training intervention for school psychologists that showed positive effects on psychologists' self-reported job satisfaction and decreased anxiety.

Kilburg, Nathan and Thoreson (1986) published an edited volume on professionals in distress that included chapters on the management of distressed psychotherapists and psychologists. Chapters addressed a range of aspects related to 'impaired psychologists' including disciplinary issues for those practitioners who had breached ethical standards, through to psychotherapy and social support. However, as O'Connor (2001) notes the response to impaired psychologists tends to emphasise code enforcement rather than prevention and education.

There is also a small literature on professional practice in psychology that is aimed at both new practitioners and at practitioners at various stages in their career development (e.g. Buckner, 1992). Strictly speaking, this is not 'stress management' material, but more a series of recommendations and guidelines based on the wisdom of experienced practitioners aimed at enhancing performance outcomes.

13.6 SOURCES AND NATURE OF STRESS IN AUSTRALIAN PSYCHOLOGICAL WORK

This study was originally developed in response to a substantive interest in rural and remote psychological work in Australia. The largely narrative commentary from Australia and elsewhere on rural work (e.g. Andrews *et al.*, 1995; Cohen, 1992; Griffiths, 1996; Griffiths and Andrews, 1995; Hargrove, 1986; Harvey and Hodgson, 1995; Schank and Skovolt, 1997) suggested that psychological practitioners in non-urban settings encounter a range of demands and challenges unique to their geographic context. On the basis of this literature, and the broader literature on work stress and psychological work, it was hypothesised that rural and remote psychologists may experience higher levels of work-related stress than their urban colleagues. In accordance with the stress–strain model proposed earlier, higher levels of stress increase the likelihood strain, which for service providers in largely under-serviced areas (see Griffiths, 2000) has social as well as professional and personal ramifications.

In addition to this original interest a further interest arose related to exploring the role of context in perceived work stress. The above review of the literature suggests that the context of work is a relevant but neglected factor in many studies. Also, of interest was the degree of comparability between the Australian experience and studies conducted elsewhere, with a particular focus on assessing the usefulness of existing stress measures for use with Australian psychologists. A final aspect of interest, that appears to be neglected in previous research, is the role and nature of the appraisal process. An understanding of appraisal is essential in determining the needs of psychologists in both proactive stress management and education and for designing appropriate interventions when strain is manifested. This exploratory study is not only focused on perceived stressors and from where they derive, but also on the meaning of the identified stressors and *why* they are appraised as stressful.

This study is conducted in two parts. The focus of part one is on the relevance of context to the perceived sources and nature of stress encountered by Australian psychologists and the appraisal process. The research questions for the first part of the study are: (1) what are the main sources of stress identified by Australian psychologists; (2) from what broad systemic domains do they derive; and (3) do rural and remote (R/R) psychologist differ from urban and metropolitan (U/M) in respect of the first and second research questions.

The second part of the study is related to the appraisal process. The focus was on identifying *why* some phenomena are appraised as stress invoking. The intention was two-fold. First, more understanding about the meaning of the stressors was required. For example, as noted earlier a stressor such as 'too much work' may mean different things to different people in relation to the stress–strain process. Second, more understanding was sought about the priorities of psychologists that would lead them to appraise some phenomena as stress-invoking yet others as benign. The research questions for this part of the study are: (1) what is the nature of work stress affecting Australian psychologists; (2) what appraisal processes underpin the experience of stress in psychological work, and (3) do rural and remote (R/R) psychologists differ from urban and metropolitan (U/M) in respect of the first and second research questions.

Fifty-one registered psychologists participated in the study. The R/R group consisted of 30 participants and the U/M group of 21 participants. Of the 30 R/R psychologists, 16 were male and 14 female. Their average age was 45 years. Two members of the R/R group reported practising on a three-year degree in psychology, 19 on a four-year degree and nine were practising with a Master of Psychology (or equivalent). The U/M group consisted of 6 males and 15 females, with an average age of 43 years. Ten members of the U/M group reported holding a four-year degree in psychology and eleven reported having a Master of Psychology. All participants had been practising for a minimum of two full-time equivalent years, the average years experience for the R/R group being 13.25 years and for the U/M group, 10.5 years. The categories of R/R and U/M were defined in accordance with the *Rural and Remote Areas Classification* (Department of Human Services and Health [DHS and H], 1994). Although not consistent with the sampling procedures of phenomenologically oriented research (Morse, 1989) an attempt was made to recruit participants that represented the diversity of settings in which Australian psychologists live and work. Because Australia is geographically and culturally diverse this was thought to be an important element in the design. All participants were engaged in clinically oriented work, either with agencies and/or in private practice. They will be referred to as 'practising psychologists' to distinguish them from registered psychologists who are employed in non-clinically oriented roles. The participants were interviewed either face-to-face or by telephone using a semi-structured interview schedule (see, Smith, 1995) that originally asked the following question:

In general terms what do you find demanding or challenging about your work?
1.   More specifically, what do you find demanding or challenging about the service that you are contracted to deliver to clients?
2.   What do you find demanding or challenging about the organisation or workplace in which you work?
3.   What do you find demanding or challenging about the geographic location in which your work is conducted?

This question was later modified to include the following emergent questions:
1.   What are the demands and challenges associated with professional, industrial and regulatory bodies? and,
2.   What are the demands and challenges associated with family and personal life?
3.   Are there times, in any circumstances, where you feel threatened?

The interviews were transcribed and analysed.

13.6.1   Analysis – Part One

A content analysis was performed (Anderson, 1997) in which all the transcribed comments directly pertaining to the stimulus question were identified. For example, comments such as 'My organisation has no funding for professional development' were identified. The comments were then assigned to one or more categories reflecting the levels of the interview question. In the above example the comment was placed under organisational level. The object was to identify each stressor and to link it to a source. Although five sources were implicit in the interview question, provision was made for the development of additional categories. The purpose of this procedure was to identify the broad systemic sources of stress affecting Australian psychologists.

Individual comments were referred to as 'content items'. Because the number of content items was so large they were organised into 'content themes'. The proportion of respondents reporting one content item or more in any content theme was calculated for both the R/R and the U/M groups. In this way proportional comparisons could be made between the R/R respondents and the U/M respondents on a number of dimensions including: (1) nature of the stressors; (2) frequency of reporting; (3) relative salience of stressors, and (4) relative salience of sources of stress. In addition to comparisons across geographic location the analysis permitted the type, range and salience of stressors affecting Australian psychologists as a whole to be identified.

13.6.2   Results and Discussion – Part One

The stressors identified by Australian psychologists derive from six domains: (1) services to clients; (2) organisation/workplace; (3) geographic location (4) professional and industrial bodies; (5) family and personal life; and (6) the broader socio-historical-political context. The first five domains make the most immediate and direct demands on the psychologist, but each component is embedded in and affected by the broader socio-historical-political context. None of the domains is discrete and interdependence was demonstrated by content themes appearing in more than one domain (see below). The model that best describes the relationship between the various domains is not inconsistent with Bronfenbrenner's (1979) ecological model of nested sub-systems.

Fifty-three content themes were identified across all domains although some were repeated with a different emphasis in different domains. It is beyond the scope of this chapter to itemise the full range of identified stressors however, they show a high degree of correspondence with areas covered in the stress inventories described earlier. Where there is a clear departure from previous research on work stress in psychological work is in relation to stressors associated with the geographic domain and the socio-historical-political context. In addition, sources of stress associated the professional and industrial context of the work and with family and personal aspects of life appear to be more refined in their definition than in previous studies although in other instances (e.g. client related stressors) the Australian data appears to show less specificity. These differences may pertain to unique features of the Australian context. However, they may also be related to the broader contextual orientation to work stress taken by the researcher.

The number of stressors (i.e. content themes) within each domain varies. In addition, some content themes within a domain attracted responses from a greater proportion of the respondents. For both rural and urban psychologists the domain with the greatest number of content themes was the organisational and workplace domain with 18 content themes. The domain that showed the highest mean reporting rate of respondents was the domain pertaining to clients.

Only one content theme differentiated the R/R psychologists from their U/M counterparts. The rural group alone identified lack of essential services (e.g. education and medical services) for family as a source of stress. Otherwise, rural and urban psychologists reported the same content themes in all domains. However, strength and rate of reporting and relative salience of the domains differentiated the two groups.

The six domains (i.e. sources of stress) were rank ordered according to the mean reporting rates of respondents for each of the two geographic groups. Although services to clients was the highest for both groups the rank order thereafter showed a different pattern (Table 13.4).

In addition to differences in rank ordering of domains rate other differences between the two groups were apparent. Of the 53 content themes, 22 were reported by 50% or more of the R/R group, whereas only ten were reported by 50% or more of the U/M group. All domains attracted reporting rates of 50% or more for some content themes from the rural respondents, but the urban respondents only responded to content themes in four domains in a similar way. It appears that although both groups identify the same stressors, a greater proportion of the R/R sample reported them.

Table 13.4 Rank order of domains from which stress derives based on mean reporting rates for content themes

Rank	Rural Group	Urban Group
1	Services to clients (62.6%)	Services to clients (45.6%)
2	Geographic location (47.3%)	Organisation and workplace (34%)
3	Organisation and workplace (40.7%)	Family and personal (28.6%)
4	Family and personal (37.9%)	Professional and industrial (26.5%)
5	Professional and industrial (36.1%)	Geographic location (23.9%)
6	Social-historical context (23.3%)	Social-historical context (22.7%)

A second comparison involved relative strength of response between the two groups. For example, a content theme may be reported by only ten per cent of the R/R group but at ten per cent still represent a higher response rate than the U/M group on the same content theme. It is interesting to note that the strength of response for the R/R group was consistently higher than for the U/M group. Of the 53 content themes, 42 were reported by a larger proportion of rural respondents. Only nine content themes overall were reported by a higher proportion of U/M respondents.

These data suggest that Australian psychologists report similar stressors regardless of their geographic location of work. However, the salience of the domains from which stressors derive distinguished rural from urban practitioners. Similarly, R/R psychologists are differentiated from U/M psychologists overall reporting rates and relative strength of response from the group. It seems that R/R are more affected by work stressors, or are more likely to identify and report stressors that affect their work.

13.6.3   Analysis – Part Two

The second part of the study used thematic and hermeneutic analysis (see Wilson and Hutchinson, 1991) and elements of both Wicker's (1989) 'substantive theorising' model and grounded theory (see Charmaz, 1995) to explore the nature and meaning of stressors identified by Australian psychologists in part one of this study. The intention of this analysis was to gain a greater understanding of the appraisal process in respect of work stress for Australian psychologists by developing an explanatory model to account for their identified stressors. As discussed earlier, an understanding of appraisal is essential for effective intervention with work stress.

For each of the content themes described in part one, the following question was asked: 'What is the nature of the stress expressed in this content theme?' In consultation with two registered and experienced psychologists (not participants in the study) tentative explanations were proposed and considered in the light of similar content themes elsewhere in the content analysis (if relevant). For example, professional development was reported in the geographic location domain, the organisation and workplace domain and then in the industrial and professional domain. In response to the question 'What is the nature of the stress?' it was proposed that a major concern underpinning this content theme was lack of professional support. Lack of professional support as an interpretation appeared consistent with all other instances of professional development related content themes arising within other domains of the content analysis. This suggested an overarching thematic category pertaining to professional support. Interview transcripts were consulted to ensure that the interpretation of the content theme remained consistent with the original statements and with the explanatory statements volunteered by the participants. The procedure was repeated until eight themes, accounting for all the material in the content analysis, were generated. Next, the material within each theme was organised into related sub-themes. For example, the professional support theme is comprised of four sub-themes: professional development, education, clinical supervision and collegiality.

To address the issue of *why* the identified themes were relevant to the respondents, theoretical relationships between the themes were considered, until a model emerged that accounted for the thematic categories and was congruent with the narrative and explanatory accounts in the original transcriptions. Again, the two registered and experienced psychologists were consulted in order to gauge the validity of interpretations. Throughout the consultation process differences of opinion were resolved through discussion and as a final validity check a summary of the results was circulated for comment to all participating psychologists (Sharpe, 1997). Of those who communicated back to the author, all said that the analysis was consistent with their experience. Their views were integrated into the final analysis.

13.6.4   Results and Discussion – Part Two

Eight themes were generated using the above analytic strategy (Table 13.5). There is a very high degree of correspondence between this list of themes and the factor analysis of clinical psychologists data on the MHPSS reported by Cushway and Tyler (1996).

Although differences identified in the content analysis between the R/R group and the U/M group relating to the strength of reporting were obscured in this analysis, qualitative differences related to the nature of the stressors became more evident. For example, in the content analysis R/R and U/M respondents alike identified stressors associated with career development. The same stressor explored in the thematic analysis highlights the unique challenges for rural psychologists in gaining access to further education, limited promotional opportunities in organisations where they may be the only psychological practitioner and an increased likelihood of promotion being associated with relocation. Although, urban practitioners also identified stressors associated with career development, their challenges were more related to organisational restructuring limiting career pathways and frustration at needing to upgrade qualifications (even though they had access to educational institutions). These are important distinctions in the nature of the stressors being reported.

Overall the R/R and U/M groups appear to experience similar stressors, but circumstances characteristic of Australia's remote and rural areas interact with the demands and challenges of psychological work creating subtle but important differences in the nature of the reported stressors. Six features of rural and remote communities were identified that contribute to the R/R experience of stress inducing phenomena. These are: (1) local resources and infrastructure; (2) distance from a major population centre; (3) population size and dispersion; (4) local population characteristics; (5) local economic base; and (6) geographic and climatic characteristics.

By exploring the relationship between the themes generated in the thematic analysis a picture begins to emerge of what it means, at a subjective level to be a psychologist in Australia. It is the meaning inherent in the professional identity of psychologists that underpins the appraisal process.

A cursory examination of Table 13.5 shows that there are two major thematic categories. The first category applies to Themes I–VII and relates to professional and work stress. The second category relates to stress in the personal and private domain and is largely captured in Theme VIII. However, these two categories are not exclusive, for example, client issues as noted in Theme I, can affect the psychologist both professionally and personally. Similarly, organisational stability as noted in Theme VII can also have personal and professional effects.

Psychological work is human work. It involves a high degree of personal involvement combined with specialist knowledge, skills and competencies. The personal nature of the work interacts with the psychologist's own quality of life and many respondents noted the reciprocal relationship between the two.

In addition the decision to enter a profession is personal in nature and as Cherniss (1980) has noted, the decision carries with it a set of expectations about work and about lifestyle. An exploration of what it means to be a psychologist, that is, how psychologists' construe their professional identity has to take account of these two elements – the professional and the personal. To facilitate this exploration each will be discussed in turn. However, the interactive nature of the two should not be overlooked.

Although it was not overtly articulated the participants in this study clearly conceive of psychology as a human service *profession* in that it constitutes a 'performance for public good' (Brooks, 1952, in Fullinwider, 1996, p. 79). In clinically oriented work 'the public' is encountered in the form of clients and at the heart of the work described by the respondents is the provision of quality services to clients. The psychologists in this study clearly desire the best outcomes for their clients and perceive as stressful those aspects of their professional and work

Table 13.5 Nature of stressors affecting Australian psychologists

Themes		Sub-themes
I	Client Encounters	Client issues
		Client resources
		Resistant clients
II	Professional Standards	Responsibility
		Competence
		Efficacy
		Accountability
		Confidentiality
III	Integrity of the Profession	Expectations
		Professional integrity
		Valuing
IV	Professional Support	Professional development
		Education
		Clinical supervision
		Collegiality
V	Professional Resources	Accommodation
		Office equipment
		Professional materials
		Personnel resources
		Transport
VI	Service Delivery	Constrained services
		Diverse services
		Peripatetic services
VII	Organisational context	Nature of employment – private practice
		Financial management
		Isolation
		Nature of employment – salaried position
		Professional recognition
		Management
		Communication
		Organisational stability
		Administrative tasks
VIII	Quality of Life	Services
		Residential location
		Social support
		Visibility
		Safety
		Spill-over
		Career
		Remuneration

situations that impede this, including client variables such as reluctance or complexity. Most of the material integrated into the thematic analysis both supports this interpretation of the data and can be understood in terms of it.

For a profession to serve the public good individual practitioners need to maintain their standards of practice and the profession must be regulated so that it maintains its autonomy and integrity (Fullinwider, 1996). It is not sufficient that individual psychologists do 'good work', the profession must also be recognisable as making a valued contribution to the community and valued by the community. To this end practitioners must engage in 'sound professional practice . . . in order to safeguard the welfare of consumers of psychological services and the profession (Australian Psychological Society [APS], 1997 p. 1). Themes II and III of the thematic analysis – *Professional Standards* and *Integrity of the Profession* reflect these aspects of professional responsibility. The sub-themes of Theme II reveal a concern for duty of care, competence to assist others and quality of client outcomes. Also included were issues of accountability and vulnerability of both the profession and individual practitioners in the face of adverse publicity. These latter concerns do not appear to be driven by self-serving factors such as fear of censure but rather by a genuine desire to provide effective and ethical psychological services and to uphold the profession so that it can continue to make a positive social contribution.

The public image of psychology is reflected again in Theme III. However, the emphasis is slightly different from that in Theme II as the content of this theme reflects social issues that bear on professional integrity. Berger and Luckman (1979) argued as early as 1966 that social roles are negotiated. That is, it is impossible to maintain or perform a social role that has no meaning for others. In Theme III the stress associated with negotiating the social viability of psychology were indicated. Respondents suggested that frequently the profession is not well understood or valued. If a constrained social contract is indeed an accurate perception it will have serious implications for individual practitioners, both in their work and in their careers. However the focus of concern, once again, was on how serving the public good is impeded by lack of social understanding and/or valuing. This content of this theme augments the usual understanding of '[safeguarding] the integrity of the profession' (APS, 1997, p. 1).

Themes IV and V, Professional Support and Professional Resources, identified the requirements for the maintenance and fulfilment of the psychologist's professional role. Without access to further training opportunities, professional standards may be difficult to maintain when the requirements of the work exceed the psychologist's expertise. In addition, on going training and development is understood to be an essential aspect of professionalism psychologists perceive their professional status as under threat when this is either not recognised or access is impeded in some way. Similarly, both standards and integrity of the profession are negatively affected when the resources essential to psychological work are unavailable or limited.

In Themes VI and VII attention is drawn to the organisational interface. Whether a psychologist works in private practice or is employed or in a salaried position, approaches to service delivery and organisational factors are going to be relevant.

Finally, as noted earlier, the individual psychologist who sets out to serve the common good by providing services to clients in accordance with established ethical codes of conduct (APS, 1997) is also engaged in work that has personal implications for the practitioner. Clinical work draws on the humanity of the practitioner, and consequently affects the psychologist both positively and negatively. In addition, the psychologist has expectations of what it means to be a professional person and the extent to which psychology meets these expectations will also have implications for the practitioner's sense of well being. The private life of a psychologist also constitutes a competing set of responsibilities. If for example, a work situation is causing distress to the psychologist's family, not only may work performance be affected, the practitioner may also come to question the value of the job and/or the profession.

In summary, the professional identity of the psychologist underpins the appraisal process. In addition those aspects of the psychologist's personal life that interact with the professional identity also underpin stress appraisals. An understanding of how psychologists' construe themselves and their work allows for a deeper understanding of *why* some phenomena are perceived as stressful. In turn, these understandings assist in identifying points of entry in terms of support and other interventions.

This study has indicated that there is no geographically related difference in terms of what it means to be a psychologist in Australia. Rural psychologists do not constitute a different professional group. They share the same priorities and values and concerns as their urban colleagues. That is, it means the same thing to be a psychologist whether one is in the urban or rural sector. However, in the course of their work R/R psychologists are exposed to contextual differences that appear to result in more stress related appraisals, differential relevance of stressors and qualitative differences related the nature of the encountered stressors. These distinctions are important for the provision of appropriate stress-management interventions, support and training experiences designed to equip psychologists for working in different contexts.

13.7 CHAPTER SUMMARY

This chapter has attempted to map the research efforts in psychological work stress using a model of occupational stress derived from the broader work stress literature. The piecemeal nature of these research efforts in addition to what is largely a substantive interest in psychological work stress results in a body of knowledge that is difficult to utilise in the service of professional support and enhancement.

Despite these limitations a growing body of work is emerging that both sheds some light onto the nature of stress in psychological work and provides a basis for future research. Using the Matteson and Ivancevich (1987); Ivancevich and Matteson, (1980) model to integrate the research there is a reasonably clear picture emerging of both the broader sources of stress and the types of stressors that affect psychologists in their professional work. The identification of stressors and the development of measures such as that published by Cushway and Tyler (1994; 1996) permit coherent and organised research efforts, whereas the acknowledgement of the role of contextual factors on work permit research efforts that can be tailored to specific subsets of practitioners.

Unfortunately, the dominant position of burnout as a strain construct tends to obscure other fruitful areas of exploration in the area of strain. Stress per sé does not have negative implications. The duration, level and/or extremity of stress required to compromise practitioners can only be determined by understanding more about the nature and diversity of compromised states. This is an important area for research. For example, in rural and remote Australia, attrition from the profession, the job and/or the geographic location of work may be one of the most damaging forms of strain outcome for an already compromised health service.

Finally, the least explored aspect of the stress–strain process is appraisal. Appraisal is a two-edged sword. On the one hand, the appraisal process lends itself well to CBT approaches in intervention with distressed practitioners. However, Griffiths' (2000) findings also show that the kinds of appraisals made by Australian practitioners are related to their core professional identity as practising psychologists and the context of their work. These findings signpost the way to more sociologically oriented analyses of work stress that may ultimately provide a more fruitful avenue of exploration in terms of intervention than psychological models that locate the problem with the person.

13.8 REFERENCES

Anderson, J., 1997, Content and text analysis. In *Educational research, methodology and measurement: An international handbook*, edited by J.P. Keeves (New York: Pergamon Press), pp. 340–344.

Andrews, H.B., Griffiths, S.P. and Loney, A.M., 1995, Confidentiality in the country. *The Bulletin of the Australian Psychological Society*, **17**, pp. 17–19.

Arvay, M.J. and Uhlemann, M.R., 1996, Counsellor stress in the field of trauma: A preliminary study. *Canadian Journal of Counselling*, **30**, pp. 193–210.

Australian Psychological Society, 1997, *Code of ethics* (Carlton, Victoria: Author).

Basson, C.J., 1988, Potential sources of work-related stress for the educational psychologist in the Republic of South Africa. *School Psychology International*, **9**, pp. 203–211.

Berger, P. and Luckman, T., 1979, *The social construction of reality* (New York: Penguin).

Bronfenbrenner, U., 1979, *The ecology of human development: Experiments by nature and design* (Massachusetts: Harvard University Press).

Buckner, M.O., 1992, New professional in private practice. *Counseling Psychologist*, **20**, pp. 10–16.

Burden, R.L., 1988, Stress and the school psychologist: A comparison of potential stressors in the professional lives of school psychologists in three continents. *School Psychology International*, **9**, pp. 55–59.

Byrne, B.M., 1993, The Maslach Burnout Inventory: Testing for factorial validity and invariance across elementary, intermediate, secondary teachers. *Journal of Occupational and Organizational Psychology*, **66**, pp. 197–212.

Charmaz, K., 1995, Grounded theory. In *Rethinking methods in psychology*, edited by J.A. Smith, R. Harre, and L. Van Langenhove (London: Sage Publications), pp. 27–49.

Cherniss, C., 1980, *Professional burnout in human service organizations* (New York: Praeger).

Cohen, D., 1992, Occupational hazards of the rural psychotherapist. *Psychotherapy in Private Practice*, **10**, pp. 13–35.

Cordes, C.L. and Dougherty, T.W., 1993, A review and integration of research on job burnout. *Academy of Management Review*, **18**, pp. 621–656.

Cushway, D. and Tyler, P.A., 1994, Stress and coping in clinical psychologists. *Stress Medicine*, **10**, pp. 35–42.

Cushway, D. and Tyler, P.A., 1996, Stress in clinical psychologists. *International Journal of Social Psychiatry*, **42**, pp. 141–149.

Cushway, D., Tyler, P.A. and Nolan, P., 1996, Development of a stress scale for mental health professionals. *British Journal of Clinical Psychology*, **35**, pp. 279–295.

Department of Human Services and Health, 1994, *Rural/remote areas classification* (Canberra, ACT: Author).

Deutsch, C.J., 1984, Self-reported sources of stress among psychotherapists. *Professional Psychology: Research and Practice*, **15**, pp. 833–845.

Dewe, P., 1991, Measuring work stressors: the role of frequency, duration and demand. *Work and Stress*, **5**, pp. 77–91.

Dunnette, M.D. and Hough, L.M. (eds) 1992, *Handbook of industrial and organizational psychology,* Volume 3 (Palo Alto, CA: Consulting Psychologists Press).

Figley, C.R., 1995, Compassion fatigue as secondary traumatic stress disorder: An overview. In *Compassion fatigue: Coping with secondary traumatic stress disorder in those who treat the traumatised,* edited by C. Figley (New York: Brunner/Mazel), pp. 1–20.

Forman, S.G., 1981, Stress-management training: Evaluation of effects on school psychological services. *Journal of School Psychology*, **19**, pp. 233–241.

Freeman, A., 1987, 'Fix it quick' - stresses of the EP and problems of effectiveness. *Educational Psychology in Practice*, **3**, pp. 44–50.

Fullinwider, R.K., 1996, Professional codes and moral understanding. In *Codes of ethics and the professions*, edited by M. Coady and S. Bloch (Melbourne, Victoria: Melbourne University Press), pp. 72–89.

Griffiths, S.P., 2000, *Occupational stress and burnout in Australian rural and remote psychological practice: An exploration.* Unpublished doctoral thesis (Perth, WA: Curtin University of Technology).

Griffiths, S.P., 1996, Issues in rural health: The utilisation and perception of psychology services. In *Psychology services in rural and remote Australia: Issues paper* edited by R. Griffiths, P. Dunn and S. Ramanathan (Wagga Wagga, NSW: ARHRI/Charles Sturt University), pp. 17–23.

Griffiths, S.P. and Andrews, H. B., 1995, *Issues in rural and remote psychological practice.* Paper presented at the Australian Psychological Society Conference, Perth, WA.

Hargrove, D.S., 1986, Ethical issues in rural mental health practice. *Professional Psychology: Research and Practice,* **17**, pp. 20–23.

Harvey, D. and Hodgson, J., 1995, New directions for research and practice in psychology in rural areas. *Australian Psychologist,* **30**, pp. 196–199.

Hellman, I.D., Morrison, T.L. and Abramowitz, S.I., 1986, The stresses of psychotherapeutic work: A replication and extension. *Journal of Clinical Psychology,* **42**, pp. 197–205.

Hellman, I.D., Morrison, T.L. and Abramowitz, S.I., 1987, Therapist flexibility/rigidity and work stress. *Professional Psychology: Research and Practice,* **18**, pp. 21–27.

Henwood, K.L. and Pidgeon, N.F., 1992, Qualitative research and psychological theorising. *British Journal of Psychology,* **83**, pp. 97–111.

Huebner, E.S., 1993, Professional under stress: A review of burnout among the helping professions with implications for school psychologists. *Psychology in the Schools,* **30**, pp. 40–49.

Huebner, E.S. and Mills, L.B., 1997, Another look at occupational stressors among school psychologists. *School Psychology International,* **18**, pp. 359–374.

Ivancevich, J.M. and Matteson, M.T., 1980, *Stress and work: A managerial perspective* (Glenview, IL: Scott Foresman).

Jackson, S.E., Schwab, R.L. and Schuler, R.S., 1986, Toward an understanding of the burnout phenomenon. *Journal of Applied Psychology,* **71**, pp. 630–640.

Kahill, S., 1988, Symptoms of professional burnout: A review of the empirical evidence. *Canadian Psychology,* **29**, pp. 284–297.

Kahn, R.L. and Byosiere, P., 1992, Stress in organizations. In *Handbook of industrial and organizational psychology,* Volume 3, edited by M.D. Dunnette and L.M. Hough (Palo Alto, CA: Consulting Psychologists Press), pp. 571–650.

Kilburg, R.R., Nathan, P.E. and Thoreson, R.W. (eds), 1986, *Professionals in distress: Issues, syndromes, and solutions in psychology* (Washington, DC: American Psychological Association).

Lazarus, R.S. and Folkman, S., 1984, *Stress, appraisal, and coping* (New York: Springer Publications).

Maslach, C. and Jackson, S., 1981, *Maslach burnout inventory manual,* 1st edn (Palo Alto: Consulting Psychologists Press).

Maslach, C. and Jackson, S., 1986, *Maslach burnout inventory manual,* 2nd edn (Palo Alto: Consulting Psychologists Press).

Matteson, K.T. and Ivancevich, J.M., 1987, *Controlling work stress: Effective human resource and management strategies* (London: Jossey-Bass).

McCann, I.L. and Pearlman, L.A., 1990, Vicarious traumatisation: A framework for understanding the psychological effects of working with victims. *Journal of Traumatic Stress*, **3**, pp. 131–149.

Mills, L.B. and Huebner, E.S., 1998, A prospective study of personality characteristics, occupational stressors, and burnout among school psychology practitioners. *Journal of School Psychology*, **36**, pp. 103–120.

Morse, J.M., 1989, Strategies for sampling. In *Qualitative nursing research: A contemporary dialogue* edited by J.M. Morse (Rockville, Maryland: Aspen Publishers), pp. 117–131.

Nash, J., Norcross, J.C. and Prochaska, J.O., 1984, Satisfactions and stresses of independent practice. *Psychotherapy in Private Practice*, **2**, pp. 39–48.

O'Connor, M.F., 2001, On the etiology and effective management of professional distress and impairment among psychologists. *Professional Psychology: Research and Practice*, **32**, pp. 345–350.

Ozanne, J., 1999, *Attrition among rural psychologists,* Unpublished master's dissertation (Perth, WA: Curtin University of Technology).

Perlman, B. and Hartman, E.A., 1982, Burnout: Summary and future research. *Human Relations*, **35**, pp. 283–305.

Rodolfa, E.R., Kraft, W.A. and Reilley, R.R., 1988, Stressors of professionals and trainees at APA-approved counseling and VA medical center internship sites. *Professional Psychology: Research and Practice*, **19**, pp. 43–49.

Schank, J.A. and Skovolt, T.M., 1997, Dual relationship dilemmas of rural and small-community psychologists. *Professional Psychology Research and Practice*, **28**, pp. 44–49.

Schaufeli, W.B., 1999, Burnout. In *Stress in health professionals: Psychological and organisational causes and interventions,* edited by J. Firth-Cozens and R. Payne (New York: John Wiley and Sons), pp. 17–32.

Schaufeli, W.B. and Buunk, B.P., 1996, Professional burnout. In *Handbook of work and health psychology,* edited by M.J. Schabracq, J.A.M. Winnbust and C.L., Cooper (New York: John Wiley and Sons), pp. 311–346.

Sharpe, L., 1997, Participant verification. In *Educational research, methodology and measurement: An international handbook,* edited by J.P. Keeves (New York: Pergamon), pp. 314–315.

Smith, J.A., 1995, Semi-structured interviewing and qualitative analysis. In *Rethinking methods in psychology,* edited by J.A. Smith, R. Harre, and L. Van Langenhove (London: Sage Publications), pp. 9–26.

Sterling, D.L., 1992, Practising rural psychotherapy: complexity of role and boundary. *Psychotherapy in Private Practice*, **10**, pp. 105–127.

Thorenson, R.W., Miller, M. and Krauskopf, C.J., 1989, The distressed psychologist: prevalence and treatment considerations. *Professional Psychology: Research and Practice*, **20**, pp. 153–158.

Wicker, A.W., 1989, Sustantive theorizing. *American Journal of Community Psychology*, **17**, pp. 531–547.

Wilson, H.S. and Hutchinson, S.A., 1991, Triangulation of qualitative methods: Heideggerian hermeneutics and grounded theory. *Qualitative Health Research*, **1**, pp. 263–276.

Wise, P.S., 1985, School psychologist's rankings of stressful events. *Journal of School Psychology*, **23**, pp. 31–41.

Wolfenden, K., Blanchard, P. and Probst, S., 1994, *Rural information for mental health staff: Final project report* (Canberra, ACT: Commonwealth Department of Human Services and Health/RHSET).

CHAPTER FOURTEEN

Volunteering Work Stress and Satisfaction at the Turn of the 21st Century

Jacques C. Metzer

14.1   INTRODUCTION

Volunteering work is unpaid work. As such, it would not be unexpected to find that volunteering shares many of the characteristics of paid work. Yet to suggest that stress may affect volunteering work in ways similar to paid work strikes one almost as a counterintuitive statement. Surely a volunteer worker experiencing significant job strain would quickly leave the work and thus the remainder of volunteer workers (who might not be experiencing such strain) should be expected to be a low strain unpaid workforce? This chapter will outline some examples of volunteer work with these questions as background.

14.2   DEFINITIONS OF VOLUNTEERING

According to Noble (1997), a volunteer is a person who provides a service to the community, does so of his or her own free will, and does not receive any monetary reward for this service. For psychology, the use of the term 'own free will' presents a challenge for operationalising it appropriately in this context, notwithstanding the literature on 'choice' behaviour.

The Australian National Agenda on Volunteering (2001, p. 5) defined formal volunteering as

'an activity which takes place through not for profit organisations or projects and is undertaken:

- To be of benefit to the community;
- Of the volunteer's own free will and without coercion;
- For no financial payment; and
- In designated volunteer positions only'.

This definition mentions again a free will component, though the inclusion of no coercion is an interesting addition. Presumably, some work for government unemployment benefits schemes, which in Australia recently have taken the form of allowing a volunteer option, are not included, since a significant monetary incentive or inducement (or threat) is assumed (i.e. not losing one's unemployment benefits). It is an interesting question whether coercion can be conceived to include non monetary incentives such as social rewards or whether coercion must always include an aversive element.

Perhaps the simplest and least ambiguous defining criterion to distinguish volunteer work from paid work is the absence of remuneration.

While single acts of help are as ubiquitous as any other commonly observed prosocial behaviour and have been studied by social psychologists (e.g. Eagly and Crowley, 1986), the term 'formal volunteering' here is reserved for planned and sustained helping behaviour conducted through an organisation. Consequently, informal volunteering work is not covered in this chapter, though its ubiquity is not in doubt.

14.3   ECONOMIC BENEFITS OF VOLUNTEERING

Over the last seven years, various estimates have been made of the economic benefits to Australia of formal volunteering. Perhaps the most trustworthy ones come from large scale surveys conducted by the Australian Bureau of Statistics (1996, 2001). In 1995, 3,189,900 people in Australia and 4,400,000 in 2000, representing 24% and 32% respectively of the population, undertook some formal volunteering activity. Australian dollar estimates for the economy range from $5.6 billion in 1996 to $42 billion in 2000 (Debelle, 1996; Warburton and Oppenheimer, 2000). In 2000, there were 704 million hours injected into the economy. Jamrozik (1996) argues that the inclusion of informal volunteer work would dramatically increase these estimates.

It seems clear, from these data, that volunteer work is an important contributor, at least in a quantitative sense, to the economy of Australia.

The increase between 1995 and 2000 appears to represent a response to increasing demands on the volunteer sector, particularly for welfare and service organisations. This is not altogether unexpected, since the 1990s saw the introduction in Australia of economic rationalist policies. Effects of these policies include the attempt to measure the worth of almost every activity in $ value, and the forcing of people in the microcosm of their own lives to do likewise. It is no coincidence that conservative governments of the day have become more aware of the efforts of the volunteer workforce, its economic value and the potential for tapping into this resource, while introducing cost-saving and downsizing measures for their own programmes. The reduction of government services has put increasing pressure on volunteer services to fill the inevitably larger gap. Consequently, the volunteer sector has come under pressure to formalise its workforce with an increased emphasis on training and management. Metzer *et al.* (1997) suggested that, despite many studies examining the quality of worklife of paid workers being reported in the occupational literature (e.g. Karasek and Theorell, 1990; Schaufeli *et al.*, 1993), there is a paucity of studies investigating the influence of stressful events on retention rates, personal wellbeing and health, job satisfaction and so on, in the volunteer workforce. It seems evident that these are becoming of greater importance at the same time as other evidence is mounting for the erosion of 'social capital', so necessary to maintain relative cohesiveness and social support structures and processes in an increasingly stressed society (Cox, 1995).

14.4   MOTIVATION FOR VOLUNTEERING

The published literature on people's reasons for commencing and for their continuation of volunteering include the findings of an Australian Bureau of Statistics surveys (ABS, 1996, 2000) and several research articles in psychological (e.g. Penner and Finkelstein, 1998; Metzer, 1996) and volunteering journals (e.g. Jamrozik, 1996). In this chapter, the emphasis will be on selected *psychological* reasons.

Findings from the ABS surveys enable some interesting inferences about underlying motivations for volunteering. In particular, Metzer (1996) concluded that the reasons given by respondents for engaging in volunteer work could be classified into intrinsically and extrinsically motivated behaviours, of approximately equal distributions. This is contrary to the commonly held assumption that altruism is the main or only motivator for volunteer work.

Clary and co-workers (e.g. Clary *et al.*, 1998) have developed the Volunteer Function Inventory (VFI), to measure different motivations in volunteers. Their six subfactors (Values, Understanding, Enhancement, Defensive, Social Adjustment, Career Advancement) emerged from factor analytic studies conducted by them. While demonstrating reliable differences on these measures for different samples of volunteer workers and indicating reliably initial intentions of volunteering, there remain questions of their comprehensiveness and of predicting long-term volunteering behaviour (e.g. Battaglia and Metzer, 2000; Elshaug and Metzer, 2001).

It may be that an analysis of underlying motivations is neither a simple nor a single issue: it may be a more relevant question to ask what the contribution of various motivational measures might be to measures of the worklife of volunteers, retention rates of volunteers in various organisations and so on.

14.5   VOLUNTEERING AND PERSONALITY

Elshaug and Metzer (2001) provide a good, though not exhaustive, review of the contribution of studies relating personality to volunteering. Penner and Finkelstein (1998), likewise give a good discussion of the relative contribution of dispositional variables to the conduct of volunteering. Both reviews conclude that there are significant and consistent effects. The latter authors concluded that the main personality orientation consistently correlating with long-term volunteering is a prosocial one. While this comes as no surprise, one dimension, Other Oriented Empathy, of the Prosocial Personality Battery emerged as correlating consistently with volunteering, whereas the other dimension of Helpfulness did not do so. Elshaug and Metzer (2001) found, using the NEO-PI-R (Costa and McCrae, 1992), that there were consistent personality similarities in two different volunteer samples (food preparers and fire fighters) and reliable differences when these samples were compared with a sample of paid food preparers. The results indicated that, after controlling for age, socioeconomic status, education level, years of service and hours worked per week, the two volunteer samples were more Extraverted and Agreeable than the paid workers. Of the Extraverted dimension's facets, the volunteers scored higher on both warmth and positive emotions. Of the Agreeable dimension's facets volunteers scored higher on trust, altruism and tender minded-ness. The two volunteer groups differed only on the facet of assertiveness, firefighters scoring higher. The authors concluded that there is limited evidence for the existence of a constellation of these traits to characterise a volunteering disposition. Interestingly, these traits did not include Openness or (low) Neuroti-cism, the latter finding suggesting that volunteer workers' affect scores may need to be considered in any studies of the effects of stress in volunteer work.

14.6   VOLUNTEER WORK AND OLDER ADULTS

In addition to the likely increasing demands and pressure on volunteer services and organisations, the pool of available volunteers has also been shrinking. There are several reasons for this. First, women, the traditional source of much volunteer work, are now an integral part of the paid workforce and have much less time to contribute to volunteering. Second, many paid workers are under greater pressures in their work to perform at high levels and consequently tend to put in longer hours in their paid work, leaving less time and energy to devote to volunteer work. Third, the greater demands of paid work together with decreased interaction with and activities in the local community have a tendency to lead to a feeling of less involvement in and support for and from the community: there is a feeling of lower 'connectedness'. The combination of these three reasons is believed by some social commentators to lead to decreased 'social capital', the decline of rural communities and the dislocation of some sectors in society (Cox, 1995; May, 1999), the very factors which are likely to contribute to an increased pool of recipients for (volunteer) welfare services.

   With the above reasoning in mind, it is not surprising to expect volunteer organisations to look to those who have retired from the paid workforce in order to meet the increasing demands for resources. Typically, these people represent an important component of the volunteer workforce: the potential for high levels of existing skills transferred from the paid workforce and now available time can be alluring for a manager of a volunteer organisation struggling to meet demand and needing to expand its volunteer workforce. Warburton (1997) though, cautions against creating a norm that retired people should be expected to volunteer. Many retirees prefer a period of leisure after a lifetime of work, whereas others are keen to continue to commit to regular volunteer work. Battaglia and Metzer (2000) in reporting the results of a survey of both young and retired volunteers, note that retirees

> ...may welcome the opportunity to be free of responsibility and obligation and may not want the commitment of regular volunteer work. This point is underscored by the results of our study which showed that those retired adults who strongly agreed with statements designed to tap a sense of obligation, e.g. '*Other people at the organisation would probably make me feel guilty if I stopped volunteering*', were less satisfied with their volunteer work than those who did not feel obligated.

Hence, a cautionary note is required here, lest satisfied retirees be pressed into (dissatisfied) service by the use of moral force. It is an arguable point whether this is an example of coercion! It is not inconceivable that even mildly unscrupulous organisations or individuals, who themselves are under performance pressure, might wish to use such techniques to expand their 'volunteer' workforce.

On the flip side of their survey, the authors reported strong perceptions of retirees having benefited substantially from their volunteer work. The survey results suggested that the benefits were:

- The maintenance of identity and a sense of self esteem.
- Meeting social needs and a feeling of connectedness with a social group.
- Affording opportunities to undertake new learning.

From an excellent review of the literature, Wilson and Musick (2000) concluded that there is now sufficient evidence to support a causal link from engagement in volunteering to good physical health and decreased mortality, and a correlation between volunteering and good mental health, these effects being particularly pronounced for older adults. In other words, the interpretation of information available from the literature about the stresses and benefits of volunteering by older adults is complex, rather than vexing. Provided (even mild?) pressure or coercion is avoided in recruitment, the benefits of engagement in volunteering might be expected to be an important aspect in later life.

14.7 STRESS FACTORS IDENTIFIED IN THE VOLUNTEERING LITERATURE

Some information regarding certain variables that may form or affect stress and satisfaction in volunteers has been provided in the literature. Some authors are directly concerned with stress and burnout in volunteers. Others focus on the motivation and management of volunteers, while further publications examine the rights and responsibilities of volunteer workers.

Using data collected from discussions with hospice volunteers, Paradis *et al.* (1987) identified four major sources of volunteer stress including: role ambiguity (referring to uncertainty regarding the volunteer's obligations and responsibilities); status ambiguity (referring to the volunteer's location in the organisational hierarchy, and their lack of influence or power over their work); stress related to patients and families (referring to the volunteer's feelings of inadequacy due to their lack of training in dealing with the complex needs of patients and their families); and stress related to personal circumstances (referring to personal circumstances in volunteers' life that encourages or dissuades them from becoming involved with the organisation).

The volunteers interviewed in the above study indicated that being unclear about their role produced job stress. Other specific sources of stress mentioned were: feelings of inadequacy, insufficient information to perform the required tasks, lack of support from management as well as inappropriate supervision, and unrealistic expectations of the volunteers' work output. These volunteers stated a need for more training so as to feel like part of the staff, and highlighted the need for increased communication with paid staff.

Bryant and Harvey (1996) studied stress in volunteer fire-fighters. Susceptibility to post traumatic stress was found to increase with previous experiences of traumatic situations. Sixty per cent of volunteers identified feelings of helplessness, exhaustion and inadequate training or equipment as main causes of stress. In studies of caregivers of people with AIDS (Guinan *et al.*, 1991) categories of stressors identified were emotional overload, client problems, lack of support, and lack of training.

Savishinsky (1992) noted that volunteering in nursing homes was an emotionally demanding experience that some people handled more successfully than others. Several factors mediated how volunteers felt about their activity, and whether they continued their volunteer work. These included: the motives people had for becoming volunteers, prior experience in doing similar work, career orientations and current family/living situation, and the image held of the kind of people the volunteer organisation helped or represented.

Capner and Caltabiano (1993) stated that potential sources of burnout in volunteers include personal, interpersonal and organisational stressors. Volunteer counsellors reported feelings of loneliness and lack of feedback as stressful. These stressors were unique to volunteers; other identified stressors were common to both volunteer and professional counsellors. This would suggest that volunteers experience similar job-related stresses to paid workers, but that some stressors are unique to the volunteer worker.

Lafer (1991) recommended that better management support, improved training and structured communication between staff and volunteers would assist in alleviating volunteer burnout. Maslanka (1996) found that staff support lowered stress levels, while a sense of efficacy acted to motivate volunteers and decrease burnout.

In an Australian study of the volunteer experience, the Australian Council of Social Service (ACOSS, 1996) emphasised the importance of training. Volunteers called for clear duty statements, references and a system of evaluation. The negative side of volunteering was identified by the following issues: lack of support; hard work; long hours; lack of acknowledgement; poor management; unclear management; lack of information; lack of reliability of other volunteers increasing the workload; personal expenses; not being confident regarding their entitlements; and lack of appreciation.

McSweeney and Alexander (1996) identified numerous promoters of stress in the volunteer workforce including: staff shortages; excess workload; poor communication and poor relationships with other workers; frequent unpleasant working environment; intrinsic hazards; conflicts of home, family and work; and poor time management. These authors recommended that there should not be too much bureaucratic interference with volunteer tasks, and that volunteering should be meaningful to the individual. Teamwork was also considered to be important in alleviating volunteer stress.

McSweeney and Alexander (1996) also noted some of the more common reasons for volunteering, including: having free time; a desire to help others; a desire for social contact or friendship; identification of an explicit need for help; a wish to gain experience prior to paid work; religious or spiritual reasons; unemployment; a wish to fill time positively; political motives; and an opportunity to tackle a social problem. These authors emphasised that training should make volunteers feel more competent and confident, and that supervisors should ensure volunteers receive ongoing support. These authors also reported that people derived satisfaction in their volunteer work through achievement and recognition, and became dissatisfied with their volunteer work when confronted with conflict with company policy and administration.

Hackman and Suttle (1977) argued that dissatisfied workers were more likely to suffer from stress, which implies that satisfaction is a moderator of stress. Salancik and Pfeffer (1978) found that those working in objectively dull and alienating jobs can and do report high levels of job satisfaction, where satisfaction has stemmed from the work's social context. Pearce (1993) wrote about the attitudes of volunteers and reviewed the psychological research on work attitudes. Pearce asserts (albeit quite hastily) that there is selective retention of volunteers with positive attitudes. Due to the fact that volunteers have no financial dependence on their work, and volunteer work is peripheral to self and societal roles, 'dissatisfied volunteers can be expected to leave their organizations virtually the moment they become unhappy' (p. 90).

14.8 MEASUREMENT OF JOB SATISFACTION AND STRESS IN VOLUNTEER SAMPLES

The requirement for reliable measures of (job) satisfaction and stress or strain has been never more apparent than today, since control and predictability over training and retention, concerns over occupational health and safety in the workplace and so on of the volunteer workforce is now much more critical than even ten years ago (Dollard *et al.*, 1999). In many areas the questions asked now about volunteers is a recurrence of similar questions asked 50 or 60 years earlier of the paid workforce. Hence the answers to these questions might be expected to affect both productivity or service delivery and the quality of worklife of both volunteer workers and those who manage them.

With these issues in mind, the author of this chapter set out to devise and test two scales which would measure reliably levels of volunteer job satisfaction and volunteer job stress (or strain) in a number of volunteer work areas. The theoretical context for this study utilised Karasek's (1979) and Karasek and Theorell's (1990) models of stress.

14.8.1   The Demand/Control Model in Occupational Stress Research

Karasek's (1979) demand/control (DC) model was considered to be an appropriate theoretical model for understanding stress in volunteers. Karasek's model is considered to present a stimulus-based approach to stress management (Dollard and Winefield, 1998), emphasising the causal effects of the organisational structure and the objective properties of the job on the onset of worker stress, rather than the role of the individual (e.g. individual differences in stress responses, coping styles, personality, etc.). Karasek assumes that occupational stress results from high demands (e.g. excessive work, work pressure, conflicting demands) and low control (e.g. low control over workload and tasks performed, and lack of autonomy and decision making authority) in the workplace (Karasek and Theorell, 1990). Basically, stress occurs when the worker cannot access the necessary controls to cope with the strain caused by high demands. Conversely, the least stressful work environment exists when demands are low and controls are high. Deriving satisfaction from one's work tasks is different, in which the highest levels of job satisfaction should occur when high levels of demands are matched with high levels of controls ('active jobs'), as such jobs allow for the development of skills, learning and competency. On the other hand, when both demands and controls are low, the result is often reduced motivation, low skill acquisition and decreased learning ('passive jobs'), resulting in low levels of satisfaction with the job (Karasek and Theorell, 1990).

A modified version of Karasek's original DC model was adopted, termed the demand/control – support (DC-S) model (Karasek and Theorell, 1990). In the DC-S sense, support is derived from within the organisation (e.g. provided by co-workers and supervisors). It is assumed that the variable of support has the potential to moderate the impact of organisational demands on stress outcome .

Considering that support can be conceptualised as a stress moderator, for the present study the DC-S model was revised to include other moderators. Other potential moderators of stress include volunteer motivators (reasons for volunteering) as measured by the VFI, external social support (e.g. support not directly provided by the organisation), feeling connected to the organisation, and values surrounding volunteering and the volunteer organisation.

Although social support has been classified as a stress moderator, for the purpose of the present study, some factors of organisational support were considered to fit better within the broader factor of controls. This is because the concept of organisational support also includes factors such as cooperation within a group of workers which should act to reduce the workload for a single worker. Such tangible support would also include variables such as good communication, ideas from workers being welcomed by the organisation, teamwork and other organisationally oriented control mechanisms. On the other hand, social support in the form of positive interpersonal contact with others in the organisation (e.g. appreciation, experiencing fun, respect and liking others) seems more appropriately categorised as support, that is as a stress moderator.

14.8.2   Sample/Materials/Procedure

The sample comprised 131 Australian volunteers (53 male and 78 female) from 18 volunteer organisations (refer Table 14.1). The mean age of respondents was 53 years ($SD = 15$ years, range: 16–81 years). A total of 40% of respondents were employed in paid work and 60% had no paid work. The mean tenure within the current volunteer organisation was 7 years ($SD = 7$, range: .08–30 years), and the mean number of other volunteer jobs engaged in was 1.3 ($SD = 1.5$, range: 0–6). The number of mean hours contributed to the respondents' main volunteer job was 8 ($SD = 5$, range: 1–50).

A questionnaire entitled *The Volunteer Experience Survey* was developed by the author to measure various work experiences of volunteers. In its design the variables that might form or affect perceived stress in the workplace were taken from the existing literature on stress in volunteering and paid work. The concepts in the literature were noted, then separated and simplified into individual variables. In accordance with Karasek and Theorell's (1990) DC-S model of stress, most of these variables could be categorised as either demand, control or support variables. The remainder could be classified as either motivators for doing volunteer work, or as some other moderator of stress.

A further division was made in the process of isolating variables within the domains of demands/controls/moderators, in which variables were classified into three groups of orientation: (1) self, (2) interpersonal, and (3) environmental/organisational.

Single variables were then converted into statements for the questionnaire, allowing respondents to indicate their degree of agreement or disagreement with each statement on a Likert-type scale. For example, *level of training* was a pertinent issue in the literature, and was categorised as an organisation-orientated control variable. Thus, a corresponding statement for inclusion in the questionnaire was worded, 'I have been adequately trained for my volunteer work.'

The Volunteer Experience Survey comprised several separate sections, which were as follows:

Section A (32 items) was labelled *demands* and included items on:
• organisational demands
• time and effort put into volunteer work; and
• role ambiguity

Section B was labelled *controls* (28 items) and measured:
• autonomy
• training and ability; and
• organisational resources

Section C (20 items) was labelled *Supports/Values* and measured:
• feeling appreciated/respected by and connected to the organisation
• beliefs in organisational values
• feelings of obligation to volunteer

Section D comprised the Volunteer Functions Inventory (30 items, 6 sub-scales) (Clary *et al.*, 1998) measuring reasons for volunteering. Section E contained the MOS Social Support Scale (19 items) (Sherbourne and Stewart, 1991).

Section F centred on volunteering outcomes (20 items) and measured:
• volunteer stress
• volunteer satisfaction; and
• determination to continue volunteering

Section G measured demographic characteristics of the respondents including:
• type of volunteer activity, age, hours contributed to volunteer activity, years spent at this volunteer job, sex, employment status, satisfaction with paid work, and stress with paid work.
• original reasons for volunteering (three options based on orientation of self, others/community and organisation/environment); and
• comments regarding the questionnaire (optional)

The General Health Questionnaire (12 items) was also included as a measure of general emotional well-being.

Two hundred and twenty questionnaires were posted out with reply paid return envelopes to 22 volunteering organisations within South Australia (Table 14.1). Of these 22 organisations, 18 responded, with a total response rate of 59% (N=131) of questionnaires returned.

14.8.3   Results and Analysis

Data gained from Section A (demands), Section B (controls), Section C (supports), and Section F (outcomes), were subjected to a principal components factor analysis with varimax rotation. Items were excluded from the analysis based on the following criteria:

- greater than 5% of cases were missing
- extreme skewness
- item loading was less than 0.4

Table 14.1 Participating South Australian Volunteer Organisations (*n*=18)

Australian Council for Health, Physical Education and Recreation	Adelaide Zoo	Aged Care and Housing Group
Amnesty International	Art Gallery of SA	Family and Community Services
Girl Guides Association	Investigator Science and Technology Centre	Julia Farr Volunteer Service
Lavender Lads and Ladies Royal Adelaide Hospital	SA Volunteer Fire Brigade Association	SA Museum
Trees for Life	Victim Support Service	Wheelchair Sports Association
Tennis SA	Volunteer Involvement Program (Office for Recreation and Support)	OMNIA

Demands

An inspection of the unrotated eigenvalues indicated the presence of three factors, jointly accounting for 51.9% of the total variance. The first three components were rotated and the resulting loadings indicated that these three factors were interpretable as:

1. *Organisational demands* (10 items, alpha = .86). This factor is defined by lack of guidance by supervisors, conflict and confusion within the organisation, high volunteer turnover, bureaucratic interference, and obstacles to performing work tasks.
2. *Time and effort* (8 items, alpha = .81), characterised by high levels of time and physical/mental effort contributed, high levels of responsibility, tight deadlines, and perceived unfair share of workload.

3. *Extreme demands* (3 items, alpha = .74), characterised by dangerous work, and dealing with conflict and distressing situations.

Controls

An inspection of the unrotated eigenvalues indicated the presence of three factors, jointly accounting for 59.3% of the total:

1. *Organisation* (6 items, alpha = .88). This factor is defined by adequate organisational resources, good working relationships and communication between staff, and adequate assistance and information provided to perform work tasks.
2. *Knowledge/Influence* (5 items, alpha = .77). Markers of this factor include volunteers being well informed and provided with feedback on their work, and whether volunteers are welcome to contribute ideas and opinions to the organisation.
3. *Autonomy* (6 items, alpha = .77) was characterised by adequate training (volunteers know their job well), and feeling in control of job tasks.

Supports

The same procedure was utilised for support items, highlighting two obvious factors, which taken together accounted for 45.2% of the total variance. Rotation of these components indicated that these two factors were interpretable as:

1. *Appreciated/Connected* (7 items, alpha = .88), characterised by feelings of performing interesting and important work, feeling appreciated and respected by the organisation/others, and feeling connected to the organisations values.
2. *Obligation* (2 items, alpha = .53). This factor was marked by feeling pressured and obligated to volunteer.

Volunteer Outcomes

Analysis of outcomes variables indicated two factors which jointly accounted for 47.9% of the total variance. These components were subjected to varimax rotation, with the resulting two factors being interpretable as:

1. *Satisfaction* (6 items, alpha = .89). Markers of this factor were feeling enthusiastic about the work, being personality fulfilled by the work, making a worthwhile contribution, feeling successful, and overall satisfaction with volunteer work.
2. *Stress* (4 items, alpha = .71), characterised by experiencing anxiety in relation to volunteering, finding difficulties in relaxing, negative health consequences, overall stress in volunteer work.

Table 14.2 Correlation Matrix for Stress, Satisfaction and Demand/Control/Support Variables

	SAT	ORD	T/E	EXD	ORC	K/I	AUT	A/C	OBL
STR	.16	.42**	.55**	.12	.33**	−.02	−.05	.00	.29**
SAT		−.10	.33**	.18*	.35**	.52**	.54**	.63**	−.05
ORD			.33**	.20*	−.57**	−.19*	−.19*	−.15	.34**
T/E				.19*	−.21*	.21*	.18*	.10	.26**
EXD					.13	.23**	.33**	.28**	−.08
ORC						.51**	.48**	.52**	−.28**
K/I							.58**	.58**	−.05
AUT								.62**	−.14
A/C									−.02

*$p < .05$ **$p < .01$ STR=Stress, SAT=Satisfaction, ORD=Organisational Demands, T/E+Time/Effort, EXD=Extreme Demands, ORC=(control) Organisation, K/I=Knowledge/Influence, AUT=Autonomy, A/C=Appreciated/Connected, OBL=Obligation.

Stress and demand/control/support variables

Pearson correlations indicated significant, positive relationships between volunteer stress and the demand factors of organisational demands ($r = .62$, $p < .01$) and time and effort ($r = .56$, $p < .01$). A significant negative correlation was found between stress and the control factor of organisation ($r = −.20$, $p < .05$) although correlations between stress and other control factors were non-significant. No significant correlations were evident between the support factor of appreciated/ connectedness and stress. A positive relationship existed between obligation to volunteer and stress ($r < .29$, $p < .01$).

Stress and VFI Sub-factors

Significant, positive correlations existed between volunteer stress and the VFI subfactors (reasons for volunteering): a career function ($r = .24$, $p < 01$); an enhancement function ($r = 36$, $p < 01$); a protective function ($r = .29$, $p < .01$); a social function ($r = .31$, $p < .01$); an understanding function ($r = .27$, $p < .01$).

Stress and Demographic Variables

Significant, positive correlations were found between volunteer stress and respondents' age ($r = .19$, $p < .05$), hours (per week) contributed to present volunteer organisation ($r = .46$, $p < .01$), and number of other volunteer activities the respondent participated in ($r = .20$, $p < .05$). There was a positive relationship between volunteer stress and engaging in paid work (outside of volunteer work) ($r = .18$, $p < .05$), where stress in one's paid work was significantly correlated with stress experienced in volunteer work ($r = .37$, $p < .01$). Stress and GHQ displayed a significant, positive correlation ($r = .38$, $p < .01$); stress and satisfaction did not correlate significantly.

Satisfaction and Demand/Control/Support Variables

Pearson correlations indicated a significant, negative correlation between satisfaction with volunteer work and the demand factor of time and effort ($r = .33$, $p < .01$), and a positive correlation with extreme demands ($r = .18$, $p < .05$). Positive correlations were found between volunteer satisfaction and all control factors. The support factor of appreciated/connectedness also correlated strongly with volunteer satisfaction ($r = .63$, $p < .001$) .

Satisfaction and VFI Sub-factors

Significant, positive correlations were found between satisfaction with volunteer work and the VFI sub-factors of: an enhancement function ($r = .30$, $p < .01$); a social function ($r = .22$, $p < .05$); a values function ($r = .38$, $p < .01$); an understanding function ($r = .37$, $p < .01$).

Satisfaction and demographic variables

Significant, positive correlations existed between satisfaction with volunteer work and length (years) of volunteering ($r = .28$, $p < .01$), and between satisfaction with volunteer work ($r = .42$, $p < .01$).

Multiple Regression Analyses

Demand/control/support variables as predictors of stress and satisfaction
A standard multiple regression was performed entering stress as the dependent variable, and another regression was performed entering satisfaction as the dependent variable, with demand, control and support variables entered as predictors in both analyses (refer Table 14.3). Three regressions were performed for both stress and satisfaction. The first included the total sample, the second included only the respondents who engaged in paid work in addition to volunteer work, and the third included only the respondents with no paid work.

Results indicated two demand factors as significant predictors of volunteer stress for the total sample. These were time and effort, and organisational demands. Significant predictors of stress for volunteers with paid work were controls, the demand factor of time and effort, and the support factor of appreciated/connectedness. In contrast, only one predictor, the demand factor of time and effort, was significant for volunteers with no paid work.

Three variables were found to predict satisfaction for the total sample. These were controls, the support factor of appreciated/connectedness, and the demand factor of time and effort. Significant predictors of volunteer satisfaction for respondents with paid work were identical to results derived from the total sample. However, significant predictors of satisfaction for volunteers without paid work were slightly different, indicating the support factor of appreciated/connected, the demand factor of time/effort, and feeling obligated to volunteer as predictors of volunteer satisfaction.

Table 14.3 Standard multiple regressions for stress and satisfaction (total sample, respondents with paid work, and respondents without paid work) using demand, control and support variables as predictors (Beta values in nonitalics)

	Volunteer Stress			Volunteer Satisfaction		
	Total sample	Paid work	No paid work	Total sample	Paid work	No paid work
Organisational demands	.18*	.11	.21	.01	.01	.03
Time and Effort	.46**	.44**	.46**	.30**	.22*	.34**
Extreme demands	.02	.02	.04	−.08	.02	−.18
Controls	−.16	−.61**	.03	.21*	.32*	.15
Appreciated/ Connected	.09	.42*	.10	.49**	.43**	.52**
Feeling Obligated	.08	.01	.11	−.09	.03	−.22*
R^2	.38	.46	.37	.51	.64	.44
(df)	(6,122)	(6,51)	(6,76)	(6,122)	(6,51)	(6,76)
F	12.69**	6.35**	6.81**	20.80**	13.40**	9.15**

*$p < .05$, two-tailed **$p < .01$, two-tailed.

Overall predictors of stress and satisfaction

Standard multiple regression analyses were performed for stress and satisfaction to indicate which variables, overall, best predicted stress and satisfaction. The choice for predictor variables for entry in the regression was based on results of previous regression analyses and significant correlations with the dependent variable.

A standard multiple regression, in which all the independent variables were entered at the same time, was performed with stress as the dependent variable and time/effort, organisational demands, controls, hours contributed to the volunteer organisation, and the VFI function of protect as predictor variables. Results indicated four significant predictors of stress (accounting for 46.0% of the variance), which were time/effort (beta = 39, $p < .001$), controls (beta = −.22, $p < .01$), hours contributed (beta =.20, $p < .05$), and volunteering to serve a protective function (beta =.20, $p < .05$). This regression was significant, F $(6, 124) = 16.89$, $p < .001$.

A standard multiple regression was performed with satisfaction as the dependent variable and time/effort, controls, appreciated/connectedness, years spent volunteering, and the VFI functions of enhancement, social, understanding and values as predictor variables. Results indicated three significant predictors of satisfaction (accounting for 52.0% of the variance), which were appreciated/ connectedness (beta = .41, $p < .001$), controls (beta = .24, $p < .01$), and time/effort (beta = .22, $p < .01$). This regression was significant, F$(8, 128) = 16.3$, $p < .001$.

Partial correlations

The original (Pearson) correlation between stress and demands was significant ($r = .56, p < .001$). Partial correlations were conducted between stress and demands, controlling for each of the proposed moderators in turn. If these variables were moderators of stress it was expected that the relationship between demands and stress would increase when controlling for each of these variables. Partial correlations between stress and demands controlling for each of the moderators yielded the following results: controlling for obligation ($r = .52, p < .001$); controlling for appreciated/connectedness ($r = .57, p < .001$); controlling for MOS social support ($r = .55, p < .001$); controlling for satisfaction ($r = .55, p < .001$); controlling for VFI career ($r = .54, p < .001$); controlling for VFI enhancement ($r-.52, p < .001$); controlling for VFI protect ($r = .55, p < .001$); controlling for VFI social ($r = .53, p < .001$); controlling for VFI understanding ($r = .53, p < .001$); controlling for VFI values ($r = .56, p < .001$).

The results of the partial correlations suggest that neither obligation, satisfaction, feeling appreciated, social support, nor any of the VFI functions appear to have a moderating effect on the relationship between demands and volunteer stress.

While further work on stress moderators needs to be conducted, interest was centred on the comparison between volunteers who had paid work in addition to their volunteer work and those who did not.

Of particular interest were the positive correlations found between firstly, volunteer stress and engaging also in paid work and secondly, stress experienced in one's paid work and stress experienced in volunteer work.

These suggest a generalisation effect of perceived stress (or strain) of engaging in both paid and volunteer work.

14.8.4   Brief Discussion of Findings

Three scales for the use in research on stress and satisfaction in volunteers were designed and used: a Stress scale (4 items), a Satisfaction scale (6 items) and a Connectedness scale (7 items). These are freely available from the author upon request, and may prove of use in research on volunteers' quality of worklife.

While the data above give an indication that Karasek and Theorell's model has some generalisation from paid work to volunteer work in predicting stress, some differences were indicated in the results.

First a distinction needs to be made on the accuracy of the model in predicting perceived stress for volunteers who have paid work also and those who do not. The model seems to perform in a similar fashion for the former group of volunteers to occupational groups of paid workers. This satisfactory performance of the model must be contrasted to its lack of predictive power for perceived stress in volunteers who do not also have paid work. Only one demand, that of time and effort, appeared to be a predictor in this volunteer group. It appears that this group is more unlike other work groups, including the other volunteer group of those who also have paid work, in perceiving stress. They seem to be more buffered from stress compared with their paid counterparts, and it may be that they are the main recipients of the improved health effects of volunteering, as noted by Wilson and Musick (2000).

Whether these beneficial effects also generalise to those volunteers with paid work needs further investigation.

On the other hand, volunteers' satisfaction appeared to be determined more similarly between both the volunteer groups and other occupational groups, a finding also supported by Robertson (2000) who further confirmed the power of prediction of the connectedness scale in both paid and volunteer workers.

Volunteers are a group of workers who seem to have some similarities to other occupational groups in the understanding of their stress and satisfaction, but also appear to have sufficient differences to warrant further investigation in their own right.

14.9 ACKNOWLEDGEMENTS

The author is indebted to Carol Elshaug and David Cann for their valuable work in assisting with the design, running, analysis and interpretation of the data of the survey reported in the latter half of this chapter.

14.10 REFERENCES

Australian Bureau of Statistics, 1996, *Voluntary Work*. (4441.0).

Australian Bureau of Statistics, 2001, *Voluntary Work*. (4441.0).

A National Agenda on Volunteering, 2001, Beyond the International Year of Volunteers. *Australian Journal on Volunteering*, **6**, pp. 5–9.

ACOSS, 1996, Volunteering in Australia (Sydney, Australian Council of Social Service).

Battaglia, A.M. and Metzer, J., 2000, Older adults and volunteering: A symbiotic association. *Australian Journal on Volunteering*, **5**, pp. 5–12.

Bryant, R.A. and Harvey, A.G., 1996, Posttraumatic stress reactions in volunteer firefighters. *Journal of Traumatic Stress*, **9**, pp. 51–62.

Capner, M.C. and Caltabiano, M.L., 1993, Factors affecting the progression towards burnout: A comparison of professional and volunteer counsellors. *Psychological Reports*, **73**, pp. 555–561.

Clarey, E.G., Snyder, M., Ridge, R., Copeland, J., Stukas, A., Haugen, J. and Miene, P., 1988, Understanding and assessing the motivation of volunteers: A functional approach. *Journal of Personality and Social Psychology*, **74**, pp. 1516–1530.

Costa, P.T. and McRae, R.R., 1992, *Revised NEO personality inventory (NEO-PI-R) professional manual* (Odessa, FL, Psychological Assessment Resources).

Cox, E., 1995, *A truly civil society* (Sydney, Australian Broadcasting Corporation).

Debelle, P., 1996, *$5.6 billion volunteers* (Adelaide, Advertiser).

Dollard, M.F. and Winefield, A.H., 1998, A test of the demand-control/support model of work stress in correctional officers. *Journal of Occupational Health Psychology*, **3**, pp. 243–264.

Dollard, M., Rogers, L., Cordingley, S. and Metzer, J., 1999, Volunteer work: Managers' conceptualisations of factors affecting volunteer quality of life. *Third Sector Review*, **5**, pp. 5–23.

Eagley, A.H. and Crowley., 1986, Gender and helping bahvior: A meta-analytic review of the social psychological literature. *Psychological Bulletin*, **100**, pp. 283–308.

Elshaug, C. and Metzer, J., 2001, Personality attributes of volunteers and paid workers engaged in similar occupational tasks. *Journal of Social Psychology*, **14**, pp. 752–763.

Guinan, J.J., McCallum, L.W., Painter, L., Dykes, J., 1991, Stressors and rewards of being an emotional-support volunteer: A scale for use by care-givers for people with AIDS. *Aids Care*, **3**, pp. 137–150.

Hackman, J.R. and Suttle, J.L. (eds), 1977, Improving life at work: Behavioral science approaches to organizational change (Santa Monica, Goodyear Publishing Company).

Jamrozik, A., 1996, Voluntary work in the 1990s: Comments on the ABS 1995 survey. *Australian Journal on Volunteering*, **1**, pp. 15–19.

Karasek, R.A. and Theorell, T., 1990, Healthy work: Stress, productivity and the reconstruction of working life (New York, Basic Books).

Lafer, B., 1991, The attrition of hospice volunteers. *Omega Journal of Death and Dying*, **23**, pp. 161–168.

Maslanka, H., 1996. Burnout, social support, and AIDS volunteers. *AIDS Care*, **8**, pp. 195–206.

May, J., 1999, Welcome to the brave new world of partnerships. *Australian Journal on Volunteering*, **4**, pp. 4–12.

McSweeney, P. and Alexander, D., 1996, *Managing volunteers effectively* (Hants, UK, Arena).

Metzer, J.C., 1996, The psychology of volunteering: External or internal rewards? *Australian Journal on Volunteering*, **1**, pp. 20–24.

Metzer, J.C., Dollard, M., Rogers, L. and Cordingley, S., 1997, Quality of worklife of volunteers. *Australian Journal on Volunteering*, **2**, pp. 8–15.

Noble, J., 1997, *Volunteers and paid workers: A collaborative approach* (Adelaide, The Volunteer Centre).

Pearce, J.L. 1993, *The organizational behaviour of unpaid workers* (London, Routledge).

Penner, L.A.F. and Finkelstein, M.A., 1998, Dispositional and structural determinants of volunteerism. *Journal of Personality and Social Psychology*, **74**, pp. 525–537.

Robertson, M., 2000, Comparison of job strain and job satisfaction between volunteer and paid staff in nursing homes. *Unpublished Honours Thesis* (University of South Australia).

Salancik, G.R. and Pfeffer, J., 1978, A social information processing approach to job attitudes and task design. *Administrative Science Quarterly*, **23**, pp. 224–456.

Savishinsky, J.S., 1992, Intimacy, domesticity and pet therapy with the elderly: Expectation and experience among nursing home volunteers. *Social Science & Medicine*, **34**, pp. 1325–1334.

Schaufeli, W., Maslach, C. and Marek, M., 1993, *Professional burnout: Recent developments in theory and research* (Washington, Taylor & Francis).

Sherbourne, C.D. and Stewart, A.L., 1991, The MOS social support survey. *Social Science & Medicine*, **32**, pp. 705–714.

Warburton, J., 1997, Older people as rich resource for volunteer organisations: An overview of these issues. *Australian Journal on Volunteering*, **2**, pp. 19–25.

Warburton, J. and Oppenheimer, M., 2000, *Volunteers and volunteering* (Federation Press).

Wilson, J. and Musick, M., 2000, The effects of volunteering on the volunteer. *Law and Contemporary Problems*, **62**, pp. 142–168.

CHAPTER FIFTEEN

Conclusion

Maureen F. Dollard

15.1 INTRODUCTION

The book, *Occupational Stress in the Service Professions* set out to explore and describe the current work context for a range of service professions: police, oncology care providers, nurses, general practitioners, teachers, academics, prostitutes, social and human service workers, clergy, psychologists and volunteers. As discussed in each of the chapters, the service professions have become increasingly concerned about the issue of work stress because of the costly negative consequences to the professional, to the efficacy of professional services, and therefore to the repute of the profession. Reduced performance and increased absenteeism and mistakes, are some organisational outcomes of work stress that have serious economic consequences for those who employ service professionals and more broadly for society. For the service professional, increased psychological distress, job dissatisfaction, and increased symptoms of burnout, reduced professional efficacy and more physical illness, can result from exposure to hazardous psychosocial work conditions.

The book did not intend to argue that work in the professions is becoming more stressful. Indeed, it is difficult to sustain this position as contributing authors commonly reported increasing stressors not only in the form of sheer contact hours and work load, but also in the range of taxing tasks, and complexity of information to be managed by modern service professionals (e.g. see Chapter 8).

15.2 EMERGING THEMES

It is difficult to draw major themes from the text and the following should be supplemented by further reading of each of the chapters. Here are some issues that have arisen.

15.2.1 Rural and Remote

A major challenge for rural and remote professionals is the separation of roles, social and family on the one hand and work on the other hand. This issue is well covered in the Chapters on psychologists (13), social workers (11) and general practitioners (7). Indeed, the whole notion of 'professional' is never so clearly under surveillance by the public as in these remote environments. Professionals are challenged to provide generic rather than specialist services for which they are rarely trained, and to cope with numerous variations in emerging client issues (including violence) without support. The barrier of anonymity available in metropolitan work, so helpful in maintaining client distance and affording stress recovery, is often not available in rural and remote work. Challenges to the maintenance of modern professions are constantly being played out in rural and remote settings, and it calls into question the relevance of metropolitan training and current support for rural and remote practitioners (Dawson *et al.*, 2002; Dollard *et al.*, 2001).

15.2.2   Management and Leadership Issues

The findings of research in relation to police work (Chapter 4) echo the conclusions of many of the other chapters. Specifically the way the professional's job is organised and managed (e.g. leader and management practices, appraisal and recognition processes, career opportunities, clarity of roles, co-worker relations, goal alignment) appears to be more important in predicting distress and morale than the operational aspects (e.g. being exposed to danger, dealing with victims) peculiar to the different professions. These findings are consistent with the long held belief by many researchers that organisational interventions rather than individual focused interventions should have the most far-reaching impact.

15.2.3   Organisational Recognition

An important theme related to management and leadership issues is (in relation to) organisational recognition (or validation). It appears to be a critical ingredient for occupational well-being when experienced not only on a day to day basis (e.g. see Chapter 8, and Chapter 12) but especially in times of great distress, for example when a worker experiences a critical incident (see Chapter 4). Interventions involving support from the supervisor may be just as important in recovery as that received from more experienced professionals in the form of debriefing (e.g. Rose *et al.*, 2000).

15.2.4   Performance

Many researchers reinforced the idea of linking aspects of the work environment to the bottom line (work performance). Evidence is emerging that there are two separate but dynamically operating pathways: one that is associated with strain on the one hand, and another with efficacy (Chapter 12) (see also Dollard *et al.*, 2001) or morale (Chapter 4) or satisfaction. The latter pathway appears to be further associated with withdrawal behaviour (Chapter 4) and performance. The correlation between job satisfaction and performance was confirmed in a recent meta-analysis to be about .52 for complex jobs (Judge *et al.*, 2001), and the size of this correlation was recently confirmed by Cotton *et al.* (2002) in a study using objective performance and satisfaction.

15.2.5 Socio-political and Economic Context

Prevailing stressors in any profession are in many ways determined by the larger socio-political and economic context. Resources to support the professions are under threat due to fiscal restraint, driven by an ideology of conservatism that fosters individualism rather than communalism (see Chapter 11). Due to legal and social imperatives the work of the professions has changed. A poignant example is in the case of social workers. Greater focus is now on child protection work, and although extremely important, due to limited resources this has led to a shift away from a historic concern with community development initiatives. A redoubling of demands occurs for human service professions as workload increases due to direct cut backs, and then less opportunity for preventive work leads to a greater work load. Examples of this kind are also clearly seen in the education (Chapter 8) and health sectors (Chapter 7).

Policy Implications

Having established that work stress is an issue for the service professions the question arises, what can be done? Chapter 1 offers a range of possible interventions: national policy, surveillance and monitoring; work redesign; and individual level interventions. In addition, many of the chapters have provided a review of interventions tried and implications from research for further intervention.

It is not the aim here to provide an exhaustive list of policies that could be implemented because each of the professions has its own challenges and nuances. Although international literature reviews were conducted, it is difficult to draw broad policy implications that are relevant across cultures and nations (although the clergy, for example, are often internationally organised). Differences if they exist cross-culturally 'may affect the development of a national strategy and the implementation of job stress reduction/ control' (Kawakami, p. 2, 2000). Nevertheless, both nationally and trans-nationally a number of policy, position and practice documents are emerging to which the reader can be referred: European Commission, Guidance on work-related stress: Spice of life or kiss of death, 2000, European Communities, Luxembourg; National Occupational Health and Safety Commission (NOHSC) Symposium on the OSH Implications of Stress, Melbourne 2001; and, The Tokyo Declaration (1998).

From these and the preceding chapters a philosophical framework rather than specific policy emerges for advancing our understanding and developing processes to deal with new stressors as they emerge. First, primary prevention/ intervention should be the first priority (La Montagne, 2001). Second, guidelines for the organisational implementation of interventions suggest that they should: be step-wise and systematic; require an adequate risk assessment; combine person and work-directed measures; use a participatory approach; involve top management support; and be evaluated for cost–benefits in terms of both health and productivity (Kompier *et al.*, 1984). Third, the participation of a range of stakeholders in dialogue and research activities (e.g. risk assessment approach, intervention) and mutual understanding is critical to the development of policy within the professions that is responsive to new insights gained from the field. Professional bodies can then act to lobby for local, national and international changes (system level), and call for changed employer/organisational practice and training standards. This process will be much less easily achieved in the case of prostitutes (Chapter 10) and volunteers (Chapter 14) who have less well developed and organised bargaining capacity (i.e. little or no professional/union representation).

Future Research

Each of the chapters has provided some insights and directions for future research of the service professions. In particular, research that promotes positive or productive aspects of work such as morale (e.g. Chapter 4), engagement (Maslach, 1998), performance and efficacy (Chapter 12), and explores emerging issues such as emotional and cognitive demands (Houkes *et al.,* 2001) and the causes and consequences of workplace violence, will be particularly valuable (see Chapter 10).

Major theories in this field such as the demands-control model and the effort-reward imbalance model, and burnout theory were reviewed in Chapter 1, and throughout the book their relevance in understanding stress in the service professions was underscored. Nevertheless, capturing the strengths of each model as well as ironing out limitations, has led to the emergence of new theory espoused here as the DISC model (Chapter 2). Fresh research is now required to validate the model empirically.

As pointed out in Chapter 1, it is difficult to propose a grand theory of work stress, and inevitably the application of theory requires some currency and meaning at the local level (i.e. including stressors of special relevance to each profession).

Most work stress theories have been developed and tested in Western countries, such as US and Europe. Job stress is however a common concern among many countries in the world, in both industrialised and emerging economies. In the last ten years several countries in East Asia, including China, Korea, and Taiwan, have rapidly industrialised and have grown economically. Among these countries there are many concerns about job stress and its adverse effects on worker health (Kawakami, 2000). We do not yet know the extent to which the findings presented here, where similar professions exist, are transferable to eastern cultures. Although research has relocated the findings in Japan and other non-Western countries in general, it is expected that there are differences in sources, and the stressor–strain relationship in non-Western cultures, at least among East-Asian countries which share common cultural characteristics, such as collectivism (Kawakami, 2000). Further research is required to discover the local relevance of contemporary theories for the professions where they exist in the Southern as well as the Northern Hemisphere and in cultures other than Western, and in rural and remote versus metropolitan contexts.

Urgent research is required to evaluate prevention and intervention efforts and their effect on well-being and performance (Chapter 3). In conducting such research the utilisation of multiple methods (interviews, observation, policy documents, surveys) and testing points (longitudinal designs) are recommended to capture and comprehend the complexity of the work stress process.

15.3 RELEVANCE

The book has obvious relevance for policy makers, unions, academics, and students. We hope it will also be relevant for prospective professionals. If potential trainees had a better idea about what was involved in a profession – the positives, challenges and drawbacks – they might be in a more informed position to make appropriate decisions about whether their characteristics and resources would be sufficient to cope.

The book will also hopefully be of interest to the public generally in their exploration about the professions, the functions they serve and their value to contemporary society. For example, in the case of clergy (Chapter 12), 'unrealistic expectations from others' (e.g. the congregation) is rated by them as one of the greatest demands. Greater clarity among social service recipients more generally about the emotional demands experienced by the range of service providers, may lead to greater reciprocity, support, acknowledgement and mutual regard.

Organisational managers and supervisors should gain some insights from the book about the cost–benefits of stress intervention, about how to construct a healthy productive work environment, and about how to monitor and continuously improve the work environment through participatory action and risk management approaches (see Chapters 3, 5 and 6 in particular). Further, Chapter 4 introduces a heuristic model of organisational health where occupational well-being is clearly linked with outcomes that affect organisational performance.

Among academics, professionally the book should provide a helpful almanac for those teaching in the areas of stress management, occupational health and safety, occupational health psychology, as well as being a salutary reminder about the contemporary and damaging levels of stress that academics themselves are experiencing (Chapter 9). The book also provides a challenge for those interested in the theory and philosophy of work stress in Chapters 1 through to 3. Specifically Chapter 2 provides a state of the art advance on current theoretical formulations of work stress, proposing an integrated model applicable to service work, called the Demand-Induced Strain Compensation (DISC) Model of job stress.

15.4 CONCLUSION

Globalisation and regional economic imperatives have no doubt led to modern work environments increasingly characterised by 'too much work', 'not enough work' and 'no work' rather than optimal 'healthy-productive' job characteristics. Besides negative implications for national economies, there is a strong belief that mental health problems and stress-related disorders are the biggest overall cause of premature death in Europe (WHO, 2001) as discussed in the Preface. Income inequality that arises from such disparate work states has negative health consequences for all members of society as social cohesion which characterises 'healthy egalitarian societies' progressively breaks down (Wilkinson, 1996). This anomic state in turn leads to increased pressure on social services provided by all of the professions discussed here as well as the volunteer work force (Chapter 14).

Changes in the work environment are taking place at an unprecedented pace. As demands for quality and productivity increase, and new demands emerge for service professionals, such as emotional and cognitive demands, and workplace violence, work management will require change. Workers will require greater education about the inherent risks of their proposed vocation, and more varied organisational responses to assist them to cope with old, new, and emerging risks as well as high performance. 'Policies and strategies for continuous monitoring and dialogue between the full range of stakeholders are imperative. Rather than relying on old models and methods, we now require active, coordinated and on-going meta-analysis of how we conceptualise stress, how we reduce negative effects and enhance positive outcomes for both the person and organisation' (see Dollard, 2001, p. 39).

We note both the high cost of stress in the modern work environment and a body of evidence to suggest that much of the stress experienced could be prevented if organisations and work arrangements were developed, constituted and constructed in more sustainable ways (Cameron, 1998). The human services which professionals seek to provide could therefore be offered with greater quality, professionalism, and continuity than might currently be the case. Change within organisations is rapidly taking place and it is incumbent upon managers, workers, and unions, to remain constantly vigilant about current work contexts that are being constructed for workers, and the impact these may have on the health and well-being of service professionals and their families.

Finally, we expect that the utilisation of multimedia methods, such as video and the WWW will become increasingly important for education and the understanding of work stress, its prevention and intervention.

15.5 REFERENCES

Cameron, I., 1998, Retaining a medical workforce in rural Australia, *The Medical Journal of Australia*, **169**, pp. 293–294.
Cotton, S.J., Dollard, M.F. and de Jonge, J., 2002, Stress and Student Job Design: Satisfaction, Well-Being, and Performance in University Students. *International Journal of Stress Management*, **9**, pp. 147–162.
Dawson, A., Dollard, M.F., Blue, I. and Wilkinson, D., 2002, CRANA Personal Support Network evaluation: Providing high quality support and services to

remote area health practitioners. South Australian Centre for Rural and Remote Health and the Work & Stress Research Group, Adelaide.

Dollard, M.F., 2001, *Work stress theory and interventions: From evidence to policy*. National Occupational Health and Safety Commission Symposium on the OHS Implications of Stress, Melbourne, pp 1– 57.

Dollard, M.F., Farrin, J. and Heffernan, P., 2001, *A model for improved health of rural and indigenous Australians through education, support and training for psychologists*. Good Health, Good Country, 6th National Rural Health Conference, Conference Monograph, National Rural Health Alliance.

Dollard, M.F., Winefield, H.R. and Winefield, A.H., 2001, Occupational strain and efficacy in human service workers (Dordrecht: Kluwer Academic Publishers).

European Commission, 2000, Guidance on work-related stress: Spice of life or kiss of death (European Communities, Luxembourg).

Judge, T.A., Thoresen, C.J., Bono, J.E. and Patton, G.K., 2001, The job satisfaction-job performance relationship: A qualitative and quantitative review. *Psychological Bulletin*, **127**, pp. 376–407.

Houkes, I., Janssen, P.P.M., de Jonge, J. and Nijuis, 2001, Specific relationships between work characteristics and intrinsic work motivation, burnout and turnover intention: A multi-sample analysis. *European Journal of Work and Organizational Psychology*, **10**, pp. 1–23.

Kawakami, N., 2000, Preface. *Job stress in East Asia: Exchanging experiences among China, Japan, Korea, Taiwan and Thailand*. In Proceedings of the First East-Asia Job Stress Meeting, Waseda University International Conference Centre Japan, pp.1–2.

Kompier, M., De Gier, E., Smulders, P. and Draaisma, D., 1994, Regulations, policies and practices concerning work stress in five European countries. *Work & Stress*, **8**, pp. 296–318.

LaMontagne, A.D., 2001, *Evaluation of occupational stress interventions: An overview*. National Occupational Health and Safety Commission Symposium on the OHS Implications of Stress, Melbourne, pp. 80– 97.

Maslach, C., 1998, A multidimensional theory of burnout. In C. Cooper (ed.). Theories of organizational stress (Oxford: Oxford University Press).

National Occupational Health and Safety Commission Symposium on the OHS Implications of Stress, 2001, Melbourne.

Rose, A., Bisson, J. and Wessely, S., 2002, Psychological debriefing fro preventing post traumatic stress disorder (PTSD). In The Cochrane Library, Issue 1, (Oxford: Update Software).

The Tokyo Declaration, 1998, *Journal of Tokyo Medical University*, **56**, pp. 760–767.

WHO, 2001, *Mental Health in Europe* (Copenhagen: World Health Organization, Regional Office for Europe).

Wilkinson, R.G., 1996, *Unhealthy societies: The afflictions of inequality* (London: Routledge).

Index

ESSENTIAL READING

Fundamentals of Health at Work
Carol Wilkinson, University of Lincoln, UK

Hbk: 0–419–24820–X
Spon Press   Pbk: 0–419–24830–7

Healthy and Productive Work, an International Perspective
Edited by **Lawrence R Murphy**, NIOSH, USA and **Cary L Cooper**, UMIST, UK

Taylor & Francis   Hbk: 0–7484–0839–8

The Burnout Companion to Study and Practice, a critical analysis
Wilmar Schaufeli and **Dirk Enzmann**, the University of Utrecht, The Netherlands

Hbk: 0–7484–0697–2
Taylor & Francis   Pbk: 0–7484–0698–0

This title is part of the Issues in Occupational Health Series, edited by Tom Cox and Amanda Griffiths

---

Workplace Health, Employee Fitness and Exercise
Edited by **John Kerr**, Nijenrode University, The Netherlands, **Tom Cox** and **Amanda Griffiths** both of The University of Nottingham, UK

Hbk: 0–7484–0142–3
Taylor & Francis   Pbk: 0–7484–0143–1

This title is part of the Issues in Occupational Health Series, edited by Tom Cox and Amanda Griffiths

---

Coping, Health and Organisations
Phil Dewe, Massey University, New Zealand, **Tom Cox**, University of Nottingham, UK and **Michael Leiter**, Arcadia University, Canada

Hbk: 0–7484–0824–X
Taylor & Francis   Pbk: 0–7484–0823–1

This title is part of the Issues in Occupational Health Series, edited by Tom Cox and Amanda Griffiths

---

Information and ordering details
For price availability and ordering visit our website
www.ergonomicsarena.com
Alternatively our books are available from all good bookshops